Advancement in Blockchain Technology and Applications

Advancement in Blockchain Technology and Applications

Editor

Hamed Taherdoost

MDPI • Basel • Beijing • Wuhan • Barcelona • Belgrade • Novi Sad • Cluj • Manchester

Editor
Hamed Taherdoost
University Canada West
Vancouver
Canada

Editorial Office
MDPI
St. Alban-Anlage 66
4052 Basel, Switzerland

This is a reprint of articles from the Special Issue published online in the open access journal *Electronics* (ISSN 2079-9292) (available at: https://www.mdpi.com/journal/electronics/special_issues/Blockchain).

For citation purposes, cite each article independently as indicated on the article page online and as indicated below:

Lastname, A.A.; Lastname, B.B. Article Title. *Journal Name* **Year**, *Volume Number*, Page Range.

ISBN 978-3-7258-1193-9 (Hbk)
ISBN 978-3-7258-1194-6 (PDF)
doi.org/10.3390/books978-3-7258-1194-6

Contents

About the Editor

Hamed Taherdoost

Dr. Hamed Taherdoost is an award-winning leader and R&D professional. He is the founder of the Hamta Group, Hamta Business Corporation, Associate Professor and Chair of RSAC at University Canada West, and Director of R&D at Q Minded, Quark Minded Technology Inc. He has over 20 years of experience in both the industry and academia sectors. He has worked at international companies from Cyprus, the UK, Malta, Iran, Malaysia, and Canada and has been highly involved in development of several projects in different industries: healthcare, transportation, residential, oil and gas, and IT. Apart from industry, he has been a university lecturer in three different parts of the world, Southeast Asia, the Middle East, and North America. He is a certified cybersecurity technologist and a senior member of IEEE, IAEEEE, IASED, and IEDRC, a fellow member of ISAC, WGM of IFIP TC11, and a member of CSIAC, ACT-IAC, and AASHE. Hamed has been an active multidisciplinary researcher and R&D specialist involved in several academic and industrial research projects. Currently, he is involved in several multidisciplinary research projects, including studying innovation in information technology, blockchain, cybersecurity, people's behavior, and technology acceptance.

Preface

Blockchain technology is changing many different industries, and as it becomes more widely used, many important questions regarding privacy, security, and trust will arise. This preface functions to provide an overview of a compilation of scholarly articles that explore the aforementioned inquiries, presenting novel perspectives, evaluations, and resolutions to the obstacles and prospects that arise from blockchain technology.

Adoption of blockchain technology, security issues and concerns, smart contract accuracy, assaults and defense tactics, consensus algorithms, data privacy protection, regulatory considerations, and a variety of other topics are covered in the extensive scope of the work. Our aim in delving into these areas is to enhance the field's comprehension of blockchain technology and provide valuable insights to both scholars and practitioners in the field.

Blockchain could revolutionize decentralized systems and digital transactions, but privacy and security need to be considered. Therefore, thorough research and analysis are needed to identify and mitigate blockchain technology adoption risks. This Special Issue supports the development of reliable, safe, and privacy-preserving blockchain solutions and contributes to the conversation. By sharing our findings in these papers, we hope to engage stakeholders and improve blockchain security and privacy understanding.

Dr. Taherdoost, the Special Issue Guest Editor, oversees contributions from academia, industry, research institutions, and policymaking. Alongside Bhadoria et al., who are well-known for their expertise in secure online transactions and cryptography, Taherdoost and Madanchian provide an organized review of how blockchain technology impacts business models. Their innovative proposal for the storage of vehicle network data involves Sangeeta and Nam; the ICT security and configuration management experts Chatziamanetoglou and Rantos; the network performance evaluation specialists Eltahlawy et al.; Du and colleagues, who bring experience in identity authentication and cloud computing; Song and colleagues, who are well known for their knowledge of carbon trading and payment channels; Qiu and colleagues, who specialize in zero-knowledge-proof technology and smart contracts; Hajian Berenjestanaki and colleagues, who concentrate on electronic voting systems; and Baldauf and colleagues, who specialize in Ethereum development. Together, these writers add insightful perspectives, thoughtful evaluations, and creative solutions to the problems and possibilities that blockchain technology raises, enhancing the conversation about blockchain security and privacy.

We appreciate everyone who helped make this scientific work possible. The writers' thoughtful contributions and dedication to blockchain privacy and security are appreciated. We also thank editors, reviewers, and colleagues for their support during publication. Their input helped shape this work and ensure its quality and applicability.

Hamed Taherdoost
Editor

electronics

MDPI

Editorial

Blockchain Innovations, Applications, and Future Prospects

Hamed Taherdoost [1,2,3]

1 Department of Arts, Communications and Social Sciences, University Canada West, Vancouver, BC V6B 1V9, Canada; hamed.taherdoost@gmail.com
2 Research and Development Department, Hamta Business Corporation, Vancouver, BC V6E 1C9, Canada
3 Q Minded | Quark Minded Technology Inc., Vancouver, BC V6E 1C9, Canada

1. Introduction

This Special Issue delves into the diverse applications of blockchain technology, spanning topics including democratic elections, business models, secure data storage, and large-scale ICT security. From MANET performance to cloud computing authentication challenges, this collection of papers covers innovative solutions for carbon trading and car insurance. These papers collectively showcase the evolving landscape and promising future of blockchain technology application across various domains.

Blockchain is a decentralized ledger system that utilizes cryptographic algorithms to guarantee tamper-proof, secure, and transparent transactions. Network integrity can be improved through the use of consensus mechanisms, and the technology's adaptability extends to smart contracts for automating agreements. Blockchain has the potential to revolutionize sectors such as healthcare, finance, supply chain management, and governance by providing a secure, streamlined, and intermediary-free method for conducting digital transactions.

This Editorial provides a complete overview of the Special Issue titled "Advancement in Blockchain Technology and Applications" by combining and contextualizing the featured papers' different contributions. This Editorial highlights trends, patterns, and important developments in blockchain's applications in improving democratic processes, business models, data security, and network performance while addressing obstacles and unanswered concerns. This Editorial also guides readers from various fields by providing insights into potential future research avenues, encouraging a deeper awareness of blockchain's cross-sector impact.

Citation: Taherdoost, H. Blockchain Innovations, Applications, and Future Prospects. *Electronics* **2024**, *13*, 422. https://doi.org/10.3390/electronics13020422

Received: 15 January 2024
Accepted: 16 January 2024
Published: 19 January 2024

2. Blockchain Applications in Various Sectors

Blockchain technology has affected several industries, with each industry having its own issues and opportunities. Bhadoria et al.'s (Contributor 1) article proposed blockchain-based traceable certificates for use in democratic elections to improve fairness and competition. The technique used a distributed digital ledger with strong encryption methods to record transactions securely, transparently, and in a tamper-proof manner, boosting democratic transparency and voter privacy. This study added to the wider discussion on using blockchain to secure democratic elections worldwide.

Beyond elections, Taherdoost and Madanchian (Contributor 2) undertook a systematic analysis of blockchain's role in creating new business models. This comprehensive study of 75 articles from the last decade showed how blockchain technologies like NFT and P2E might revolutionize corporate strategies and models. The study examined blockchain-based business models and identified research gaps and interesting possibilities. The study also shed light on blockchain's commercial applications by focusing on journals and utilizing particular selection criteria.

A decentralized InterPlanetary File System (IPFS) and blockchain-based solution designed by Sangeeta and Nam (Contributor 3) addresses vehicle network data storage issues. Recognizing the importance of CCTV cameras and black boxes in road safety, the authors proposed a cost-effective solution that coupled blockchain security with IPFS's decentralized file-sharing protocol. The proposed system provided transparency and data integrity, using keyword searches for sensitive data retrieval. This paper addressed vehicle network security and data integrity issues, offering a decentralized and efficient blockchain storage system.

Chatziamanetoglou and Rantos (Contributor 4) stressed the importance of security configuration management in the design of ICT systems. The article proposed a permissioned blockchain-based mechanism to maintain system configuration integrity throughout its lifecycle. The authors provided smart-contract-based and role-based access control and examined permissioned blockchain models' security configuration management benefits and problems. This article highlighted the need for common techniques and blockchain solutions to safeguard large-scale ICT infrastructures and systems in many sectors.

3. Blockchain Technologies and Performance

Within the domain of network performance evaluation, Eltahlawy et al. (Contributor 5) methodically examined the obstacles encountered in Mobile Ad Hoc Networks. Given the absence of centralized infrastructure, these networks are distinguished by their dynamic node formations, requiring a nuanced understanding of environmental parameters. The study scrutinized 50 recent publications to showcase the widespread use of the NS-2 simulator in MANET studies. The study illuminated the critical factors impacting performance, offering a thorough analysis of simulation environments to facilitate reliable assessments of MANET efficacy, especially in adversarial environments.

Du et al. (Contributor 6) presented the hyperledger fabric identity authentication (HIDA) protocol as a solution for security concerns inherent in conventional authentication approaches used in cloud computing authentication. To accommodate the revolutionary effects of cloud computing on resource accessibility, secure authentication channels within trusted domains are required. HIDA implements zero-knowledge-proof technology and federated chain technology, thereby enhancing the security of user data and access efficiency. The protocol's effectiveness was confirmed by performing formal semantic analysis and simulations, providing novel approaches for identity authentication in cloud computing applications.

4. Innovations and Challenges in Blockchain Implementation

Song et al. (Contributor 7) discussed novel approaches for enhancing the efficiency of high-frequency carbon-trading procedures with regard to carbon trading and payment channels. The utilization of blockchain's intrinsic characteristics in their multi-factor routing payment Scheme (MFPS) increased transaction success rates and decreased processing costs. The proposed asymmetric time-lock contract (ATLC) protocol exhibited superior computational verification and safeguarded against potential assaults, thereby ensuring security and privacy.

To rectify the inefficiencies inherent in conventional automobile insurance, Qiu et al. (Contributor 8) introduced an innovative approach that integrated smart contracts, blockchain, and zero-knowledge-proof technology. Privacy preservation was prioritized during the design process by incorporating private smart contracts for insurance creation and revocation, as well as public contracts utilized for authorization and validation. The effectiveness of the ZoKrates technical implementation approach for off-chain zero-knowledge proofs in terms of minimizing blockchain data storage and computation was highlighted.

Regarding electronic voting systems, Hajian Berenjestanaki et al.'s article (Contributor 9) offered an exhaustive analysis of the effects of blockchain technology on elections. Although the study emphasized key advantages such as transparency and security, it also detected deficiencies in some areas, including usability and accessibility. This segment examined the obstacles associated with and consequences of using blockchain technology in electronic voting, providing valuable perspectives regarding the present status of scholarly inquiry and possible future research directions. Baldauf et al.'s (Contributor 10) final contribution concerned Ethereum development strategies, specifically focusing on the compilation of smart contract programming best practices that ensure security and efficiency. It is crucial to prioritize code quality and security to successfully navigate the ever-changing Ethereum landscape.

5. Perspectives

This Special Issue comprises a diverse array of viewpoints regarding blockchain technology, as each paper provides distinct and valuable contributions regarding its practical implementations and advancements. This collection encompasses various topics, including an analysis of blockchain technology's effects on democratic elections, business models, data storage, network performance, and authentication in cloud computing. Additionally, our scholarly articles explore the potential of blockchain technology to streamline financial transactions, as demonstrated by recent advancements in carbon-trading payment channels and privacy-preserving automobile insurance claims. This discourse is enhanced by a pragmatic approach to Ethereum development and a critical evaluation of blockchain-based electronic voting systems. Collectively, these contributions will expand our understanding of blockchain technology's adaptability while also offering pragmatic resolutions and strategic counsel, thereby emphasizing the potential of this technology to revolutionize entire industries and sectors.

Conflicts of Interest: The authors declare no conflict of interest.

List of Contributions:

1. Bhadoria, R.S.; Das, A.P.; Bashar, A.; Zikria, M. Implementing Blockchain-Based Traceable Certificates as Sustainable Technology in Democratic Elections. *Electronics* **2022**, *11*, 3359.
2. Taherdoost, H.; Madanchian, M. Blockchain-Based New Business Models: A Systematic Review. *Electronics* **2023**, *12*, 1479.
3. Sangeeta, N.; Nam, S.Y. Blockchain and Interplanetary File System (IPFS)-Based Data Storage System for Vehicular Networks with Keyword Search Capability. *Electronics* **2023**, *12*, 1545.
4. Chatziamanetoglou, D.; Rantos, K. Blockchain-Based Security Configuration Management for ICT Systems. *Electronics* **2023**, *12*, 1879.
5. Eltahlawy, A.M.; Aslan, H.K.; Abdallah, E.G.; Elsayed, M.S.; Jurcut, A.D.; Azer, M.A. A Survey on Parameters Affecting MANET Performance. *Electronics* **2023**, *12*, 1956.
6. Du, Z.; Jiang, W.; Tian, C.; Rong, X.; She, Y. Blockchain-Based Authentication Protocol Design from a Cloud Computing Perspective. *Electronics* **2023**, *12*, 2140.
7. Song, Y.; Xiong, A.; Qiu, X.; Guo, S.; Wang, D.; Li, D.; Zhang, X.; Kuang, Y. A Blockchain-Based Method for Optimizing the Routing of High-Frequency Carbon-Trading Payment Channels. *Electronics* **2023**, *12*, 2586.
8. Qiu, Z.; Xie, Z.; Jiang, X.; Ran, C.; Chen, K. Novel Blockchain and Zero-Knowledge Proof Technology-Driven Car Insurance. *Electronics* **2023**, *12*, 3869.
9. Hajian Berenjestanaki, M.; Barzegar, H.R.; El Ioini, N.; Pahl, C. Blockchain-Based E-Voting Systems: A Technology Review. *Electronics* **2024**, *13*, 17.
10. Baldauf, M.; Sonnleitner, E.; Kurz, M. Exemplary Ethereum Development Strategies Regarding Security and Gas-Saving. *Electronics* **2024**, *13*, 117.

electronics

MDPI

Article

Exemplary Ethereum Development Strategies Regarding Security and Gas-Saving

Manfred Baldauf , Erik Sonnleitner * and Marc Kurz

Department for Smart and Interconnected Living, University of Applied Sciences Upper Austria, Softwarepark 11, 4232 Hagenberg, Austria; manfredbaldauf@gmail.com (M.B.); marc.kurz@fh-hagenberg.at (M.K.)
* Correspondence: erik.sonnleitner@fh-hagenberg.at; Tel.: +43-50804-22823

Abstract: Ethereum is a rapidly evolving blockchain with new features as well as new vulnerabilities being introduced regularly. Interaction with the network is costly compared to other blockchains or traditional software systems. When starting to develop on Ethereum, a supported smart contract programming language needs to be learned, most notably Solidity. Having various pitfalls raises the question of what the best practices for the safe and efficient usage of Ethereum are. This study primarily aims to combine knowledge from existing research resources, while also introducing new approaches learned from practical smart contract development analysis and inquiry, which are subsequently compiled into lists of best practices. The most important findings are that code quality and security should be prioritized. Moreover, some simple gas-saving strategies can help to decrease interaction costs with little effort.

Keywords: smart contracts; gas; gas saving; Ethereum; solidity; NFTs; research synthesis; best practices

Citation: Baldauf, M.; Sonnleitner, E.; Kurz, M. Exemplary Ethereum Development Strategies Regarding Security and Gas-Saving. *Electronics* **2024**, *13*, 117. https://doi.org/10.3390/electronics13010117

Academic Editor: Hamed Taherdoost

Received: 29 November 2023
Revised: 18 December 2023
Accepted: 25 December 2023
Published: 27 December 2023

1. Introduction

The problem with Ethereum's smart contract security issues stems from multiple sources. As summarized by Luu et al. [1], smart contracts operate in a permissionless network where anyone can join and they can hold large monetary value. On top of this, the code is publicly available and fixing errors is difficult due to the immutability of the network. Secondly, Ethereum is the largest Turing-complete blockchain by market capitalization. Interacting with the network is costly because of a maximum throughput limit and the high price of Ether. Ethereum commonly gets congested and experiences high gas prices [2]. The importance of best practices is also heightened, due to multiple reasons. The execution environment is more unfamiliar to developers. The code is run by a global network of anonymous nodes. The Ethereum stack is under constant, fast-paced development. New security issues are therefore found regularly [3–5]. Lastly, the saying "move fast and break things" that is sometimes used in software development is not applicable due to the immutability of Ethereum and its limited patching functionality [6].

From ten blockchain-specific vulnerabilities that were found by Chen et al. [7], nine can be avoided by applying best practices. They also note a mismatch between high attack efforts and trailing defence efforts. In interviews conducted by Zou et al. [8], 70% of interviewees answered that guaranteeing the security of smart contracts is hard. One reason that was mentioned is the lack of missing best practices. This study seeks to combine best practice knowledge by leveraging previous research, the knowledge of the Ethereum community and own findings. The aim is to create a best practice list that is easy to follow and that contains the most important information.

1.1. Research Questions

To improve the current situation, a set of research questions was formulated. These tackle parts of Ethereum development where an improvement could have a positive impact.

The following are the research questions that this study tries to answer, which are answered in Section 6.

- *RQ1:* What are general best practices that should be followed for developing on Ethereum?
- *RQ2:* What best practices help to defend against major attack vectors used to hack smart contracts?
- *RQ3:* Which coding techniques help save gas? Are there negative side effects when using those techniques?

1.2. Summary

The results of this study are a best practice list for the sections of general philosophy, security and gas saving. Each section contains the most important parts, and the entries for each topic range from four to ten recommendations. An important finding was the priority of security over other best practices. Without a security focus, a smart contract is vulnerable to many attacks inexperienced developers are not aware of. The gas savings were separated into two lists, as not all are always preferable and depend on the context. If the list can improve the knowledge of other Ethereum developers and how they interact with the network will be seen in the future. A limitation is that it is hard to define the list as complete, as new features are continuously introduced and new security issues are discovered. The best practices could be further extended to contain best practices for other programming languages of Ethereum and focus on other parts of the blockchain, like decentralized finance (DeFi) or gaming. Another extension could be to include other blockchains and compare the best practices to the ones explained here.

1.3. Paper Outline

Section 2 gives some information on the background of Ethereum. In Section 3 the related work is introduced. Publications from different areas of best practices are shown and their differences and contributions to this study are explained. Section 4 focuses on the methodology used for finding further best practices through hands-on experience. Afterwards, the best practices from the research and the methodology have been combined into the best practice list. This list is depicted in Section 5 and its points are elaborated on. Lastly, the research questions are revisited in Section 6 and implications, limitations and future work are presented in Section 7.

2. Background

The background section is a collection of some important Ethereum information related to the findings of the paper.

2.1. Gas Fee

Since the Ethereum Virtual Machine is Turing-complete, code can be executed that includes nondeterministic code. Infinite loops could be utilized by malicious users or unintentionally occur due to a programming error. The whole network could collapse, as the computation would not come to a stop. The problem is known as the halting problem [9], which says it is not possible to predict if a given piece of code will terminate or not. To tackle this problem, a fee for using the network is introduced [10].

The following list contains the most important fee-related terms [11]:

- *Gas:* this fee is a unit of measurement for computational resource consumption and is referred to as gas.
- *Gas limit:* the user sets a maximum amount of steps that the code is allowed to run with the gas limit. If the code does not terminate before all gas is used, the transaction is reverted, but the fee still has to be paid.
- *Gas price:* this is the price per unit of gas used during a transaction. Most commonly, Gwei is used to price gas.

- *Base fee and priority fee:* the fee that is required to be paid to be able to be included in a block. The current base fee is calculated by the previous block and it can only grow a maximum of 12.5% per block. The base fee set for the block is getting burned and only the priority fee is being paid to the validators. The higher the tip, the more likely a transaction is included before other transactions.

Per block, there is a maximum throughput that the network can handle. If many participants want to interact at the same time, the network gets congested and the base fee is increased. The users outbid each other to get included in the network [11].

2.2. Solidity Optimizer

There are different types of optimizer programs that reduce the size of the code and make it more efficient. Optimizer tools are trying to find optimizations on Solidity code, bytecode or an intermediate language like Yul [12–16]. The tools then try to optimize the respective code representations and create a smaller code for deployment. Solidity has its own optimizer included in the Solidity compiler. There are two optimizer modules titled "old" optimizer and "new" optimizer. The "old" optimizer is opcode-based and has simplification rules to change opcodes. Moreover, dead code is removed, and equal code sets are merged. The "new" optimizer is Yul-based which allows it to optimize across functions, according to the Solidity documentation. The optimizations improve the cost of calling a function and deploying the contract to the blockchain [16].

2.3. Blockchain Security

Security on Ethereum is crucial for the safe usage of the network. Security is required in multiple parts of interacting with the network. It requires users and developers to be security aware. Problematic are the two attributes permissionless and immutability, according to Chen et al. [7]. Those features allow malicious users to attack at will, and no one can normally change the immutability of the blockchain. The DAO hack [17] was an extraordinary circumstance where the Ethereum Foundation overrode a smart contract, which led to a fork in Ethereum [18]. More on overcoming immutability can be seen in Section 2.4 and the trade-offs involved are reviewed in Section 5.2.2.

Alchemy has created a simulator API that can simulate the changes that would happen when authorizing a transaction. This makes it easier for users to see changes without having to look at the code. Not only users have to be careful, but also developers—they need to secure their code against attacks [19] and bugs. A more thorough survey regarding smart contract vulnerabilities, exploits and countermeasures has been proposed by Kjiam et al. [20]. The steps recommended for developers to improve the safety of Ethereum are covered Section 5.

2.4. Immutability and Proxy Contracts

Smart contract code that has been deployed on Ethereum is immutable. This makes bug fixes on deployed code impossible. This is a negative for security because fixing bugs is seen as a standard development practice. However, one can get around the immutability by deploying multiple smart contracts containing the code and having a proxy contract that links to the implementations of the contracts. The addresses linking to the other contracts are stored in the modifiable storage of the proxy contract [10,11].

Amri et al. [21] say that the number of proxies increased in the last couple of years and the transparent proxy increased significantly. They are unsure whether overcoming immutability has a positive or negative impact. Salehi et al. [18] report a growth in proxy upgradability contracts on Ethereum as well. Salehi et al. further identified changes with the terms retail changes and wholesale changes. Retail changes are made as small changes to components inside a smart contract. Those contracts need to be set up beforehand to be able to handle changes. The simplest retail change is a parameter that can be updated with a setter method. The second retail change is when a contract has a function that calls another contract with the logic for that function. Here, the contract can store the address in a changeable variable and therefore the logic of the function can be changed or bugs can be fixed. On the other hand, wholesale changes enable the modification of entire contracts. The simplest wholesale change is a so-called "social upgrade" or contract migration, where a new contract is deployed and everyone is informed about that change. Another way to upgrade a whole contract is via CREATE2-based Metamorphosis. This allows for deleting a contract and redeploying a new one at the same address. However, the required opcode SELFDESTRUCT is expected to be deprecated in a future version [22]. Lastly, the DELEGATECALL-based or CALL-based data separation is another pattern found by them. Those two patterns both consist of a proxy contract, a storage contract and a logic contract. Here, the logic contract is changed to handle updates.

2.5. Ethereum 1.0 vs. 2.0

Most publications related to gas-saving or smart contracts in general refer to Ethereum 1.0. It is, however, important to note that the core virtual machine which eventually runs compiled smart contract code, the Ethereum Virtual Machine (EVM), is used by numerous other blockchains e.g., Binance Smart Chain, Polygon (formerly Matic) and Avalanche. As a general rule, all open permissionless blockchains allowing the execution of smart contracts always rely on some kind of metric for making execution steps costly in order to prevent overtaxing or exploiting the network [23]. For EVM-based blockchains, gas is almost exclusively used for such a metric.

Ethereum itself is currently undergoing a far-reaching and architecturally challenging change, which is commonly referred to as Ethereum 2.0. This change includes significant modifications to the overall network design, most notably switching from the energy-intensive proof-of-work consensus to the much more economic proof-of-stake scheme. This also includes introducing a new blockchain solely for coordination, the Beacon chain, as well as sharding concepts for scalability. The transformation towards Ethereum 2.0 is happening in multiple phases. It is still in progress as of the end of 2023 and is expected to continue for another 5–10 years [24].

The Ethereum organization claims that around 99.95% less energy is consumed by the network now. This is in line with Kapengut and Mizrach [25], as they state that the consumed energy by the Ethereum network was reduced by 99.98% with the switch. Some misconceptions were that the network would be faster afterwards and that the gas fee would become lower. However, the change was only for how to reach consensus and not for improving the performance or capacity of Ethereum [11]. Generally, the amount of gas required for any particular smart contract to run will remain similar, even though the variability of effective costs for such transactions still remains a significant factor and is subject to change.

2.6. Oracles

While smart contracts are self-executing, deterministic entities, they do not have direct access to data outside the blockchain ecosystem they are running in. Oracles are third-party services that can provide smart contracts with such information as, e.g., stock data or game results [26]. Even though Oracles may drastically widen the potential areas of application regarding smart contracts, they are typically represented by a central authority, disrupting the claim of trustlessness of many blockchains and introducing a single point of failure or manipulation [27]. From a security point of view, an attack known as Oracle manipulation aims at manipulating Oracle answers in order to invoke unintended execution flows in smart contracts. Kjiam et al. [20] discuss countermeasures against this attack, which primarily revolve around using time-weighted average values when processing Oracle data, and instantiate an M-of-N Oracle approach rather than only relying on a single Oracle provider. Oracles can directly (service fees) or indirectly (transaction fees) increase the amount of gas required, but they are not inherently relevant to the topic of gas savings and hence disregarded for this study.

3. Related Work

This section presents an overview of the related literature. Each work is briefly described and its differences compared to this paper are highlighted. They are sorted into the different categories of the best practices list. First, the publications for general best practices are presented and we continue with security recommendations and gas-saving patterns. Next, related works for tools assisting with security and gas optimization are introduced. The last part is about token standards, especially NFT standards.

3.1. General Best Practices

The related works in this section are two-sided, with software engineering recommendations and suggestions specifically tailored to blockchain development and Solidity.

Jones [28] examined over 600 companies as well as government organizations and created best practices for software engineering, ranked by an average score. The biggest expenditures were located and put into a list, with the four most expensive ones being bug fixes, cancelled projects, documentation and security flaws. In 6th place is coding and in 14th place is avoiding security flaws. Comparing it to smart contract development, the security aspects could be even ranked higher, as they have the highest emphasis, according to Zou et al. [8]. Jones, moreover, argues about two important points. Firstly, that software varies greatly in size and therefore the best practices vary as well. Smaller projects might gain more from some practices than larger projects and vice versa. Secondly, software engineering often does not have a universal way of proceeding with different types of software. For instance, open-source applications, military applications and games all have different requirements. This is equally true for Solidity, where the range of functionality can vary greatly. The ConsenSys best practice list [29] uses a similar philosophy and gives examples of when to use different patterns over others. Comparing the resulting best practice list of Jones [28] with the results of the best practices for Solidity from other works [8,29,30] indicates some relations. Reusability is ranked at the first position in the list and it is also an important part of smart contracts. ConsenSys [29] states that maximizing the reuse of code wherever reasonable is desirable and the safest way is to reuse own code.

Martin [31] argues in his book Clean Code about how to write better code. He starts with general clean code and continues with discussing naming conventions, functions, classes, comments and other topics of his best practices. He further urges testing the code heavily, which is highly recommended for Ethereum as well [29,32]. James O. Coplien says in Clean Code [31]:

> "Code is never perfect."

This quote should alert Ethereum developers of the implications for blockchain development. ConsenSys [29] predicts something similar and says that there will be bugs in any non-trivial contract. ConsenSys recommends having the option to pause a contract, manage rate limits and have upgrade paths for enhancements and fixing bugs. In general, Clean Code is a helpful general advisor for the best practice list.

ConsenSys [29] documentation for Ethereum smart contract best practices explains how to write good and safe Solidity code according to their standard. The documentation is set up as a git repository and currently has more than 6600 stars, making it their most starred repository, followed by their developer tool list for Ethereum with around 4600 stars. Their recommendations are security-related and they target intermediate Solidity programmers [29]. The collection begins with general security mindset, development recommendations and code patterns. Furthermore, known attacks, how to avoid them and security tools helping to detect vulnerabilities are explained. Their work is a great resource for developers looking for best practices, as can be seen by the popularity of their GitHub repository. ConsenSys's work still differs from this paper, as they do not go into detail on gas savings and do not contain all recommendations given in this paper.

Antonopoulos and Wood [32] present a great starting resource for Ethereum development. They start with an introduction to Ethereum for beginners and afterwards explain more technical topics suited for developers. The more advanced topics contain Ethereum clients, wallets, transactions, the Ethereum Virtual Machine and consensus algorithms. They also explain smart contracts with Solidity and security-related best practices. Their work is a great resource for Ethereum developers, but it is partly more focused on giving a broad overview of blockchain technology while neglecting more specific topics like gas savings and the Solidity optimizer. Furthermore, the development of Ethereum is fast-paced and some vulnerabilities and facts are not up-to-date anymore. For instance, the consensus mechanism of Ethereum changed. However, they remark that the consensus mechanism could be changing. Therefore, this paper will be an extension of points made in their work.

3.2. Security

The importance of security is emphasized by Zou et al. [8]. In their work, they interviewed 20 people with Solidity experience and did a validation survey with 232 respondents to see challenges and opportunities in the field. A total of 75% of the respondents agreed that the code security required for smart contracts is a lot higher than for traditional software. The three major reasons are the sensitive information being handled, the irreversibility of transactions on the blockchain and that the code is not modifiable after deployment. In addition, around 70% of the developers answering the survey said that it is hard to guarantee security. They mention the ConsenSys [29] documentation but argue that following this guide still is not enough for the requirements of developing smart contracts. This is where this publication tries to improve the situation by creating and collecting best practices for developing with Solidity on Ethereum.

Kushwaha et al. [33] conducted a review of Ethereum vulnerabilities. In their review, they present different attacks, their underlying causes and state which tools were able to find them. They did not state which tool is best to use, but are more concerned with improving the tools, especially when new features are introduced. In their work, they do not focus on any of the other best practices categories like gas saving or ERC standards. Furthermore, they do not discuss the security implications of proof-of-stake, as their publication was before the transition to proof-of-stake. Therefore, this study is a great guide for the security part of our proposal, with additional information being added.

Chen et al. [7] provide a survey for the system security of Ethereum. The thorough survey shows vulnerabilities, their possible attacks and lastly the recommended defences. They classified 40 vulnerabilities at different layers of Ethereum. They found some insights which are valuable and can be compared to the work of others. They argue about the importance of domain-specific best practices, because 10 out of 14 types of vulnerabilities are non-existent in traditional software. Of those 10 types, 9 can be prevented by following best practices. They further state that due to Ethereum being permissionless and immutable, writing safe code is harder compared to traditional software and it is even a security barrier. The attackers can hence attack at will and the often-used mechanism of vulnerability-patching in conventional software is by default not possible. These findings are backed by the earlier presented findings of Zou et al. [8] when interviewing Ethereum developers on why it is hard to write safe code and guarantee security. Code reuse potentially inflicts a greater risk, compared to traditional development systems. Marchesi et al. [34] write that one should use libraries but also remark on potential security issues. ConsenSys [29] states that OpenZeppelin's library for Solidity tries to provide secure code and that reusing one's own code is the safest option. The work of Chen et al. [7] helps improve the security best practices of this proposal.

Wohrer and Zdun [30] elaborate on six smart contract security patterns to mitigate typical attack scenarios. Their patterns try to improve the execution control after a smart contract is added to the blockchain. The patterns focus on allowing the owner to stop smart contract execution or by adding different types of so-called speed bumps to slow down the execution time. Their patterns are a great start to writing more secure code but are not exhaustive. They also observe the fact that a substantial part of the knowledge and the research are scattered and that it can be found in the grey literature and blog articles.

Wang et al. [35] provide a blockchain-based smart contract overview. They show the operating mechanisms, use cases, challenges and basic framework. The two most discussed blockchains in their paper are Ethereum and Hyperledger Fabric. They categorize the problems and challenges, similar to Praitheeshan et al. [36], with contract vulnerabilities and blockchain limitations, but also add legal and privacy concerns. They remark that it is difficult to keep data private and critical methods safe. This study tries to extend the written challenges from Wang et al. [35] and elaborate on security issues.

Destefanis et al. [37] provide a case study on the Parity [38] library for smart contracts In their investigation, they highlight that the vulnerability was already revealed and discussed in earlier literature. They, therefore, call for following best practices and standards in blockchain development, which this work is trying to improve by creating a detailed summary of best practices.

The work of Atzei et al. [39] is, according to them, the first systematic structured composition of Solidity and Ethereum vulnerabilities. Their work aims to help smart contract developers and researchers improve their verification and analysis methods. They created a taxonomy of vulnerabilities with common programming pitfalls and further showed corresponding actual attacks that had been carried out on contracts deployed on the Ethereum Mainnet. Their taxonomy is one of the most established ones, say Dika and Nowostawski [40], who also used it for their vulnerability list. Their work further warns of the security vulnerability leading to the Parity attack before it happened, said Destefanis [37]. Lastly, other scientific works have extended their work, for example Kushwaha et al. [33].

Marchesi et al. [41] provide a security checklist for developers containing 32 best practices and 16 abstract security patterns. The best practices are split into the design phase, the coding phase and the testing phase. This publication further tries to improve their security lists with further best practices from other parts of Ethereum and Solidity development.

Chen et al. [42] collected smart contract defects from more than 17,000 StackExchange posts and defined 20 vulnerabilities for five different aspects. The aspects are availability, maintainability, performance, reusability problems and security. Furthermore, they defined five impact levels ranging from IP1 to IP5, with IP1 being the most critical and IP5 the lowest, where IP4 and IP5 are similar to code smells from Fowler [43] or Martin [31]. These code smells do not cause critical behaviour on their own, but over time can decrease development speed and enhance the risk of bugs being added later on. Their categorization of the impact of defects highlights that vulnerabilities and tips have varying impacts and not all are equally important.

Zhang et al. [44] investigated 516 unique smart contracts deployed to the Mainnet. They created a bug model for the simplification of finding future bugs. They have earned USD 102,660 with their findings in bug bounties. Some of their findings are that around 80% of bugs are not auditable by a machine and that most exploitable bugs are difficult to find. This shows that only using tools is not a feasible option if one is aiming for high security. This serves as a reminder in the best practice list to not trust tools blindly.

3.3. Gas Savings and Design Patterns

This section contains general gas-saving related publications and scientific research that tried to find design patterns that are good for gas savings and antipatterns that can be gas costly.

Marchesi et al. [34] showed in another paper a collection of design patterns for saving gas. In total, 24 patterns were presented in the following five different categories: external transactions, storage, saving space, operations and miscellaneous. Although they also elaborate on many areas for gas savings, they do not explain and emphasize the consequences of some of their presented patterns. For instance, the introduced proxy pattern does well in saving money if one or multiple contracts had needed to be redeployed. However, they do not address the required additional complexity and potential security issues arising from proxy patterns [29].

Research on the mutability and upgradeability of smart contracts on Ethereum was conducted by Salehi et al. [18]. They classified six upgradeability patterns into two types, which they named wholesale changes and retail changes. Different patterns for upgrading are explained in Section 2.4. Furthermore, they developed a framework to detect the number of smart contracts on Ethereum that use a certain type of upgrade pattern. Around 1.4 million contracts with proxy patterns were found by their framework. Interestingly, with 8225 contracts, the number of unique proxy contracts is only a fraction of that. Their work helped identify different proxy contract possibilities.

Kong et al. [45] showed an approach for the detection and optimization of six inefficient patterns on the source code level. The focus therefore is on development issues caused by programmers. A total of 160,000 smart contracts from the Ethereum blockchain were analyzed and their findings were applied. They found that 52.75% of the contracts showed a minimum of one gas inefficiency. They state, without providing a concrete number, that a lot of money could have been saved if gas optimizations had been applied to all 16 million smart contracts that were available at that time. Their work highlights that small improvements can have huge effects on a large scale. Their work is only focused on gas savings and not on other best practices.

4. Methodology

This section contains the project that was set up to test the different aspects of smart contract programming and to gain knowledge about the domain by hands-on practice. First, the conception phase is explained and it is followed by the implementation. Eventually, the results of the project are presented.

4.1. Project Concept

Figure 1 visualizes a comprehensive overview of the methodology process. Each step is discussed in more detail in the section below. The first step was to find the related work, which was explained in Section 3. This provides a solid understanding of Ethereum's fundamentals and highlighted the parts that were less researched. Less researched parts were a comparison of the different NFT standards including their gas efficiency and a best practice collection containing as much about Ethereum as possible, including security, gas savings, development tips and other insights. Following the research, the outcome of this project was set to investigate NFT standards and combine best practices from all domains of Ethereum. The findings were presented on a token-gated website that allows for minting different NFTs. After owning one of the tokens, the findings are accessible.

4.1.1. Learning Tools and Resources

After researching scientific publications, further research was done by visiting Ethereum learning platforms to find further Solidity and Ethereum-related best practices. The following resources are a valuable addition to the Ethereum space. The Ethernaut [46] by OpenZeppelin currently contains 30 Solidity levels that are intended to be hacked. Cryptozombies is another interactive learning platform to study different blockchains with. At the time of writing, there are six free courses available. Solidity by Example is another website to introduce Solidity to new developers. All concepts and ideas are presented with a smart contract example.

4.1.2. Collection of NFT Standards

The most prominent token standards are the ERC-721 and ERC-1155. They have often been mentioned in scientific publications [47–54]. Therefore, the preferred tokens for analysis were selected by popularity, as these tokens are the ones being used and detailed research of their costs was deemed helpful. The standards ERC-721, ERC-721A, ERC-721R, TinyERC-721, ERC-998 and ERC-1155 [11] were chosen for a more thorough investigation, containing use cases, deployment costs, mint costs and transfer costs. Some contracts were again dropped from the list due to time constraints, missing popularity or bugs in their implementation. Lastly, the final selection consisted of the following four standards ERC-721, ERC-721A, item ERC-1155, ERC-721 from Zora. Zora's ERC-721 is not a standard on its own, but they have a website that allows users to create and deploy an NFT collection with little knowledge. Additionally, the contract the user creates is not a standard ERC-721 contract, but a proxy contract containing the user's modified settings and pointing to another contract with the required logic. The main logic of minting, transferring and other parts are deployed on another contract that the proxy contracts call. Hence, minting and transfer costs are interesting, as the underlying mechanism works a bit differently.

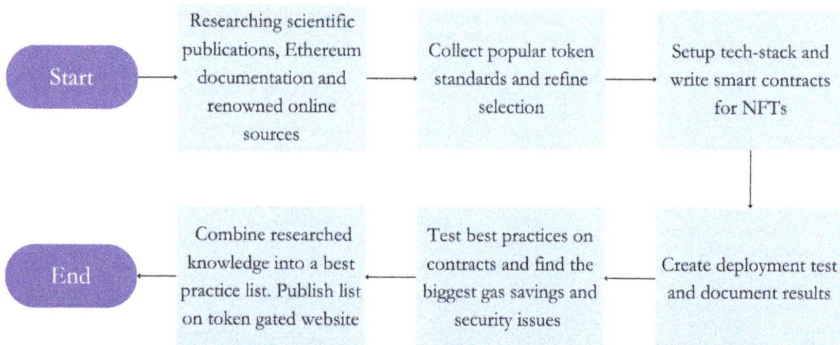

Figure 1. Overview of the methodology.

4.2. Implementation

In this section, the implementation details concerning the smart contracts, the test cases and the audits are explained. The contracts were created to be as similar as possible, with a very limited set of functionalities. Each contract contains at least a mint function that enables minting and a constructor.

The tests of the NFT standards are a continuation of the work of the Azuki team [55] and Hu [56]. The standards were tested with HardHat on gas efficiency and functionality in different domains. These tests contained minting and transferring. The deployment costs were checked on Remix, as it is easily possible to deploy to a testnet from there. The deployment was done twice, first with the default settings, where the Solidity optimizer is turned off, and a second time with the optimizer turned on. The optimizer was used with the default settings of 200 runs.

Applying Best Practices to Deployed Contracts

The collected best practices were not only applied to the standard ERC implementations but also to a few contracts deployed on the Mainnet. The contracts were reviewed manually and then locally redeployed. Each test was adapted towards the contract and the parts that were optimized. All contracts were tested on deployment costs.

4.3. Project Results

4.3.1. Deployment Costs

The deployment costs for the NFT standards are depicted in Table 1. A significant reduction in gas costs can be achieved with the optimizer turned on.

The amount of effort required, in comparison to the gain in these cases, is minimal. In Remix, the optimizer can be turned on with the check of a button. In a setup with Visual Studio Code and HardHat, one can turn on the optimizer by setting a variable to true in the hardhat.config.js file. Lastly, the expected runs of the code can be adapted to easily modify the optimizer. More background explanation can be seen in Section 2.2 and more advanced information on optimizer settings in Section 5.3.1. The ERC-721 contract created by Zora was not adaptable, as one cannot change optimizer settings on their website, albeit, they have the optimizer turned on and the created contract is optimized for 5000 runs. Its small deployment size can be explained by the minimal proxy that was deployed, while the ERC-721 requests are forwarded to their main smart contract containing that logic. This information results in Box 1:

Box 1. Optimizer savings

Insight 1: The optimizer decreases the gas required for the deployment of a basic ERC-721 or ERC-1155 contract between 43.01% and 46.81%.

Table 1. Results of ERC smart contract deployments with and without the Solidity optimizer in Wei.

Contract Type	Deployment Costs without Optimizer in Wei	Deployment Costs with Optimizer in Wei	Savings in Percent
ERC-721 Enumerable	2,782,296	1,501,689	46.03
ERC-721A	1,685,088	960,354	43.01
ERC-1155	2,482,077	1,394,607	46.81
ERC-721 by Zora		662,983	Not modifiable

4.3.2. Deployment Time

There is a significant change in gas prices when the network is congested, state Cong et al. [57]. The fluctuation can also be seen clearly on Etherscan [2], where the average daily gas price in Gwei is shifting heavily over the whole time. Gas prices also fluctuate daily. On the 6th of July 2023, the gas price fluctuated between 19 and 138 Gwei [2]. Interaction with the blockchain, for instance, a transaction, could cost around 725% of the lowest fee. The computation for the maximum price fluctuation can be seen in calculation (1). Notably, this scenario is about the greatest gap during the day, but significantly lower differences in the gas price still have a great impact.

$$138 \text{ Gwei} \div 19 \text{ Gwei} \times 100 = 726.32\% \quad (1)$$

Formula (2), shows how to calculate the gas fee for a transaction. It takes the units of gas used by the desired interaction and multiplies it by the current base fee plus an optional priority fee. By setting a priority fee, the validators are incentivized to include the transaction earlier into the blockchain, as they get to keep the tip [11]. The interaction can be a simple Ether transaction, which requires 21,000 units of gas [58]. There are also more costly contract deployments, for instance, the Bored Ape Yacht Club, the biggest Ethereum NFT project in total trading volume on OpenSea [59], costs 3,893,600 units of gas.

$$\text{Units of Gas Used} \times (\text{Base Fee} + \text{Priority Fee}) = \text{Total Fee} \quad (2)$$

It is possible to set the optional parameter maxFeePerGas. The adapted formula can be seen in Formula (3). Participants then never pay more for the transaction than declared and a surplus is returned to the user [11].

$$\text{Max Fee} - (\text{Base Fee} + \text{Priority Fee}) = \text{Returned Fee} \quad (3)$$

Akiyoshi [60] explains a technique for how to save gas with this parameter. If the network is currently congested and the current base fee is high, the max fee should be set lower to a more normal network status. How low and what a normal network status is can be inferred from Etherscan's Gas Tracker [2]. These gas fees vary at all times. Therefore, stating a value that is best for all times is infeasible. The transaction will stay with the pending status and will be conducted once the network is less busy. It is important to note that the maxFeePerGas ought not to be set too low, as this could end in the transaction never being executed. Furthermore, if the transaction is time-critical and should be added to the blockchain as soon as possible, the strategy should be inverted, meaning a high max fee should be set so that that transaction is preferred by validators over others. This leads to the next finding (Box 2):

Box 2. Network state implications

Insight 2: The cost of deployment and interaction on Ethereum is heavily influenced by the congestion state of the network. Non-urgent interactions with the Ethereum blockchain should be set with a low max fee.

4.3.3. Applied Best Practices

The audits on ERC-721 contracts showed that other gas saving tips [34,45,61] lead to significantly smaller savings. This is consistent with previous work from Kong et al. [45], where 52.75% of researched contracts could be improved by USD 0.30. However, they argue that applying it to all contracts can save a lot of money. The fact that small savings can compound to larger savings is true, yet this paper argues that easily attainable savings should be prioritized first. This is still not done by some, as can be seen on Blockscout [62], where at the time of writing some still did not use the optimizer on deployed contracts.

Moreover, compared to insight 1 and insight 2, other gas savings tips can have a bad impact on readability. In particular, more advanced gas-saving tips can reduce readability. It is not the case that any gas saving automatically decreases the security or adds bugs. However, Martin [31] cautions that obscured domain logic reduces quality, as it allows bugs to hide more easily. These findings are similar to those of related work. Zou et al. [8] write about there being a trade-off between code readability and gas optimizations. The code smells, found by Di Sorbo et al. [61], were generally agreed upon by developers of smart contracts, but the developers raised awareness that some optimizations could harm readability. Small gas savings should not be the reason why a bug or an attack is possible. This leads to the next insight (Box 3):

Box 3. Security vs. gas-saving

Insight 3: Security should be prioritized over other aspects that can harm it, such as gas-saving techniques.

If a contract is, however, used by many users, gas savings should be utilized as much as possible. Similar to ConsenSys [29], this study wants to emphasize that different projects and smart contracts need to be coded differently. A one-time mint ERC-721 NFT contract has different requirements than an OpenZeppelin library. Small gas savings can make a big impact if they are saved in a library that is often used. This shows the next observation (Box 4):

Box 4. Size vs. functionality

Insight 4: Not all smart contracts should be treated the same when it comes to best practices. Smart contracts benefit differently from the best practices, depending on their size and functionality.

4.3.4. Project Limitations

A manual analysis of smart contracts by only one reviewer leaves space for error or missed potential. However, the results are strengthened by their similarity and consistency to other publications mentioned. The application of the best practices on more contracts could have been beneficial, but again were similar to other related work and were aborted due to time constraints.

5. Best Practice Analysis

This section presents the best practices found through research on related work from Section 3 and practical tests discussed in Section 4. The various points are grouped into different categories and together form the whole recommendations list. Each section explains the individual parts and demonstrates the similarities and contrasts to other works. The category structure of the best practices was partly taken from other related works [7,29,41].

5.1. General Blockchain Development Philosophy

The composed best practice list for general development on Ethereum is analysed here. The list is depicted in Table 2 and contains the most important general findings.

Table 2. The list of general principles to follow when developing on Ethereum.

General Principles		
Name	**Description**	**Reference**
Favour Simplicity and Security	Simplicity improves security, gas-saving and testing. Complex smart contracts offer more ways to attack and can create undesired behaviour. Finding less complex ways and requiring fewer features is the recommended approach. More compact contracts are smaller and therefore less expensive to deploy and bugs are more easily detected. Clarity should be prioritised over performance, meaning sometimes gas-saving best practices should be neglected for clearness.	[11,29,31,32]
Code Quality	Similar software development methodologies should be adopted as are used in aerospace engineering. Code in smart contracts is by default immutable and fixing problems after launching is not trivial. Setting up bug bounties and starting to test in early phases, can help find vulnerabilities. It is not advisable to apply the motto "move fast and break things" [6].	[32]
Prepare for Failure	Non-trivial smart contracts should be expected to contain bugs and vulnerabilities. Adding a function to pause contracts and including buffers like minimum wait time for execution can help to be able to respond to problems gracefully.	[29,30,32]
Adapt the Strategy	The size, functionality and predicted usage of a contract are crucial for determining what to optimize it for. A short-lived contract can be less concerned about gas usage and upgradeability than a library.	[28,29]

5.1.1. Simplicity

The first recommendation is to prefer simple contracts over complex ones. This should be done with clean code and small contracts. Antonopoulos and Wood assert [32]:

"Complexity is the enemy of security."

This quote explains why one should strive for simplicity. Simplicity helps to make code more secure, by making it easier to understand the program flow, the intention behind the code and making it harder for bugs to hide [31]. Security is a primary concern for programming on Ethereum, claim Tikhomirov et al. [6]. Alchemy [63] adds to this rule by reminding that it does not imply to avoid creating smart contracts with many features. Its core message is to start from a simple standpoint and build more complex functionality with high code quality from there.

5.1.2. Expectation of Failures

The next principle is to prepare the code for failure. Wang et al. [64] say that it is impossible to claim that code has no bugs. They further state that even highly experienced teams that write contracts and audit them are not exempt from that rule. In conventional software, it is normally easier to stop, pause or remove your service when it is malfunctioning. On Ethereum, that is by default not possible, which is yet another reason why preparing for failure is important. To get some functionality to assist with this problem, one of the following mechanisms could be used:

- An upgrade mechanism for smart contract;
- A pause functionality of critical functionality;
- Some limitation rate to reduce maximum usage or withdrawal.

Upgrade mechanisms help improve security and can help save gas. Therefore, this tip is further analyzed in the security Section 5.2 and the gas-saving Section 5.3. A pause functionality can be helpful to pause withdrawals or other important parts of a smart contract in the event of an unexpected behaviour occurring on the smart contract. Lastly, limiting the usage is another solution to have more time to respond to undesired interactions.

5.1.3. Strategy Adaptation

The recommendation to adapt the strategy is true for general software development and for Ethereum development. There is often no one-size-fits-all technology, technique or "silver-bullet", as stated by various sources [7,28,29,43,65]. The strategy should be first adapted from conventional software engineering to blockchain development. Secondly, adaptations between different smart contracts should be undertaken as well. Some unconventional programming patterns should be used on Ethereum, which is due to its blockchain characteristics [30]. The quote from Maslow explains what happens if one does not have different strategies to choose from [66]:

> "I suppose it is tempting, if the only tool you have is a hammer, to treat everything as if it were a nail."

Different requirements require different solutions. Adapting the strategy also means implementing different functionalities at different times.

5.2. Security

The problem with security on Ethereum and other public blockchains is a combination of circumstances. Generally, security [8] and bug-free code [64] are hard to guarantee. Furthermore, attackers can attack whenever they want, since Ethereum is permissionless [36]. Moreover, contracts can hold high monetary value, making it attractive for adversaries to attack [1]. The situation is improved in permissioned blockchains, but according to Staples et al. [67], not even private blockchains are potentially private enough. They argue that private blockchains share information between their nodes and if competitors are in the same network, they have the potential to attain sensitive information about their competitors. Developers need to therefore prioritize security and make Ethereum safer. Table 3 contains guidelines to improve safety. Although it is not feasible to get rid of all security issues, since new attack vectors can arise at any time, it is at least possible to protect oneself against known attacks.

5.2.1. Code Reuse

Reusing code is a common practice and recommended by many in the Ethereum domain [11,29,41,68]. Antonopoulos and Wood [32] emphasize that reusing trusted code is possibly the most fundamental principle for security. They further warn that reinventing the wheel with new code is likely going to be less secure than extensively tested and used libraries. There is a saying in cryptography for that: "Don't roll your own crypto". However, it is noteworthy that Chen et al. [7] remark that the reuse of code on Ethereum possibly carries a higher risk than in traditional systems and therefore reused contracts should be security audited. In summary, it can be said that well-tested and widely used libraries can and should be used. OpenZeppelin [69] is a prominent example that offers a variety of such smart contracts that should be reused.

Table 3. The security best practices list.

Security Best Practices		
Name	**Description**	**Reference**
Reuse Code and Libraries	Well-tested and widely used libraries are good for reuse and most often outweigh the benefits of creating everything from scratch. The more complex the code is, the better it is to rely on audited libraries.	[11,29,32,41]
Contract Upgrades	For more complex and long-lived smart contracts, a proxy pattern or another upgradeable pattern is advisable to fix critical flaws that could render a contract useless or insecure. This reduces gas costs, as otherwise, multiple contracts would need to be redeployed. The security implications are two-sided. It enables bug fixing but also introduces new security vulnerabilities.	[18,41,69]
Check-Effect-Interact	It is difficult to check if code could be used differently than anticipated. As a defensive safety mechanism, the general order of the Solidity code should be, to check the condition first. Afterwards, update the checked state and only, as a last step, should an external call happen. Otherwise, reentrancy attacks are possible.	[7,16,32]
Pull Over Push	There are two possible ways to send ETH. A smart contract can be sent to other accounts or have the ability for other accounts to actively pull the resource. For external calls, it is advisable to implement a pull mechanism. They should be separated into their own function that the user calls. Otherwise, a denial of service with an unexpected revert could happen.	[7,29,40,70]
Readability	Clear and simple code is easier to check for bugs and better for audits. Contracts should be built with open-source methodologies and collaborations. Documenting code and following the styling and naming conventions further help.	[16,29,32]
Tokenization	Using token standards is highly recommended and each has different benefits and side effects. Prominent tokens are explained in Section 5.2.5.	[32,41,51,71]
Testing	High test coverage is recommended as smart contracts are executed in a public environment and anyone can interact with them. Make use of the test networks and test over multiple stages during the development.	[6,11,29,32,37,41]
Stay Up-To-Date	Always use the current stable versions of Solidity. All tools and the libraries used for development should be up-to-date. When new security techniques are introduced and vulnerabilities are discovered, check if they concern your smart contracts.	[29,37]
Interact With Ethereum Securely	Safe general interaction with Ethereum is key to avoid pitfalls. These include key management and being mindful of scams. Due to Ethereum's decentralized nature, it is not possible for a centralized authority to help in the case of key loss.	[11]

5.2.2. Upgradeability Trade-Off

Generally, blockchain-based software benefits from the immutability aspects, but some degree of modification is necessary for feature improvements and fixing problems, says OpenZeppelin [69]. If a vulnerability is found on production smart contracts, the desired action is to fix the problem. However, circumventing immutability and adding upgradeability bring about their own risks [29]. OpenZeppelin [69] report that upgradable proxies are difficult to use securely and correctly. They discuss the problem that both the proxy contract and the contract containing the logic access the same state variables and the same risk of overwrites. Furthermore, a programming overhead and increased complexity is added when including proxy logic.

The recommendation is for bigger projects and contracts that are used over a longer period should make use of upgradable patterns. There is a variety of possible solutions. For instance, Meisami and Bodell III [72] found 11 different patterns for upgradable proxies. Small and short-lived projects might work better without the added overhead.

5.2.3. Reentrancy and Check-Effect-Interaction Method

Reentrancy means that during the execution of the code, the executed smart contract is called multiple times by an attacker contract during one transaction. The first reentrancy attack was used during the DAO hack [29]. One possible scenario of reentrancy is to drain the balance of a smart contract. Reentrancy is the vulnerability that has been investigated most extensively, with 17 defence mechanisms developed, says Chen et al. [45]. Many defence mechanisms aid with this problem. A simple one is the recommendation "Check-Effect-Interact" of Table 3.

5.2.4. Pull over Push

There are two options to make an external call for a ETH transaction. A smart contract can try to push the ETH or have a pull mechanism that is activated by the other account that should receive the ETH. If a transaction is required to go through, an implementation that uses a push mechanism is at risk. Either maliciously or unintentionally, users can bring the smart contract to a halt. If the receiving account is a contract, it could, as an example, revert when receiving ETH. A real-world example is the Akutar auction contract [73]. It had a push mechanism to refund ETH back to bidders who did not win the auction. A hacker used it and locked it to prove the point of audits and security. Since the hacker had no ill intentions, they reopened it again. This could have been avoided by implementing a pull mechanism.

5.2.5. Token Standards

Each token standard has its benefits and disadvantages and should be used for the right use case. The ERC-20 is often used for simple interchangeable tokens. ERC-721 and its modified versions allow for differentiating between the tokens. If multiple different tokens are planned or multiple tokens are expected to change owner, the ERC-1155 is recommended to use.

5.2.6. Readability

The importance of readability of code is often mentioned [8,31,32,61]. Readability is a lot about clear and simple code, as it helps developers and auditors to quickly understand the purpose of a function or a smart contract. One additional thing to consider with Ethereum, is the trade-off with some gas savings, as they can hurt readability. Therefore, if there is no strong reason for a specific gas saving, especially for a small saving, one should deter from adding it. This tip was a finding from Section 4 and has been similarly mentioned by others [8,61].

5.2.7. Testing and Staying Up-to-Date

Over time, new token standards have been introduced to Ethereum. It is important to keep up to date, as they help with the development and make it safer to interact with them. Testing should be done often and early and can be seen as a minimum security requirement. There are different tools to test smart contracts. One can have automated testing with unit tests, integration tests, property-based tests and static and dynamic analysis tools. Furthermore, smart contracts can and should be tested manually. This includes testing on a local blockchain, testing on a currently active testnet, having bug bounties and audits. The recommended approach is to use a mix of all possible tools and to adjust the effort relative to the project size and functionality [11,32].

Dannen [74] also concludes that it can be challenging to acclimate to the fast pace of blockchain development. The best practice of staying up-to-date leads to the next finding (Box 5):

Box 5. Occurrence of breaking changes

Insight 5: The development of Ethereum and its accompanying tools is fast-paced. Breaking changes happen continually.

5.2.8. Secure Ethereum Usage

With Ethereum, there is no centralized authority that can intervene due to the underlying technology. The user has to keep their private key safe or use a strong password for their wallet. It is possible to lose access to one's wallet if the user forgets their private key. On top of the required key management, the users need to be aware of scams. A sceptical mindset is recommended. Common scams include asking for the private key or telling someone to send ETH to an address with the promise of huge returns. Moreover, scams also occur when interacting with smart contracts [11].

Vitalik et al. [75] have proposed the ERC-4337, which allows users to have smart contract accounts instead of externally owned accounts. It would allow for more user-friendly keys and key recovery according to Alchemy [76]. Alchemy acknowledges that it also possibly introduces new security risks and that interaction with using account abstraction would lead to more expensive interactions due to the overhead.

5.3. Gas Savings

The gas-saving parts of the best practices are separated into a general and an advanced part. The reason for that is that the general section is mostly a selection of low-cost, high-impact tips that provide great benefits with little effort. On the other hand, the more advanced savings require more in-depth knowledge about Solidity and yield smaller rewards. While smaller changes combined can also make a huge difference, the effort is higher. These tips can also change with newer versions of Solidity.

5.3.1. Solidity Optimizer

As found in Section 4, the Solidity optimizer improved simple ERC-721 and ERC-1155 contracts by more than 40%. The findings on the optimizer were interesting because a reduction of more than 40% for projects should be emphasized more than it is and it should also be turned on by default. It is, however, noteworthy that using an optimizer does not come without some concerns or past bugs. There are multiple occasions where issues with the Solidity optimizer were found [77]. Moreover, the Solidity documentation [16] warns against using some experimental features. Lastly, the improvements found for simple ERC-721, ERC-721A and ERC-1155 token contracts cannot automatically be assumed for other contracts. Checking the Solidity optimizer against other types of contracts is a future work task. According to a blog post by the Solidity team [78], they admit that earlier versions of the optimizer were complicated and contained bugs, but since late 2020, they recommend to always use the optimizer. They state that only if someone does not care about gas costs at all can they then disregard the optimizer.

An important setting is the runs setting. It gives the optimizer a criterion on what to optimize for. If the value is set low, one tells the optimizer that the code will not be executed often and therefore a small deployment size is important. On the other hand, a high value means that the contract will often be executed, which sets the focus on cheaper execution costs. The standard value is 200 [16]. Finding the optimal value is not trivial and needs to be tested individually. One can do some manual tests or run some test cases and see what works best.

5.3.2. Deployment Time and Network Congestion

The time when one wants to deploy has a huge impact on the cost required to do so. Furthermore, the urgency with which the interaction has to happen can also impact the cost. In Section 2.1, the mechanism of the network that is dealing with congestion is explained.

To begin with, a clear trend can be seen for when the network is normally congested and when it is more idle, according to Marchioro [79]. The best time for interacting is at night, using Coordinated Universal Time. Interestingly, the hour of the day has a stronger dependency on the network state than the day of the week. Marchioro expects the trend to flatten out over time, as more bots for interacting with Ethereum might be used in the future. Additionally, it can be very beneficial for the cost required to interact with the network if the interactions are time-flexible. As mentioned in the methodology in Section 4.3.2, by setting the maxFeePerGas to a lower level than the current base fee, one can decrease the required cost sharply. Yet, two matters have to be taken into account. If a transaction is time-critical, one cannot utilize this. Secondly, the fee should not be set too low, otherwise the transaction might never go through. The effect of this technique is conditional on the network state. It enables the usage of the network at the lower end of the fee range. The exact execution time is not predictable.

5.3.3. Storage

All Ethereum full nodes need to store all data of the network. It would be hard for network participants to run a full node if the blockchain were to be multiple TBs large. The chain is continuously growing. Currently, it is a bit over 1.2 TB, with the Ethereum client Go-Ethereum (GETH) [80]. They state that around 14 GB are added to the database per week. The GETH client [80] has a pruning feature to keep the disk size lower. This is one of the reasons why storage on Ethereum is relatively expensive and developers have to adapt their strategy accordingly. External storage options like IPFS [81], are a solution to extract bigger data from the blockchain.

5.3.4. Solidity Gas Improvements by Versions

Newer versions of Solidity improve safety but can also help reduce gas costs. With the included check of overflow and underflow in Solidity 0.8.0, the SafeMath library has become obsolete [69]. This saves gas because the library does not have to be added. It is possible to use unchecked blocks to tell the compiler to not check those parts for overflows or underflows. This saves a bit of gas as well. However, it is not advisable to use it in circumstances where the value can overflow or underflow. Another improvement came with Solidity 0.8.4 [16], where custom errors were introduced. They can be used to improve the explanation of why a smart contract reverted. Furthermore, custom errors decrease runtime and deployment gas costs.

5.3.5. Solidity Idiosyncrasies

This recommendation should be considered more, with growing experience. It may be less feasible for a beginner who is learning Solidity to study small gas-saving details. However, over time and if the project requires it, more time should be invested in learning and utilizing more advanced tricks for saving gas.

5.4. Advanced Solidity Gas Cost Improvements

This section discusses the best practices from Table 4. As mentioned in Table 2, adapting the strategy of programming is favourable. These tips can be used to save gas, but their effectiveness depends on the various factors of a contract and its surroundings. These are the size, the expected lifetime of the contract, the expected frequency of it being used, the experience of the developer team and whether it is a library or a standard contract.

5.4.1. Gas Saving Trade-Offs

Not all developers recommend advanced gas-saving methods, since they sometimes come with reduced readability or can create misunderstandings [8,31,61]. Further, not all improvements are valid at all times nor come without side effects. Fravoll [82] provides three patterns for saving gas packing variables, memory array building and string equality comparison. These are tips to lower cost, but they also bring about problems. Packing variables increases gas cost if implemented wrong. Next, the complexity can increase when using memory array building. Lastly, string equality comparison has not been used in a production environment according to Fravoll.

According to Valverde [83], the lesser sign (<) and greater sign (>) should be favoured over the less than or equal sign ($\leq$) and the greater than or equal sign ($\geq$). This tip is for a slightly more gas-efficient code. However, this could lead to confusion, when the contract for example contains a check if the maximum number of NFTs per transaction is exceeded. The code for both smart contracts is presented in Figure 2. If one only checks the contract quickly, they could think that the maximum amount to mint is 11. If the user then tries to mint 11 pieces in one transaction, it would revert due to the maximum only being 10.

Table 4. Advanced Solidity improvements, recommended for libraries and other frequently used smart contracts.

Advanced Solidity Improvements		
Name	**Description**	**Reference**
Minimal Proxy Contract	Using a minimal proxy contract allows cheap contract functionality cloning. The contract only implements some parts and delegates to another contract holding the logic. Therefore, the deployment cost can be reduced.	[84]
Cache Storage Values	Before reading storage values multiple times from storage, it is better to copy them to memory once and then read from memory.	[85]
Use Bitmap	Bitmaps are useful to save data in a compact and gas-efficient way. Each entry only takes one bit instead of 8. The cost of writing to storage depends on the value stored before overwriting it.	[86]
Variable Packing	The state variables are stored in 32-byte slots. If smaller state variables are declared next to each other in a smart contract, the Solidity compiler packages them together and therefore saves storage and gas.	[11,34]
Use UINT256	If variables are not packaged, it is better to use uint256. Otherwise, the variable has to be converted to a uint256 variable every time.	[34]

```
// SPDX-License-Identifier: MIT
pragma solidity ^0.8.17;

contract ReadableERC721A is ERC721A {
    uint256 public constant MAX_TRANSACTION_AMOUNT = 10;

    function mint(uint256 quantity) external payable {
        require(quantity <= MAX_TRANSACTION_AMOUNT, "Max amount exceeded");
        _mint(msg.sender, quantity);
    }
}

contract GasEfficientERC721A is ERC721A {
    uint256 public constant MAX_TRANSACTION_AMOUNT = 11;

    function mint(uint256 quantity) external payable {
        require(quantity < MAX_TRANSACTION_AMOUNT, "Max amount exceeded");
        _mint(msg.sender, quantity);
    }
}
```

Figure 2. The first contract is a contract with a readability focus. The second one focuses on gas efficiency. Constructors and other parts have been omitted to reduce size and to emphasize the main point of this code.

5.4.2. Minimal Proxy Contract

If one has to issue smart contracts with similar functionality, using a minimal proxy contract could be beneficial. Manifold states that deployment costs are low due to using a delegate proxy implementation [87]. This tip is only utilizable under certain requirements. One requirement is that similar code needs to be required by multiple contracts. However due to its gas savings, it should be kept in mind. The minimal proxy contract standard ERC-1167 by Murray et al. [84] has nothing to do with upgradeable contract patterns neither is it a replacement for it.

5.4.3. Storage vs. Memory

There is a gas discrepancy between interactions with memory and storage on the EVM. The loading and writing to memory is cheaper compared to doing the same thing on storage variables [85,88]. It is recommendable for variables that are used multiple times to load them first to memory, interact with them and lastly store them back into the storage This means that a temporary variable is introduced that keeps the variable in memory. This tip should be used with caution, as it increases complexity slightly and can seem confusing for developers who do not know the gas implications for memory and storage interactions

5.4.4. Bitmaps

OpenZeppelin [69] provides a bitmap library that maps uint256 to boolean, meaning that each bit of the 256 bits is an entry. The gas savings come from writing a variable that is already non-zero. Writing from non-zero to non-zero costs 5000 units of gas, while zero to non-zero is 22,100 units of gas [89].

5.4.5. Variable Packing

The EVM operates on 256-bit elements. This can be used to save on storage space by putting multiple smaller data types into one uint256 variable. This is done automatically by the compiler if the smaller variables are placed next to each other in the code. However, if, for instance, a uint16 variable on its own has to be converted to a uint256 every time it is used, It will be more expensive than when the variable is a uint256. Furthermore, if the variables are not used together, the cost of writing is also increased, because the new value is not just replacing the old, but has to be combined with the data in the same slot [11].

5.5. Tools for Auditing and Developing

When using additional tools, two things should be kept in mind. Firstly, using a security tool does not guarantee security. Zhang et al. [44] claim that 80% of security-related bugs cannot be audited by a tool. Secondly, using a gas optimizer can also bring about unexpected issues or can even introduce vulnerabilities. Therefore, one has to consider the possible side effects and decide for themselves, if they want to accept them or not. This study recommends staying with well-tested and maintained tools.

The agreement on defects between tools is low, therefore the tools should be used complementary to each other, say Di Angelo et al. [90]. Similarly, Ivanov et al. [91] found that the coverage of vulnerabilities in security tools is very different and many defects from the SWC registry [92] are not covered by any tool. Durieux et al. [93] suggest using Mythril and Slither in combination for the best results. Both tools are well maintained, which improves their appeal further.

6. Results

This section summarizes the findings of this study and recalls the research questions and checks how they have been answered.

RQ1: Multiple best practices have been found and put together into a list. Table 2 shows what general best practices should be followed. Its main points are to favour simplicity and security over other things. Program failure and bugs are a natural part of the process and should be taken into consideration from the beginning. Code quality is always preferable, but with blockchain development, high code quality is indispensable, since code faults can lead to irretrievable damage. Smart contracts can have different sizes, functionality, use cases and lifespans. Adapting the strategy on how to implement them is preferable. Last but not least, it is beneficial to do one's own research. Up-to-date information changes regularly and some recommendations might not be applicable for every scenario. Insight 4 from Section 4 influenced the answer to this research question.

RQ2: To secure smart contracts, a lot of publications were found. Some pointed out that hacks still occur even long after defense mechanisms are developed and best practices are published. Simple best practice principles to defend against security issues are presented in Table 3. Developers should build upon well-tested libraries and code. Since protection against failure cannot be easily guaranteed, proxy patterns for updating should be applied, especially for larger and more complex projects. The two coding flows "Check-Effect-Interact" and "Pull Over Push" can boost security. Readability is important for keeping it easy to read and understand the code. Using token standards helps build upon the experience of the open-source community. Lastly, staying up-to-date is highly recommended on Ethereum. New features are introduced and vulnerabilities are discovered. The answer to this research question has been influenced by insight 3 from Section 4 and insight 5 from Section 5.

RQ3: Gas savings can have a negative influence on readability and complexity. Through the methodology in Section 4, it was visible that some techniques have fewer or no side effects and can improve gas savings heavily. These techniques can be seen in Table 5. Not all found practices are in any way new, but too little advertised for their effectiveness, for example, using the Solidity optimizer and utilizing the network when it is less busy. Furthermore, storage costs should be kept in mind when developing on Ethereum. On the other hand, there are more advanced gas savings that either are less effective or increase complexity and could harm readability. These savings are not unhelpful but should be used in the right environment. The advanced gas savings are visible in Table 4. The answer to this research question has been influenced by insights 1 and 2 in Section 4.

The research, the collection of knowledge and the answering of the research questions happened to the best of the author's knowledge and belief. Still, it cannot be ruled out that the best practices are incomplete, faulty or will lose relevance.

Table 5. The list of general gas-saving best practices.

Gas Saving Best Practices		
Name	**Description**	**Reference**
Solidity Optimizer	The optimizer is recommended to be used with smart contract development. The runs variable can be adapted and it modifies the way the optimizer improves the code. A low number of runs means that the smart contract is optimized for deployment. Choosing a large number results in a less optimized contract for deployment, but lowers the cost of calling functions.	[16,94]
Regard the Network State	The cost of interaction with Ethereum depends heavily on the state of the network. Both urgency and deployment time have an impact on the gas fee. Utilize maxFeePerGas for time uncritical deployments and transactions. This should be done by limiting the maxFeePerGas to an average gas fee below the higher fee of the congested network.	[11,60]
Be Mindful of Storage	Do not store any data that does not have to be on the blockchain. A popular solution for tokens is to use an IPFS to store additional potentially large data. For example, the metadata is linked with the tokenURI in an ERC-721 project.	[32,54,69,95]
Utilize Newer Solidity Versions	Newer Solidity versions do not only bring security updates but can save gas. With Solidity version 0.8.0 an overflow and underflow check was added, making the SafeMath library obsolete on new smart contracts.	[11,69]
Learn Peculiarities of Solidity	Gas costs for operations performed in Solidity can be reduced by small changes. Be wary of how storage and its read and write operations work on Ethereum. The changes can be small, but compound when used by many network participants. These changes should never, without a good reason, reduce readability or the security of the code.	[32,34,45,94,96]

7. Discussion

This is the section where the results are critically discussed and their implications for others are evaluated. Limitations of this work are disclosed, the project creation is presented and an outlook for future works is given.

7.1. Practical Implications

The found best practices should be seen as a guide for beginner blockchain developers and can function as a good lookup resource for more experienced developers. With four different lists, the best practices introduce each category and highlight the most important aspects. The lists also help get a feeling for which principles are more relevant and should be prioritized. The most important recommendations are to prioritize code quality, security and testing. A security breach should be avoided. This includes being aware of possible attack vectors and their corresponding defense mechanisms. Moreover, software projects have disparate challenges and needs. This is the same for projects on Ethereum and means that the strategy should be adapted to the smart contract's functionality and size. Gas savings should be utilized where useful, but should not be traded for simplicity and security Lastly, the quick progress of Solidity and Ethereum should be considered.

7.1.1. Relevance for Other Blockchains

Since multiple other blockchains make use of the EVM, some findings should be relevant to alternative blockchains as well. Some alternatives are Polygon PoS, BNB Smart Chain or Avalanche. These blockchains likely benefit the most from general best practices. However, gas savings should be less important since the fees required to be paid are much lower than on Ethereum. For example, high transaction speed and low gas fees are desired properties for Polygon [97]. In 2023, Aschauer et al. proposed a blockchain-based implementation for an Uber-like ride-sharing platform which compared execution costs of the required contracts on Ethereum, Polygon and HarmonyOne. Since all those blockchains are directly or indirectly EVM-based, the resulting gas values are similar even though the actual costs differ significantly due to vastly different transaction fees [98].

7.2. Limitation

It is hard to make the best practice list exhaustive as new vulnerabilities are discovered and other features are introduced. Hence, this study cautions the readers against blindly trusting the results. To reduce the problem of missing best practices, a wide range of popular lists, tips and helpful resources were checked to get as many diverse insights as possible. The best practices are generally not holistically applicable to other blockchains without adaptations, since many use different programming languages and architectures. Nevertheless, Section 7.1.1 summarizes the expected gains for similar blockchains. The fast-paced development of blockchains and Solidity results in some best practices becoming less relevant over time if not adapted.

7.3. Credibility of Sources

Although Ethereum was initially published in 2015 and a lot of research has been made on blockchains and Ethereum, a vast amount of knowledge and current news are first found on different blog entries, the Ethereum documentation [11] or the platform X. This point is backed by multiple papers in this domain. Oliveira et al. [99] claim that the literature on tokens is based strongly on blog posts, news articles and social media channels. Wohrer and Zdun [30] also state that best practices are found across Ethereum blogs and the Ethereum community. Ultimately, the gas-saving patterns by Marchesi et al. [34] are based partly on discussion lists and Ethereum blogs.

7.4. Future Work

Future work is imaginable in multiple directions. It is possible to add best practices for other languages on Ethereum or research other blockchains. The most apparent next language to check would be Vyper. It is the second most popular programming language on Ethereum. A security comparison between Vyper and Solidity has already been made by Kaleem et al. [70]. Such a comparison could be supplemented with further investigations into differences in tools, gas-saving techniques and standards. Another possible next step would be to research best practices for other blockchains. Checking how much these guidelines help similar technologies would be worth researching. Moreover, different blockchains, such as Solana [100] could be compared. It would be interesting to see which best practices overlap and which are completely different. There should be differences between the different architectures. Other differences like low transaction fees probably make gas savings less important.

Author Contributions: Conceptualization, M.B.; methodology, M.B.; software, M.B.; writing—original draft preparation, M.B.; writing—review and editing, E.S.; supervision, E.S.; funding acquisition, M.K. All authors have read and agreed to the published version of the manuscript.

Funding: This research received no external funding

Data Availability Statement: The entire source code associated with this publication is available at https://github.com/GitHub-Fred/Ethereum-Strategies-For-Security-and-Gas-Saving (accessed on 18 December 2023).

Conflicts of Interest: The authors declare no conflicts of interest.

References

1. Luu, L.; Chu, D.H.; Olickel, H.; Saxena, P.; Hobor, A. Making smart contracts smarter. In Proceedings of the 2016 ACM SIGSAC Conference on Computer and Communications Security, Vienna, Austria, 24–28 October 2016; pp. 254–269.
2. Ethereum Gas Tracker | Etherscan. Available online: http://etherscan.io/gastracker (accessed on 2 August 2023).
3. Schwarz-Schilling, C.; Neu, J.; Monnot, B.; Asgaonkar, A.; Tas, E.N.; Tse, D. Three attacks on proof-of-stake ethereum. In Proceedings of the International Conference on Financial Cryptography and Data Security, Grenada, Grenada, 2–6 May 2022; Springer: Berlin/Heidelberg, Germany, 2022; pp. 560–576.
4. Neu, J.; Tas, E.N.; Tse, D. Two attacks on proof-of-stake GHOST/Ethereum. *arXiv* **2022**, arXiv:2203.01315.
5. Neu, J.; Tas, E.N.; Tse, D. Two more attacks on proof-of-stake GHOST/Ethereum. In Proceedings of the 2022 ACM Workshop on Developments in Consensus, Los Angeles, CA, USA, 7 November 2022; pp. 43–52.
6. Tikhomirov, S.; Voskresenskaya, E.; Ivanitskiy, I.; Takhaviev, R.; Marchenko, E.; Alexandrov, Y. Smartcheck: Static analysis of ethereum smart contracts. In Proceedings of the 1st International Workshop on Emerging Trends in Software Engineering for Blockchain, Gothenburg, Sweden, 27 May 2018; pp. 9–16.
7. Chen, H.; Pendleton, M.; Njilla, L.; Xu, S. A survey on ethereum systems security: Vulnerabilities, attacks, and defenses. *ACM Comput. Surv. (CSUR)* **2020**, *53*, 1–43. [CrossRef]
8. Zou, W.; Lo, D.; Kochhar, P.S.; Le, X.B.D.; Xia, X.; Feng, Y.; Chen, Z.; Xu, B. Smart contract development: Challenges and opportunities. *IEEE Trans. Softw. Eng.* **2019**, *47*, 2084–2106. [CrossRef]
9. Turing, A.M. On computable numbers, with an application to the Entscheidungsproblem. *J. Math* **1936**, *58*, 5.
10. Buterin, V. A Next-Generation Smart Contract and Decentralized Application Platform. 2014. Available online: https://finpedia.vn/wp-content/uploads/2022/02/Ethereum_white_paper-a_next_generation_smart_contract_and_decentralized_application_platform-vitalik-buterin.pdf (accessed on 18 December 2023).
11. Ethereum Documentation. Available online: https://ethereum.org/en/developers/docs/ (accessed on 1 August 2023).
12. Tsankov, P.; Dan, A.; Drachsler-Cohen, D.; Gervais, A.; Buenzli, F.; Vechev, M. Securify: Practical security analysis of smart contracts. In Proceedings of the 2018 ACM SIGSAC Conference on Computer and Communications Security, Toronto, ON, Canada, 15–19 October 2018; pp. 67–82.
13. Canfora, G.; Sorbo, A.D.; Laudanna, S.; Vacca, A.; Visaggio, C.A. GasMet: Profiling Gas Leaks in the Deployment of Solidity Smart Contracts. *arXiv* **2020**, arXiv:2008.05449. Available online: http://xxx.lanl.gov/abs/2008.05449 (accessed on 18 December 2023).
14. Albert, E.; Gordillo, P.; Hernández-Cerezo, A.; Rubio, A. A Max-SMT superoptimizer for EVM handling memory and storage. In Proceedings of the International Conference on Tools and Algorithms for the Construction and Analysis of Systems, Munich, Germany, 2–7 April 2022; Springer: Berlin/Heidelberg, Germany, 2022; pp. 201–219.
15. Albert, E.; Correas, J.; Gordillo, P.; Román-Díez, G.; Rubio, A. Gasol: Gas analysis and optimization for ethereum smart contracts. In Proceedings of the International Conference on Tools and Algorithms for the Construction and Analysis of Systems, Dublin, Ireland, 25–30 April 2020; Springer: Berlin/Heidelberg, Germany, 2020; pp. 118–125.
16. Śliwak, K.E.A. Solidity Documentation. Available online: https://docs.soliditylang.org/en/v0.8.20/index.html (accessed on 14 July 2022).

17. Dhillon, V.; Metcalf, D.; Hooper, M.; Dhillon, V.; Metcalf, D.; Hooper, M. The DAO Hacked. In *Blockchain Enabled Applications: Understand the Blockchain Ecosystem and How to Make it Work for You*; Apress: New York, NY, USA, 2021; pp. 113–128.
18. Salehi, M.; Clark, J.; Mannan, M. Not so immutable: Upgradeability of smart contracts on ethereum. *arXiv* **2022**, arXiv:2206.00716.
19. Charoenwong, B.; Bernardi, M. A Decade of Cryptocurrency 'Hacks': 2011–2021. *SSRN* **2021**. [CrossRef]
20. Tjiam, K.; Wang, R.; Chen, H.; Liang, K. Your smart contracts are not secure: Investigating arbitrageurs and oracle manipulators in Ethereum. In Proceedings of the 3rd Workshop on Cyber-Security Arms Race, Virtual, 19 November 2021; pp. 25–35.
21. Amri, S.A.; Aniello, L.; Sassone, V. A Review of Upgradeable Smart Contract Patterns based on OpenZeppelin Technique. *J. Br. Blockchain Assoc.* **2023**, *6*, 1–8. [CrossRef]
22. Ballet, G.; Vitalik Buterin, D.F. EIP-4758: Deactivate SELFDESTRUCT. Available online: https://eips.ethereum.org/EIPS/eip-4758 (accessed on 21 July 2023).
23. Jangid, H.; Meel, P. Blockchain Protocols: Transforming the Web We Know. In Proceedings of the International Conference on Information and Communication Technology for Intelligent Systems, Ho Chi Minh, Vietnam, 7–8 December 2023; Springer: Berlin/Heidelberg, Germany, 2023; pp. 545–557.
24. Foundation, E. The Ethereum Roadmap. Available online: https://ethereum.org/en/roadmap/ (accessed on 12 December 2023).
25. Kapengut, E.; Mizrach, B. An event study of the ethereum transition to proof-of-stake. *Commodities* **2023**, *2*, 96–110. [CrossRef]
26. Al-Breiki, H.; Rehman, M.H.U.; Salah, K.; Svetinovic, D. Trustworthy blockchain oracles: Review, comparison, and open research challenges. *IEEE Access* **2020**, *8*, 85675–85685. [CrossRef]
27. Beniiche, A. A study of blockchain oracles. *arXiv* **2020**, arXiv:2004.07140.
28. Jones, C. *Software Engineering Best Practices: Lessons from Successful Projects in the Top Companies*; McGraw-Hill Education: New York, NY, USA, 2010.
29. Ethereum Smart Contract Security Best Practices. Available online: https://consensys.github.io/smart-contract-best-practices/ (accessed on 1 August 2023).
30. Wohrer, M.; Zdun, U. Smart contracts: Security patterns in the ethereum ecosystem and solidity. In Proceedings of the 2018 International Workshop on Blockchain Oriented Software Engineering (IWBOSE), Campobasso, Italy, 20 March 2018; IEEE: Piscataway, NJ, USA, 2018; pp. 2–8.
31. Martin, R.C. *Clean Code: A Handbook of Agile Software Craftsmanship*; Pearson Education: London, UK, 2009.
32. Antonopoulos, A.M.; Wood, G. *GitHub—Ethereumbook/Ethereumbook: Mastering Ethereum, by Andreas M. Antonopoulos, Gavin Wood*; O'Reilly Media: Sebastopol, CA, USA, 2021.
33. Kushwaha, S.S.; Joshi, S.; Singh, D.; Kaur, M.; Lee, H.N. Systematic review of security vulnerabilities in ethereum blockchain smart contract. *IEEE Access* **2022**, *10*, 6605–6621. [CrossRef]
34. Marchesi, L.; Marchesi, M.; Destefanis, G.; Barabino, G.; Tigano, D. Design Patterns for Gas Optimization in Ethereum. In Proceedings of the 2020 IEEE International Workshop on Blockchain Oriented Software Engineering (IWBOSE), London, ON, Canada, 18 February 2020; pp. 9–15. [CrossRef]
35. Wang, S.; Ouyang, L.; Yuan, Y.; Ni, X.; Han, X.; Wang, F.Y. Blockchain-enabled smart contracts: Architecture, applications, and future trends. *IEEE Trans. Syst. Man Cybern. Syst.* **2019**, *49*, 2266–2277. [CrossRef]
36. Praitheeshan, P.; Pan, L.; Yu, J.; Liu, J.; Doss, R. Security analysis methods on ethereum smart contract vulnerabilities: A survey. *arXiv* **2019**, arXiv:1908.08605.
37. Destefanis, G.; Marchesi, M.; Ortu, M.; Tonelli, R.; Bracciali, A.; Hierons, R. Smart contracts vulnerabilities: A call for blockchain software engineering? In Proceedings of the 2018 International Workshop on Blockchain Oriented Software Engineering (IWBOSE), Campobasso, Italy, 20 March 2018; IEEE: Piscataway, NJ, USA, 2018; pp. 19–25.
38. Parity. Available online: https://www.parity.io/ (accessed on 12 June 2023).
39. Atzei, N.; Bartoletti, M.; Cimoli, T. A survey of attacks on ethereum smart contracts (sok). In Proceedings of the Principles of Security and Trust: 6th International Conference, POST 2017, Held as Part of the European Joint Conferences on Theory and Practice of Software, ETAPS 2017, Uppsala, Sweden, 22–29 April 2017; Springer: Berlin/Heidelberg, Germany, 2017; pp. 164–186.
40. Dika, A.; Nowostawski, M. Security vulnerabilities in ethereum smart contracts. In Proceedings of the 2018 IEEE International Conference on Internet of Things (iThings) and IEEE Green Computing and Communications (GreenCom) and IEEE Cyber, Physical and Social Computing (CPSCom) and IEEE Smart Data (SmartData), Halifax, NS, Canada, 30 July–3 August 2018; IEEE: Piscataway, NJ, USA, 2018; pp. 955–962.
41. Marchesi, L.; Marchesi, M.; Pompianu, L.; Tonelli, R. Security checklists for Ethereum smart contract development: Patterns and best practices. *arXiv* **2020**, arXiv:2008.04761. Available online: http://xxx.lanl.gov/abs/2008.04761 (accessed on 18 December 2023).
42. Chen, J.; Xia, X.; Lo, D.; Grundy, J.; Luo, X.; Chen, T. Defining Smart Contract Defects on Ethereum. *IEEE Trans. Softw. Eng.* **2022**, *48*, 327–345. [CrossRef]
43. Fowler, M. *Refactoring*; Addison-Wesley Professional: Boston, MA, USA, 2018.
44. Zhang, Z.; Zhang, B.; Xu, W.; Lin, Z. Demystifying Exploitable Bugs in Smart Contracts. In Proceedings of the Demystifying Exploitable Bugs in Smart Contracts. ICSE, Melbourne, Australia, 14–20 May 2023.
45. Kong, Q.P.; Wang, Z.Y.; Huang, Y.; Chen, X.P.; Zhou, X.C.; Zheng, Z.B.; Huang, G. Characterizing and Detecting Gas-Inefficient Patterns in Smart Contracts. *J. Comput. Sci. Technol.* **2022**, *37*, 67–82. [CrossRef]
46. OpenZeppelin. Ethernaut. Available online: https://ethernaut.openzeppelin.com/ (accessed on 26 July 2023).

47. Agarwal, U.; Singh, K.; Verma, R. An Overview of Non-Fungible Tokens (NFT). *Int. J. Adv. Res. Sci. Commun. Technol. (IJARSCT)* **2022**, 2. [CrossRef]
48. Guidi, B.; Michienzi, A. Sleepminting, the brand new frontier of Non Fungible Tokens fraud. In Proceedings of the 2022 ACM Conference on Information Technology for Social Good, Limassol, Cyprus, 7–9 September 2022 ; pp. 75–81.
49. Guidi, B.; Michienzi, A. From NFT 1.0 to NFT 2.0: A Review of the Evolution of Non-Fungible Tokens. *Future Internet* **2023**, *15*, 189. [CrossRef]
50. Malashetti, A.P. Impact of Non-Fungible Token (NFT) on World. *Int. J. Res. Eng. Sci. Manag.* **2022**, *5*, 219–221.
51. Olsson, O. A Taxonomy of Non-Fungible Tokens: Overview, Evaluation and Explanation. Master's Thesis, Department of Informatics and Media, University of Uppsala. 2022. Available online: https://www.diva-portal.org/smash/get/diva2:1672740/FULLTEXT01.pdf (accessed on 18 December 2023).
52. Tan, Y.; Wu, Z.; Liu, J.; Wu, J.; Zheng, Z.; Chen, T. Bubble or Not: Measurements, Analyses, and Findings on the Ethereum ERC721 and ERC1155 Non-fungible Token Ecosystem. *arXiv* **2023**, arXiv:2301.01991.
53. von Wachter, V.; Jensen, J.R.; Regner, F.; Ross, O. NFT wash trading: Quantifying suspicious behaviour in NFT markets. *arXiv* **2022**, arXiv:2202.03866.
54. Wang, Q.; Li, R.; Wang, Q.; Chen, S. Non-fungible token (NFT): Overview, evaluation, opportunities and challenges. *arXiv* **2021**, arXiv:2105.07447.
55. Azuki. ERC721A. Available online: https://www.azuki.com/erc721a (accessed on 30 July 2023).
56. Hu, A. ERC721 vs. ERC721A: Batch Minting NFTs. Available online: https://www.alchemy.com/blog/erc721-vs-erc721a-batch-minting-nfts (accessed on 7 July 2023).
57. Cong, L.W.; Tang, K.; Wang, Y.; Zhao, X. *Inclusion and Democratization through Web3 and Defi? Initial Evidence from the Ethereum Ecosystem*; National Bureau of Economic Research: Cambridge, MA, USA, 2023. [CrossRef]
58. Wood, G. Ethereum: A secure decentralised generalised transaction ledger. *Ethereum Proj. Yellow Pap.* **2014**, *151*, 1–32.
59. Opensea. Available online: https://opensea.io/ (accessed on 8 July 2023).
60. Akiyoshi, M. Gas Tutorial: How to Set Your Own Gas Prices. Available online: https://medium.com/@maimai816/advanced-metamask-gas-tutorial-how-to-set-your-own-gas-prices-236d59f563b7 (accessed on 6 July 2023).
61. Di Sorbo, A.; Laudanna, S.; Vacca, A.; Visaggio, C.A.; Canfora, G. Profiling gas consumption in solidity smart contracts. *J. Syst. Softw.* **2022**, *186*, 111193. [CrossRef]
62. Blockscount Verified Contracts. Available online: https://eth.blockscout.com/verified-contracts (accessed on 8 July 2023).
63. Team, A. A Developer's Guide to Securing Ethereum Smart Contracts. Available online: https://alchemy.com/blog/a-developers-guide-to-securing-ethereum-smart-contracts (accessed on 15 August 2023).
64. Wang, Z.; Jin, H.; Dai, W.; Choo, K.K.R.; Zou, D. Ethereum smart contract security research: Survey and future research opportunities. *Front. Comput. Sci.* **2021**, *15*, 152802. [CrossRef]
65. Brooks, F.; Kugler, H. *No Silver Bullet*; Addison-Wesley: Reading, MA, USA, 1987.
66. Maslow, A.H. *The Psychology of Science: A Reconnaissance*; HarperCollins: New York, NY, USA, 1966.
67. Staples, M.; Chen, S.; Falamaki, S.; Ponomarev, A.; Rimba, P.; Tran, A.; Weber, I.; Xu, X.; Zhu, J. *Risks and Opportunities for Systems Using Blockchain and Smart Contracts. Data61*; CSIRO: Sydney, Australia, 2017.
68. Fröwis, M.; Böhme, R. In code we trust? Measuring the control flow immutability of all smart contracts deployed on Ethereum. In Proceedings of the Data Privacy Management, Cryptocurrencies and Blockchain Technology: ESORICS 2017 International Workshops, DPM 2017 and CBT 2017, Oslo, Norway, 14–15 September 2017; Springer: Berlin/Heidelberg, Germany, 2017; pp. 357–372.
69. OpenZeppelin Documentation. Available online: https://docs.openzeppelin.com/ (accessed on 1 August 2023).
70. Kaleem, M.; Mavridou, A.; Laszka, A. Vyper: A security comparison with solidity based on common vulnerabilities. In Proceedings of the 2020 2nd Conference on Blockchain Research & Applications for Innovative Networks and Services (BRAINS), Paris, France, 28–30 September 2020; IEEE: Piscataway, NJ, USA, 2020; pp. 107–111.
71. Xu, X.; Pautasso, C.; Zhu, L.; Lu, Q.; Weber, I. A Pattern Collection for Blockchain-Based Applications. In Proceedings of the 23rd European Conference on Pattern Languages of Programs, New York, NY, USA, 4–8 July 2018; [CrossRef]
72. Meisami, S.; Bodell III, W.E. A Comprehensive Survey of Upgradeable Smart Contract Patterns. *arXiv* **2023**, arXiv:2304.03405.
73. Langston, T. 34M Locked in a Smart Contract. Was the Akutars Exploit Avoidable? Available online: https://nftnow.com/features/akutars-exploit-34-million-locked-in-smart-contract/ (accessed on 20 July 2023).
74. Dannen, C. *Introducing Ethereum and Solidity*; Springer: Berlin/Heidelberg, Germany, 2017; Volume 1.
75. Vitalik, B.; Yoav, W.; Dror, T.; Shahaf, N.; Alex, F.; Kristof, G.; Tjaden, H. ERC-4337: Account Abstraction Using Alt Mempool. Available online: https://eips.ethereum.org/EIPS/eip-4337 (accessed on 15 December 2023).
76. Team, A. Introduction to Account Abstraction. Available online: https://docs.alchemy.com/docs/introduction-to-account-abstraction (accessed on 14 December 2023).
77. Ethereum Foundation Blog. Available online: https://blog.ethereum.org/ (accessed on 17 July 2023).
78. Team, S. Ask the Solidity Team Anything Nr.1 Recap. Available online: https://blog.soliditylang.org/2020/11/04/solidity-ama-1-recap/#why-do-you-think-people-are-generally-suspicious-of-the-optimizer-and-are-they-right-to-be (accessed on 7 March 2023).

79. Marchioro, M. Ethereum: How to Save Even More on Gas Price with a Weekly Plan. Available online: https://medium.com/dextf/ethereum-how-to-save-even-more-on-gas-price-with-a-weekly-plan-c6689ac09fe6 (accessed on 7 July 2023).
80. Go-Ethereum. Available online: https://geth.ethereum.org/ (accessed on 2 August 2023).
81. Benet, J. Ipfs-content addressed, versioned, p2p file system. *arXiv* **2014**, arXiv:1407.3561.
82. Fravoll. Solidity Patterns. Available online: https://github.com/fravoll/solidity-patterns (accessed on 10 August 2023).
83. Valverde, J.X. Advanced Gas Optimization Tips for Solidity. Available online: https://coinsbench.com/advanced-gas-optimizations-tips-for-solidity-85c47f413dc5 (accessed on 8 July 2023).
84. Murray, P.; Welch, N.; Messerman, J. ERC-1167: Minimal Proxy Contract. Available online: https://eips.ethereum.org/EIPS/eip-1167 (accessed on 8 July 2023).
85. Li, A.; Choi, J.A.; Long, F. Securing smart contract with runtime validation. In Proceedings of the 41st ACM SIGPLAN Conference on Programming Language Design and Implementation, London, UK, 15–20 June 2020; pp. 438–453.
86. Openzeppelin-Contracts. Available online: https://github.com/OpenZeppelin/openzeppelin-contracts (accessed on 1 July 2023)
87. Manifold Documentation. Available online: https://docs.manifold.xyz/v/manifold-for-developers/ (accessed on 5 August 2023).
88. Achiando, H. 40 Tips to Optimize Smart Contract Gas Cost. Available online: https://www.linkedin.com/pulse/optimizing-smart-contract-gas-cost-harold-achiando (accessed on 5 August 2023).
89. An Ethereum Virtual MachineOpcodes Interactive Reference. Available online: https://www.evm.codes/?fork=shanghai (accessed on 5 August 2023).
90. Di Angelo, M.; Durieux, T.; Ferreira, J.F.; Salzer, G. Evolution of automated weakness detection in Ethereum bytecode: A comprehensive study. *arXiv* **2023**, arXiv:2303.10517.
91. Ivanov, N.; Li, C.; Sun, Z.; Cao, Z.; Luo, X.; Yan, Q. Security Threat Mitigation for Smart Contracts: A Survey. *arXiv* **2023**, arXiv:2302.07347.
92. Smart Contract Weakness Classification. Available online: https://swcregistry.io/ (accessed on 17 July 2023).
93. Durieux, T.; Ferreira, J.A.F.; Abreu, R.; Cruz, P. Empirical Review of Automated Analysis Tools on 47,587 Ethereum Smart Contracts. In Proceedings of the ACM/IEEE 42nd International Conference on Software Engineering, New York, NY, USA, 27 June–19 July 2020; pp. 530–541. [CrossRef]
94. Chance. The Gas-Efficient Way of Building and Launching an ERC721 NFT Project for 2022. Available online: https://nftchance.medium.com/the-gas-efficient-way-of-building-and-launching-an-erc721-nft-project-for-2022-b3b1dac5f2e1 (accessed on 2 May 2023).
95. Nyaletey, E.; Parizi, R.M.; Zhang, Q.; Choo, K.K.R. BlockIPFS–blockchain-enabled interplanetary file system for forensic and trusted data traceability. In Proceedings of the 2019 IEEE International Conference on Blockchain (Blockchain), Seoul, Republic of Korea, 14–17 May 2019; IEEE: Piscataway, NJ, USA, 2019; pp. 18–25.
96. Chen, T.; Li, Z.; Zhou, H.; Chen, J.; Luo, X.; Li, X.; Zhang, X. Towards Saving Money in Using Smart Contracts. In Proceedings of the 40th International Conference on Software Engineering: New Ideas and Emerging Results, New York, NY, USA, 27 May–3 June 2018; pp. 81–84. [CrossRef]
97. Polygon Wiki. Available online: https://wiki.polygon.technology/ (accessed on 18 August 2023).
98. Aschauer, G.; Sonnleitner, E.; Kurz, M. Cost Efficiency Evaluation of an On-Chain, Decentralized Ride-Sharing Platform. *Sustainability* **2023**, *15*, 6230. [CrossRef]
99. Oliveira, L.; Zavolokina, L.; Bauer, I.; Schwabe, G. To token or not to token: Tools for understanding blockchain tokens. In Proceedings of International Conference of Information Systems, San Francisco, CA, USA, 13–16 December 2018.
100. Yakovenko, A. Solana: A new architecture for a high performance blockchain (v0.8.13). *White Paper*; 2018. Available online: https://solana.com/solana-whitepaper.pdf (accessed 18 December 2023).

electronics

Article

Novel Blockchain and Zero-Knowledge Proof Technology-Driven Car Insurance

Zhuoliang Qiu [1], Zhijun Xie [1,*], Xianliang Jiang [1], Chuan Ran [1] and Kewei Chen [2]

[1] Faculty of Electrical Engineering and Computer Science, Ningbo University, Ningbo 315000, China; 2111082374@nbu.edu.cn (Z.Q.); jiangxianliang@nbu.edu.cn (X.J.); 2211100275@nbu.edu.cn (C.R.)

[2] Faculty of Mechanical Engineering and Mechanis, Ningbo University, Ningbo 315000, China; chenkewei@nbu.edu.cn

* Correspondence: xiezhijun@nbu.edu.cn

Abstract: It is crucial to ensure the privacy and authenticity of the owner's information in car insurance claims. However, the current traditional car insurance claims scenario suffers from inefficiency, complex service, unreliable data, and data leakage. Therefore, considering the privacy and sensitivity of insurance information and car owner data, we can use blockchain, smart contracts, and zero-knowledge proof technology to improve the current problems. This paper proposes a novel car insurance claim scheme based on smart contracts, blockchain, and zero-knowledge proof. Our scheme focuses on preserving privacy in the car insurance authorization and claim process. We design a private smart contract for the creation and revocation of car insurance and public smart contract for the authorization and validation of car insurance. By using ZoKrates, generating zero-knowledge proofs off chain and verifying the proofs on chain reduces the amount of data storage and computation on chain and provides privacy protection for sensitive information. Experimental results confirm the efficacy of our scheme in terms of security and performance.

Keywords: car insurance; blockchain; smart contract; zero-knowledge proof

Citation: Qiu, Z.; Xie, Z.; Jiang, X.; Ran, C.; Chen, K. Novel Blockchain and Zero-Knowledge Proof Technology-Driven Car Insurance. *Electronics* **2023**, *12*, 3869. https://doi.org/10.3390/electronics12183869

Academic Editor: Hamed Taherdoost

Received: 24 August 2023
Revised: 10 September 2023
Accepted: 11 September 2023
Published: 13 September 2023

1. Introduction

Insurance is one of the most widely used forms of protection worldwide [1]. An insurance policy represents an agreement between individuals or entities and an insurance company, providing financial assistance or reimbursement in the event of a loss. A digital transformation is currently underway in the insurance industry to adapt to the needs of modern society [2]. In the car insurance industry, car insurance claims management is facilitated through the collaboration of various entities from different fields, such as the police, county administrators, insurance agents, and healthcare professionals [3]. This collaborative sharing of multi-source information is crucial for insurance companies to make accurate decisions regarding policyholders' claims.

While insurance plans are prevalent, settling and processing insurance claims can be challenging and error free [4]. Insurance companies often manipulate terms and conditions to avoid paying policyholders, while fraudulent claims can pose problems for insurers [5]. The advantages of blockchain and smart contracts can make insurance contracts more transparent, efficient, and resistant to fraud [6]. Several blockchain-based solutions have been proposed [7,8], with the core idea of using blockchain to establish a trust mechanism between customers and insurance companies, thus effectively confirming the content of car insurance payouts. Automated smart contracts can speed up claims processing and reduce insurers' operating costs.

Indeed, using blockchain-based car insurance plans still presents two significant challenges. Firstly, the identities and insurance details of users participating in the insurance are public, which could lead to privacy breaches and information misuse [9]. Attackers could access all transaction data by downloading a copy of the ledger or trace relationships between transactions and accounts by analyzing the transaction data in the ledger. Secondly, the car insurance claims process relies on the automatic execution of smart contracts which may require sensitive information to be received on the blockchain to invoke smart contracts [10], such as the vehicle owner's identity. Since these inputs are publicly transparent, they could expose the vehicle owner's privacy. To address these challenges, further advancements in privacy-preserving techniques and data encryption on the blockchain are necessary.

Zero-knowledge proofs are interactive verification protocols [11]. In this protocol, based on predefined actions, a verifier can be convinced that a prover possesses specific secret data without revealing any private information, including the prover's data, the verifier's identity, and the prover's identity. The verifier only knows that the prover has access to this data. The application of zero-knowledge proofs technology in the blockchain-based insurance sector not only helps to protect the privacy of owners and reduce the risk of information asymmetry but also optimizes the execution process of insurance contracts [12–14] Our research aims to address the problem of insurance and user information leakage in the blockchain-based car insurance industry by incorporating zero-knowledge proofs.

This paper proposes a solution based on blockchain, smart contracts, and zero-knowledge proofs to address privacy issues in traditional blockchain-based car insurance systems. In this solution, the blockchain ensures the integrity and immutability of insurance data, while smart contracts enable the decentralized execution of insurance claims processes. Additionally, zero-knowledge proofs are used to maintain the privacy of insurance data and user identities.

The contributions of this paper can be summarized as follows:

1. We propose a hybrid smart contract proxy model. Using a private smart contract for creating car insurance protects insurance data from third-party access. A public smart contract is employed for insurance verification, achieving identity authentication without revealing sensitive user information.
2. The utilization of ZoKrates enables zero-knowledge authorization and verification for car insurance. This avoids the exposure of privacy attributes' ownership in a publicly transparent distributed ledger, ensuring non-linkability between vehicle owners and their insurance details.
3. The paper includes a thorough security analysis, demonstrating the privacy and security of our proposed solution. Additionally, comprehensive performance evaluations were conducted to showcase the effectiveness of the proposed approach.

The remaining sections of this paper are organized as follows. In Section 2, we present the background to blockchain-based insurance schemes. In Section 3, we present the preliminary knowledge of the methods used. In Section 4, we propose a model for a vehicle insurance scheme based on blockchain and zero-knowledge proof. In Section 5, we perform a security and performance analysis of the proposed system. Finally, we conclude the paper in Section 6.

2. Related Work

The use of blockchain as a system service to design distributed platforms to support the execution of transactions in insurance processes is a core concept to solve the problems of traditional insurance platforms [15]. The insurance industry has adopted blockchain to automate insurance operations by transforming various policies into smart contracts [16].

Many efforts have been made to address car insurance registration using blockchain. Yadav et al. [17] proposed a blockchain-based framework for car insurance to simplify the submission of accident reports and insurance claims. Nizamuddin et al. [18] provided a decentralized blockchain and IPFS-based framework for the auto insurance industry to regulate auto insurance claims and automated payment activities. Lamberti et al. [19] presented a blockchain and sensor-based framework for car insurance that uses smart contracts and sensor data to implement an on-demand insurance system. Bader et al. [20]

proposed a blockchain smart contract-based ecosystem for simple and transparent car insurance, using smart contracts to automate the insurance process. Chiu et al. [21] used blockchain to decentralize data and services and smart contracts as insurance products for insurance companies for bicycle insurance systems. Nanda et al. [22] designed a decentralized system model for the car insurance process based on blockchain technology using Ethereum and smart contracts, with a decentralized application (DApp) for the car insurance purchase and claim process.

Blockchain is also present in other areas of insurance. Kumar et al. [23] proposed a blockchain-based trusted fire brigade service and insurance claim framework to provide immediate fire brigade service to enterprises and prevent insurance fraud while proposing a sensor network and connectivity model to detect real fires and send emergency service requests to monitoring stations. Pawar et al. [24] proposed a blockchain-based insurance system to share health insurance information between hospitals, patients, and insurance companies. Iyer et al. [25] built a decentralized peer-to-peer crop insurance system to cover the risk of excessive rainfall. Jha et al. [26] proposed a blockchain-based crop insurance system to ensure that farmers benefit from the insurance on time.

However, after analyzing existing blockchain-based car insurance schemes, we find that the authors mainly focus on achieving a decentralized insurance system, ignoring the blockchain's information transparency and privacy leakage issues. Therefore, there is a need to expand the scope of policyholder privacy and consider the privacy protection of real and sensitive data during the insurance approval and verification process.

3. Preliminary

This section reviews some of the technical preparations required for this paper.

3.1. Blockchain and Smart Contract

The advent of cryptocurrencies has profoundly impacted conventional finance ever since the inception of Bitcoin in 2009 [27]. Positioned as a distributed ledger technology, blockchain facilitates data exchange among designated participants. By aggregating and disseminating transaction data from various data sources, blockchain is structured as a sequence of blocks, each encapsulating the information of multiple transactions interconnected through cryptographic algorithms to form an immutable chain [28]. Diverging from conventional centralized databases, blockchain data are distributed across numerous network nodes, wherein each node possesses a complete ledger copy, necessitating a consensus mechanism for validating and appending new blocks [29]. This decentralized nature endows blockchain with heightened security and resilience against attacks, empowering it to establish a robust trust system within a distributed and untrusted environment.

A smart contract represents an automated contract that utilizes blockchain technology and is expressed as a computer program, facilitating autonomous execution and producing irreversible outcomes [30]. Employing a distributed ledger to store these contracts ensures transactional accuracy without reliance on intermediaries, given the assured reliability of the blockchain [31]. Smart contracts empower users to implement personalized code logic on the blockchain, enabling the establishment of decentralized systems. The key features of smart contracts, including decentralization, autonomy, observability, verifiability, and information sharing, significantly contribute to developing decentralized systems.

3.2. Cryptographic Primitives

3.2.1. Non-Interactive Zero-Knowledge Proof

The fundamental concept of zero-knowledge proofs revolves around proving a statement through interactive protocols [32]. In this process, the prover presents a set of information to the verifier, enabling the verifier to validate the accuracy of this information and gain confidence in its truthfulness without acquiring knowledge of how the prover obtained the information. This information may pertain to the prover's knowledge of the original image of a hash or awareness of the members within a Merkle tree with known

Merkle roots. Practical structures for non-interactive zero-knowledge proof (NIZK) have been demonstrated in Ethernet [33]. The formal definition of NIZK is described below:

- $KeyGen\ (1^{\lambda}) \rightarrow crs$: The input is the safety parameter λ; the output is the common reference string crs.
- $Prove(crs, u, w) \rightarrow \pi$: The inputs are the instance u of some NP-language L_R and the witnesses w; the output is a zero-knowledge proof π.
- $Verify(crs, u, \pi) \rightarrow 1/0$: The input is the proof π; the output is 1 for acceptance or 0 for rejection.

3.2.2. Fiat–Shamir Heuristic

The Fiat–Shamir heuristic is a technique employed to transform an interactive zero-knowledge proof protocol into a non-interactive version [34]. In conventional interactive zero-knowledge proofs, the prover and verifier engage in multiple rounds of interaction to accomplish the proof process, potentially incurring significant communication costs. Conversely, the Fiat–Shamir heuristic mitigates the communication overhead by converting the interactive protocol into a one-way non-interactive form, achieved through a hash function and a random number generator. The overall process of the Fiat–Shamir heuristic is as follows:

1. The prover runs $Prove(crs, u, w)$ and generates the proof π. He/she hashes the (crs, u, π) to e and sends π and e to the verifier.
2. The verifier checks if the equation $e = H(crs, u, \pi)$ holds and runs $Verify(crs, u, \pi)$ to decide whether to accept.

3.3. ZoKrates

ZoKrates [35] is an open-source tool set extensively utilized in the blockchain and cryptocurrency domains for developing and deploying zero-knowledge proofs. It offers a processing model and features a user-friendly domain specific language (DSL), allowing developers to describe intricate computational tasks and generate corresponding zero-knowledge proofs succinctly. These proofs enable the verification of computational results without necessitating an understanding of the specific computations involved. Furthermore, ZoKrates supports diverse zero-knowledge proof systems, including zk-SNARKs (zero-knowledge extensible non-interactive parameters), which streamline the generation and verification of proofs, rendering the process highly efficient and swift. The details of the implementation of the proofs of zero knowledge in ZoKrates are given below:

- Compile: To prove specific computations, circuit designs need to be developed. ZoKrates utilizes a domain specific language (DSL) to describe these circuits. Additionally, ZoKrates provides libraries for commonly used circuits, such as SHA256 and elliptic curve computation.
- Setup: Before generating a proof for each circuit, a one-time setup is required to create a common reference string (CRS).
- Compute witness: ZoKrates automatically computes the corresponding witness based on the circuit when private or public inputs are provided.
- Generate proof: This step involves generating proof information for the given computation.
- Export verifier: ZoKrates allows the exporting of proof-verifier contracts, which can be deployed on Ethereum.

4. Proposed System

In this section, we present an overview of our proposed system, which aims to safeguard the privacy of vehicle owners and their car insurance through NIZK and an Ethereum-based distributed ledger. The notations employed in this paper are presented in Table 1.

Table 1. Notation setting.

Notations	Description
C	Insurance company
U	Vehicle owner
pk_c, pk_u	Public keys for insurance company and vehicle owner
sk_c, sk_u	Private keys for insurance company and vehicle owner
$addr_c, addr_u$	Blockchain addresses for insurance company and vehicle owner
A	Unique asset identifier for insurance
R_A	Authorization record for insurance A
ε	Random number
H	Cryptographic hash function

Our new framework focuses on two different scenarios: the car insurance authorization phase and the car insurance claim authentication phase. In the car insurance authorization phase, the insurance company invokes a private smart contract to apply a hash function to the insurance information to generate an asset identifier. Subsequently, using zero-knowledge proof technology, the asset identifier, the owner's public key, and random numbers are hashed to generate an authorization record, and the private authorization of car insurance is achieved by verifying the zero-knowledge proof in the public contract, thus effectively hiding the asset identifier and the owner's information.

Moving on to the car insurance claim authentication phase, the insurance company devises a secret function and transmits it with a proof key to the vehicle owner. The vehicle owner then solves the secret function to generate a witness, which, in conjunction with the proof key, is utilized to produce proof, demonstrating awareness of the secret function as originally drafted by the insurance company. The vehicle owner subsequently interacts with the public smart contract, submitting the generated proof. Once the contract successfully verifies the proof, it can ascertain the legitimacy of the vehicle owner as the rightful policyholder, all without divulging any specific information about the vehicle owner. This process ultimately enables the realization of the insurance claim.

4.1. System Overview

Our system involves six main entities: the blockchain, vehicle owner, insurance company, service providers, smart contracts, and zero-knowledge proof tool. The architecture and workflow of our proposed system, as follows, are shown in Figure 1:

- Blockchain: The blockchain is responsible for deploying carefully designed smart contracts. Our design choice is to reduce computational overhead and avoid using complex cryptographic tools, such as zero-knowledge proofs, on chain.
- Vehicle owner: In the blockchain, the vehicle owner, as a signatory to the insurance policy, owns the identity attributes stored in the blockchain and receives insurance claims by proving ownership of his identity identifier and insurance attributes.
- Insurance company: An insurance company is an organization that provides insurance products and services. Their main responsibility is to issue car insurance policies to vehicle owners, and process claim payments in the event of an accident. By utilizing blockchain technology and smart contracts, insurance companies can create accounts on the blockchain to streamline subsequent insurance operations and improve efficiency and transparency.
- Service providers: Service providers are other entities related to the insurance business, such as vehicle workshops and emergency service providers, responsible for providing specific services to vehicle owners, such as vehicle repairs and emergency assistance. In blockchain, service providers verify the legitimacy of the vehicle owner's identity before providing services.
- Smart contracts: We designed private and public contracts, where private contracts are used to create and revoke car insurance, and public contracts are used for insurance authorization and vehicle owner authentication. We incorporated the zero-knowledge

proof verification contracts in the public contract that enable the vehicle owner to prove the validity of his identity by providing proofs and public parameters as inputs.

- Zero-knowledge proof tool: We use ZoKrates as our tool to implement zero-knowledge proofs. It performs off-chain calculations of zero-knowledge proofs and on-chain verification of their correctness.

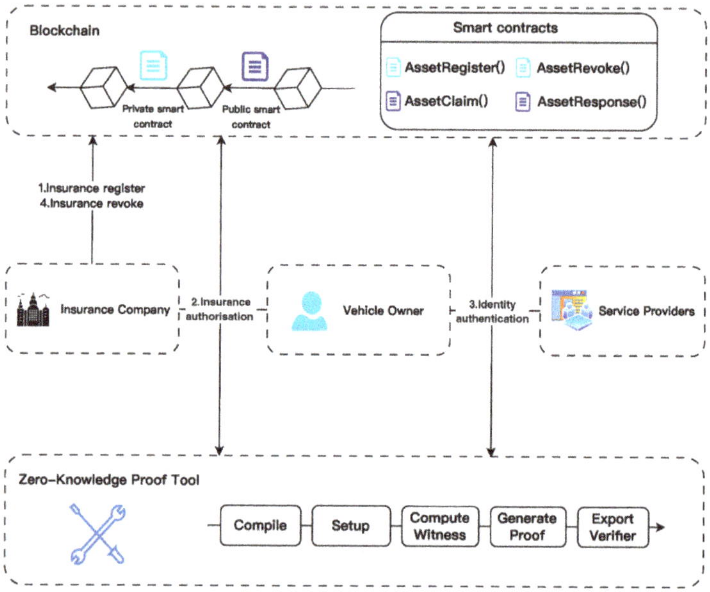

Figure 1. Proposed system model.

4.2. Insurance Register Phase

In the proposed system, the insurance company C possesses pk_c and sk_c, and the vehicle owner U possesses pk_u and sk_u. C must first complete the registration process for the insurance assets by the specific agreement to perform subsequent authorization operations on them. The private smart contract is deployed under C's blockchain address $addr_c$, and only C can invoke it. Algorithm 1 for $AssetRegister()$ is as follows.

Algorithm 1 AssetRegister

Require: $Value, Information$
Ensure: A
$A = sha256(abi.encodePacked(Value, Information, msg.sender, block.number));$
$Insurance[A].value = Value;$
$Insurance[A].information = Information;$
$Insurance[A].creator = msg.sender;$
$Insurance[A].exist = true;$
$Insurance[A].claimed = false;$
emit $LogAssetRegister(A, Value, Information, msg.sender, block.number);$
return $A;$

When C calls $AssetRegister()$ in the private smart contract, a unique asset identifier A for insurance is generated. This A is utilized to retrieve and store the mapping from A to the insurance attributes. To prevent the disclosure of their original values after a potential cyber attack, the actual inputs are concatenated with A. This ensures that sensitive information remains protected and secure. The generation of A is abstracted into the following equation:

$$A = H(Value, Information, msg.sender, block.number) \quad (1)$$

The *Value* field is utilized to record the insurance's worth. This field is specified by C during the insurance registration process, ensuring that relevant financial compensations can be promptly confirmed in the event of accidents or insurance claims. The immutability and transparency of the blockchain guarantee the accuracy and credibility of the *Value* field, eliminating the possibility of human tampering with the value. The *Information* field is employed to record the specific content and terms of the insurance. During the insurance registration process, the creator provides information regarding the insurance plan, scope of liability, and compensation conditions. These detailed pieces of information are permanently recorded, ensuring the immutability of the insurance contract and mitigating the risks of information loss or tampering. The *msg.sender* field serves to identify the address of the insurance contract creator, also known as $addr_c$. By registering this field in the smart contract, we confirm and record the identity of the insurance creator, which aids in the subsequent authorization process for verifying identities. The *block.number* field is used to increase the security and unpredictability of hash operations, which is included in the hash input data during the transaction preparation stage. This does not change the basic way the blockchain works, but it does allow for a specific hash to be generated based on the block number in which the transaction was created. The reason for using the block number instead of the timestamp is that it is determined by a consensus algorithm on the blockchain network and cannot be changed by miners. In contrast, timestamps are set by the miners and can therefore be artificially manipulated by miners. The certainty of the block number makes it safer to use the block number in a contract.

Simultaneously, we incorporated the *exist* field and the *claim* field in our design to determine the registration status and authorization of an insurance policy within the system. The *exist* field is utilized to ascertain whether an insurance policy has been successfully registered in the system. Upon registration of a vehicle insurance policy, the value of this field is set to *True*, indicating that the insurance policy has been effectively added to the blockchain. Conversely, if the registration is unsuccessful, the value of this field will remain *False*, thereby preventing duplicate or invalid insurance contracts. On the other hand, the *claim* field serves to identify whether the insurance has been authorized by the owner. Once the insurance is active, the owner of the insured vehicle will be authorized in this field and will be entitled to make a claim in case of an accident. By recording the authorization status through the smart contract, we ensure that only the legitimate vehicle owner can receive insurance compensation, thereby enhancing the security and credibility of the insurance system.

4.3. Insurance Authorization Phase

During the insurance authorization stage, C can authorize the registered insurance asset A to U through an anonymous authorization process. This process effectively establishes the mapping between A and pk_u on the blockchain while ensuring that this sensitive information remains concealed from public access. By employing this approach, we safeguard the privacy of both U's identity and the specific insurance assets allocated to them, bolstering the overall security and confidentiality of the insurance system. Algorithm 2 for $AssetClaim()$ is as follows.

Algorithm 2 AssetClaim

Require: $A, input, proof$
Ensure: $True$ **or** $False$
$result = verifyTx(input, proof);$
require ($result$, "The proof has not been verified by the contract.");
require ($creatorQuery(A) == msg.sender$, "You are not the creator of A.");
require ($claimedQuery(A) == false$, "This A has been claimed.");
require ($existQuery(A) == true$, This A has been revocationed.);
$claims[A] = A;$
$Insurance[A].claimed = true;$
emit $LogAssetClaim(A, msg.sender);$
return $True;$

To facilitate the transfer of ownership attributes generated during the registration phase to U, we incorporate a hash operation along with the introduction of a random number ε. This approach prevents attackers from cracking hash records by enumerating insurance asset identifier registered on the blockchain and also avoids replay attacks, increasing the security of the system. The process for insurance authorization is as follows:

- Step 1: C formulates a circuit **C** in accordance with the insurance authorization requirements, defining the logic for A's authorization operation within **C**.
- Step 2: C inputs A, pk_u and ε, the algorithm computes $R_A = (A, pk_u, \varepsilon)$ within **C**, where R_A and A are set as the public inputs as well as pk_u and ε are set as the private inputs. Following this, the witness value *witness* is calculated, representing a valid assignment to a variable that encompasses the computation result.
- Step 3: The algorithm generates zero-knowledge proof key pairs pk and vk based on the *witness*, employing a random source commonly referred to as "toxic waste". For generating these zero-knowledge proof key pairs, we employ the efficient Groth16 algorithm, which ensures a balance between the size of the generated proof data and the speed of operation.
- Step 4: C inputs pk, A, pk_u, ε and R_A, the algorithm produces zero-knowledge proof π.
- Step 5: During the verification phase, the smart contract automatically assesses the correctness of the provided inputs. The zero-knowledge proof π undergoes verification using vk. The insurance policy is deemed authorized to the owner only if the above validation holds true. This process guarantees secure and accurate authorization of insurance ownership while preserving privacy and confidentiality.

The validation contract is generated and deployed on the public smart contract via ZoKrates and is named $AssetClaim()$. After passing zero-knowledge verification, it is also necessary to determine whether the account address invoking the contract is the same as the one used to register A and whether A has been registered and authorized. Once all the above operations have been passed, R_A is recorded in the authorization record, and the *claim* of A is changed to *True*.

In this way, we can effectively authorize insurance to U while concealing the ownership relationship through zk-SNARK, safeguarding both U's privacy and the security of the insurance assets.

4.4. Identity Authentication Phase

During an insurance claim for a vehicle involved in an accident, the vehicle owner must provide sufficient information to establish their legal ownership of the insurance. However, relying on the traditional blockchain-based vehicle insurance claim process may expose the owner's information through the input of smart contracts, posing a risk of privacy leakage. To address this problem, we implement owner authentication with privacy-preserving features based on the Fiat–Shamir heuristic. Algorithm 3 for $AssetResponse()$ is as follows.

Algorithm 3 AssetResponse

Require: A,$input'$,$proof'$
Ensure: *True* **or** *False*
1: $result = verifyRes(input', proof')$;
2: **require** $(result,$ "The proof has not been verified by the contract.");
3: **require** $(existQuery(A) == true,$ This A has been revocationed.);
4: **require** $(calims(A) == A,$ This A has not been claimed.);
5: **return** *True*;

To authenticate as the generator of the insurance record R_A on the blockchain, U needs to provide a zero-knowledge proof to demonstrate the truth of the following two statements:

1. R_A can be recomputed: First, U has the sk_u, which allows the generation of pk_u by a hash function. Second, by combining pk_u with the unique hash value of the insurance A and a random number ε, the insurance record R_A can be recalculated.
2. The recomputed R_A is saved on the blockchain.

The first point is essentially proof of the existence of a specific computational process, demonstrating that U possesses the necessary information to generate the insurance record. The second point is essentially proof of the existence of a specific element in a set, which, in this case, is the insurance record R_A saved on the blockchain. However, the proof must maintain the confidentiality of information regarding the specific element being referred to, ensuring that sensitive details about R_A are not disclosed. In the above proof process, both ε and U's identity (pk_u) are kept confidential, ensuring privacy protection. The authentication process proceeds as follows:

- Step 1: A new circuit $\mathbf{C}'$ is designed, the logic of which is for U to prove to the service provider that he/she is the rightful owner of R_A.
- Step 2: U inputs sk_u, A, pk_u and ε, the algorithm computes $pk'_u = H(sk_u)$ and $R'_A = H(A, pk_u, \varepsilon)$, where pk'_u, R'_A and A are set as the public inputs as well as sk_u, pk_u and ε being set as the private inputs. Following this, the witness value $witness'$ is calculated, representing a valid assignment to a variable that encompasses the computation result.
- Step 3: The algorithm generates zero-knowledge proof key pairs pk' and vk' based on $witness'$.
- Step 4: U inputs pk', sk_u, A, pk_u, ε, pk'_u and R_A, and the algorithm produces zero-knowledge proof π'.
- Step 5: During the verification phase, the smart contract automatically assesses the correctness of the provided inputs. The zero-knowledge proof π' undergoes verification using vk'. And the algorithm compare whether pk'_u is the same as pk_u recorded in R_A and whether R'_A is the same as R_A. If all the above proofs are valid, the algorithm returns *True*.

The validation contract is generated and deployed on the public smart contract via ZoKrates and is named *AssetResponse*(). After passing the zero-knowledge verification, it is also necessary to determine whether R_A is the same as the one recorded on the blockchain and whether A is registered. Once all of these operations have been passed, U is determined to be the legal owner of R_A without revealing any identifying information about the vehicle owner in the process. The insurance claim process can then be carried out.

4.5. Insurance Revoke Phase

To revoke A, the *AssetRevoke*() function is called by C within the private smart contract. It is important to enforce that the account address initiating the revoke operation matches the account address used to register A. Once this condition is satisfied, the smart contract updates the *exist* status of A to *False*, effectively stopping any subsequent calls to *AssetClaim*() and *AssetResponse*(). In doing so, the smart contract ensures that the entitlement of A is revoked and prevents any further interaction with it. Algorithm 4 for $AssetRevoke()$ is as follows.

Algorithm 4 AssetRevoke

Require: A
Ensure: $True$ **or** $False$
$creator = creatorQuery(A);$
if $msg.sender == creator$ **then**
    $Insurance[A].claimed = true;$
    emit $LogAssetRevocation(A);$
    return $True;$
end if
return $False;$

The process for asset revoke is as follows:

- Step 1: The algorithm determines whether the address of the account that initiated the undo operation is the address of the account that created A.
- Step 2: The algorithm sets the *exist* field of A to $False$.

5. Analysis of System

In this section, we analyzed the proposed system in various ways.

5.1. Privacy and Security Analysis

- Security of zero Knowledge: ZoKrates offers several alternative zero-knowledge proof schemes, among which Groth16 [36] is a typical and proven secure scheme.
- Unlinkability of identity: The insurance data are stored on the private smart contract, which remains inaccessible to anyone except the insurance company. The authorization process for insurance is implemented through zero-knowledge proofs. To attempt to reveal the owner's private information through ZoKrates, an attacker would need to perform a brute-force attack on the private token within the hash statement. However, given the current computing power, calculating 2^{256} hashes is practically impossible.
- Prevention of replay attack: By adding additional data ε to the computation and incorporating it into the hash calculation, the result of each computation becomes unique even if the same A and pk_u are used. This prevents replay attacks because ε is different each time, making it impossible for an attacker to reuse previous proofs.
- Security of data transmission: All private data transmission is secured through digital signatures and hash encryption. Vehicle owners, insurance companies, and service providers can verify each other's communications through digital signatures. Ensuring the security of the certificate authority that issues the digital signatures and symmetric keys prevents attackers from executing man-in-the-middle attacks by eavesdropping on messages.

5.2. Efficiency Analysis

The performance evaluation of blockchain primarily encompasses two crucial metrics: transaction throughput and latency. Transaction throughput refers to the number of transactions processed within a specific time frame, while latency signifies the response and processing time of transactions. Low throughput may be influenced by factors such as block capacity limitations. Latency is closely associated with algorithm efficiency, and while network bandwidth constraints can also impact latency, its core reasons lie in algorithmic effectiveness rather than inherent issues of the blockchain itself. To enhance throughput, increasing block generation speed is one approach, but this could lead to blockchain forks, thereby compromising system security. To achieve increased block throughput without compromising system security, zero-knowledge proofs present an optimal solution. Exceptional zero-knowledge proof algorithms can significantly minimize latency while simultaneously ensuring the integrity and correctness of remote computation processes without divulging any private information.

Our approach employs the Groth16 algorithm from zk-SNARK to achieve privacy protection. This algorithm relies on the security of solving the elliptic curve discrete logarithm problem. We compared Groth16 with several other common zk-SNARK solutions. As there is no unified benchmark for each construction, we analyzed them based on proof size, benchmark metrics from the respective papers, or estimates from data provided by the inventors. Partala et al. [37] made statistics, and the comparative results are presented in Table 2. In the table, **C** denotes the circuit, | **C** | represents the number of gates in **C**, and **N** indicates the length of computed inputs and outputs. It is evident from the table that each solution has notable strengths and weaknesses, but Groth16 still stands out in terms of proof data size and speed.

Table 2. Time complexity of different algorithms.

	Compiling	Sizes	Prover	Verifier
Groth16	$O(\|C\|^2)$	$O(1)$	$O(\|C\|^2)$	$O(\|C\|)$
Stark	*No*	$O(log^2\|C\|)$	$O(\|C\|log^2\|C\|)$	$O(\|C\|)$
Aurora	*No*	$O(log^2\|C\|)$	$O(\|C\|log\|C\|)$	$O(\|C\|)$
Marlin	$O(\|C\|log\|C\|)$	$O(\|C\|)$	$O(\|C\|log\|C\|)$	$O(N + log\|C\|)$
Sonic	$O(\|C\|log\|C\|)$	$O(1)$	$O(\|C\|log\|C\|)$	$O(N + log\|C\|)$
SuperSonic	$O(\|C\|log\|C\|)$	$O(log\|C\|)$	$O(\|C\|log\|C\|)$	$O(log\|C\|)$

5.3. *Performance Analysis*

To accurately assess the feasibility of the solution, we tested the number of constraints and proof sizes for zero-knowledge proofs, the time consumption for zero-knowledge proofs, the gas consumption caused by smart contract operations and zero-knowledge proof operations, and compared them with other work.

We chose Ethereum as the smart contract platform and used Solidity0.8.0 [38] for smart contract development. The experiments are based on Remix0.34.1, which supports the testing, debugging, and deploying of smart contracts on Ethereum. The consensus algorithm implemented is PoS [39]. In addition, we utilized Web3.js1.10.0 to interact with Ethereum nodes. To simulate the Ethereum network environment, we used Ganache7.9.0 [40] as a personal blockchain for Ethereum development and created a test system using Truffle5.11.2 [41], the most popular development framework for Ethereum. We deployed smart contracts on Truffle and used the Truffle console to simulate data and test smart contracts. Ethereum is the most reliable and widely available blockchain and can develop and execute advanced and customized smart contracts using the Solidity programming language. All zero-knowledge proof operations were implemented on ZoKrates0.8.4.

5.3.1. Number of Constraints and Key Size

In the setup phase of the algorithm, the number of computational constraints and key results obtained by compiling two specific computations in ZoKrates are shown in Table 3. The more computational constraints that are generated, the more complex the specific computations become, resulting in larger key sizes.

Table 3. Results of particular computation pairs.

	Constraints	Proving Key (Mbytes)	Verification Key (bytes)
AssetClaim	104,486	41.6	2000
AssetResponse	131,042	50.1	3000

5.3.2. Time Cost

In the local client, we conducted performance evaluations by generating witnesses and proofs for two specific computations, and the recorded times are presented in Figure 2. Each result in the figure represents the average of 100 test runs, ensuring the accuracy and reliability of the measurements. With this configuration, the time taken to generate the proofs is deemed acceptable, while the time required for generating zk-SNARK proofs depends on various factors, including the computational resources allocated by the prover, the logic of the code, and the complexity of the computation.

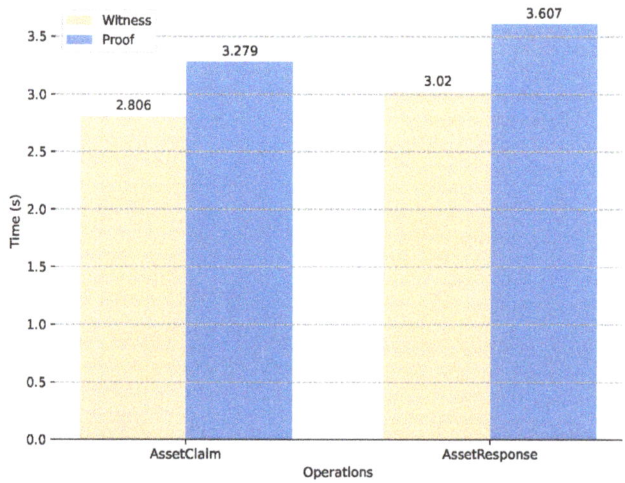

Figure 2. Average latency of witness and proof.

Meanwhile, we also tested the time consumption of key function operations in public and private smart contracts, and the results are shown in Figure 3. Each result in the figure represents the average of 100 test runs, ensuring the accuracy and reliability of the measurements. *AssetRevoke*() is the least time consuming, as this function only contains one compare and one change operation. *AssetResponse*() is the most time consuming because of this function's complex zero-knowledge proof operation.

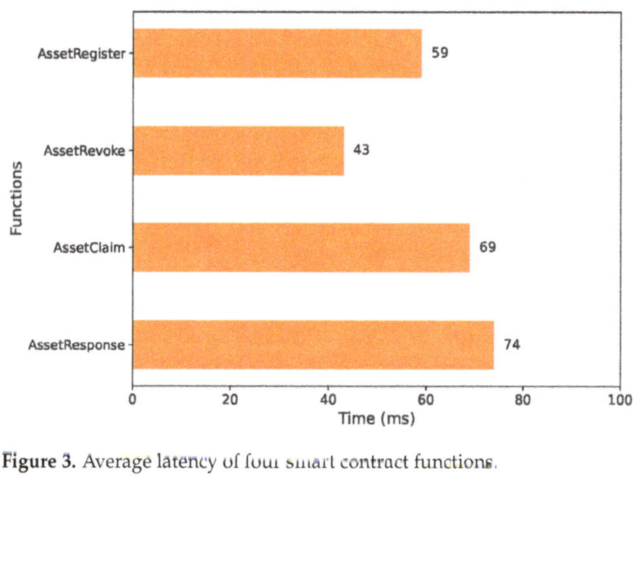

Figure 3. Average latency of four smart contract functions.

We further conducted experiments to evaluate the time consumption of implementing the *AssetClaim*() method using three distinct zk-SNARK algorithms within the ZoKrates framework as depicted in Figure 4. Groth16 is the fastest in compilation settings, witness computation, and proof generation.

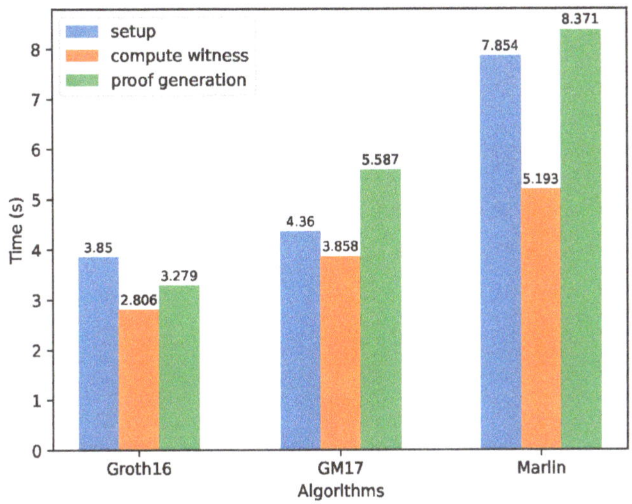

Figure 4. Average latency of three zk-SNARK algorithms.

5.3.3. Gas Consumption

In our proposed scheme, various operations, such as insurance creation, insurance revocation, insurance privacy authorization, and verification of the owner's identity, require transactions to be sent to the smart contract for execution. Insurance authorization and identity verification also necessitate submitting public inputs and zero-knowledge proofs. It is important to note that invoking smart contracts on the blockchain incurs a significant amount of gas consumption as depicted in Table 4. We set the gas price to the average of 20 Gwei (0.00000002 Eth). Gas consumption costs vary depending on the specific smart contract operations being performed. As can be seen from the table, the gas consumption and ether price for each functional operation are perfectly acceptable.

Table 4. Transaction fee statistics.

Contract Operations	Gas_{used}	Fee_{eth}
AssetRegister	98,210	0.00196420
AssetRevoke	23,071	0.00046142
AssetClaim	319,284	0.00638568
AssetResponse	298,255	0.00596510

5.3.4. Characteristic Comparison

In this section, we compare our proposed scheme and other similar insurance schemes in Table 5. Our scheme stands out by meeting the requirement for legal car insurance claims while ensuring the authenticity and privacy of the insurance authorization and the owner authentication processes, a combination not fully achieved by other related schemes. Y means yes, and N means not available.

Table 5. Comparison with other related schemes.

	Demir [42]	Roriz [43]	Liu [44]	Bhadra [45]	Our Scheme
Blockchain based	Y	Y	Y	Y	Y
Identity protection	Y	N	Y	Y	Y
Privacy authorization	N	N	N	N	Y
Data authentication	N	Y	Y	Y	Y

We also performed a performance comparison of similar solution [46], and the results are shown in Figure 5. We merged the *AssetRegister* function and the *AssetClaim* function into the *CreateInsurance* function. Except for the higher gas consumption of the revoke insurance operation, the scheme proposed in this paper outperforms other schemes in the rest of the metrics because it improves the algorithmic process of policy creation and claim verification by smart contracts.

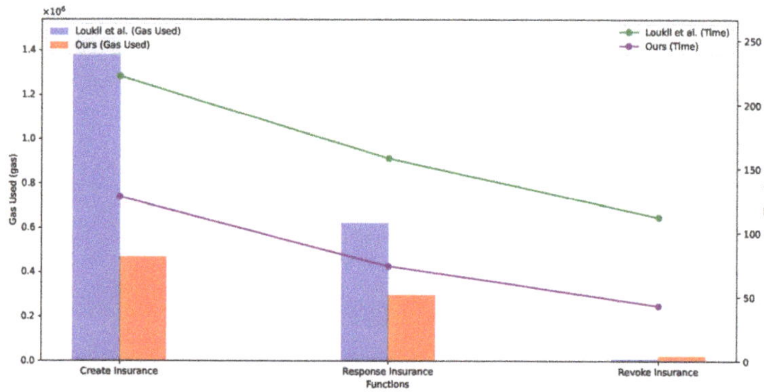

Figure 5. Comparison of time and gas used with similar solution [46].

6. Conclusions

This paper proposes a decentralized zero-knowledge proof-based car insurance claim framework to address the privacy leakage problem in car insurance schemes under the traditional blockchain framework. Currently, in the popular blockchain car insurance schemes, the contents of the insurance, ownership, and transfer records are fully public to all nodes in the chain. During identity verification at the claim stage, the vehicle owner must enter private information into the smart contract to verify the legitimacy of their identity, and this process also carries the risk of privacy leakage. Our goal is to achieve secret authorization during the insurance authorization process and secret verification of the vehicle owner's identity at the claim stage. Compared to traditional car insurance schemes in the blockchain framework, our proposed scheme achieves privacy protection by adding zero-knowledge proof technology on top of decentralization. We design both private and public smart contracts, where insurance authorization and identity verification processes are implemented on public smart contracts. Proofs are optimized using ZoKrates to reduce the size of the proof, which reduces on-chain overhead and provides privacy features. Experimental results show that the scheme performs well in terms of security and performance. A comprehensive comparative analysis with other schemes proves that our scheme achieves both secret authorization and privacy protection.

Our proposed solution increases standardization and reliability within the processes of the car insurance industry. However, the scalability of blockchain technology and the efficiency of zero-knowledge proof algorithms may present new challenges. For the future work, we will further optimize the performance of the zero-knowledge proof algorithm and focus on implementing the model in an automated manner.

Author Contributions: Conceptualization, Z.Q. and Z.X.; methodology, Z.Q. and X.J.; software, Z.Q.; validation, Z.Q., Z.X. and X.J.; formal analysis, Z.Q.; investigation, Z.Q. and Z.X.; resources, Z.Q.; data curation, Z.Q.; writing—original draft preparation, Z.Q.; writing—review and editing, Z.Q. and Z.X.; visualization, Z.Q. and C.R.; supervision, Z.Q. and Z.X.; project administration, Z.X.; funding acquisition, Z.X. and K.C. All authors have read and agreed to the published version of the manuscript.

Funding: This work was supported by the National Natural Science Foundation of China (Grant No. U20A20121); Ningbo public welfare project (Grant No. 202002N3109, 2022S094); Natural Science Foundation of Zhejiang Province (Grant No. LY21F020006); The international cooperation project of Ningbo (Grant No. 2023H012, 2023H007); Science and Technology Innovation 2025 Major Project of Ningbo (Grant No. 2019B10125, 2019B10028, 2020Z016, 2021Z031, 2022Z074, 2022Z241, 2023Z132, 2023Z133, 2023Z216); Ningbo Fenghua District industrial chain key core technology "unveiled the commander" project (Grant No. 202106206).

Data Availability Statement: The data presented in this study are available on request from the corresponding author.

Acknowledgments: The authors would like to acknowledge Ningbo University for its lab facilities and necessary technical support.

Conflicts of Interest: The authors declare no conflict of interest.

References

1. Ewold, F. Insurance and risk. *Foucault Eff. Stud. Gov.* **1991**, *197210*, 201–202.
2. Satuluri, R.K. Digital transformation in Indian insurance industry. *Turk. J. Comput. Math. Educ. (TURCOMAT)* **2021**, *12*, 310–324.
3. Catlin, T.; Lorenz, J.T.; Nandan, J.; Sharma, S.; Waschto, A. *Insurance beyond Digital: The Rise of Ecosystems and Platforms*; McKinsey & Company: New York, NY, USA, 2018.
4. Huang, C.; Wang, W.; Liu, D.; Lu, R.; Shen, X. Blockchain-assisted personalized car insurance with privacy preservation and fraud resistance. *IEEE Trans. Veh. Technol.* **2022**, *72*, 3777–3792. [CrossRef]
5. Derrig, R.A. Insurance fraud. *J. Risk Insur.* **2002**, *69*, 271–287. [CrossRef]
6. Gatteschi, V.; Lamberti, F.; Demartini, C.; Pranteda, C.; Santamaría, V. Blockchain and smart contracts for insurance: Is the technology mature enough? *Future Internet* **2018**, *10*, 20. [CrossRef]
7. Wang, K.; Safavi, A. *Blockchain is Empowering the Future of Insurance*; TechCrunch AOL Inc.: San Francisco, CA, USA, 2016; Volume 7.
8. Bhamidipati, N.R.; Vakkavanthula, V.; Stafford, G.; Dahir, M.; Neupane, R.; Bonnah, E.; Wang, S.; Murthy, J.; Hoque, K.A.; Calyam, P. Claimchain: Secure blockchain platform for handling insurance claims processing. In Proceedings of the 2021 IEEE International Conference on Blockchain (Blockchain), Melbourne, Australia, 6–8 December 2021; pp. 55–64.
9. Qi, H.; Wan, Z.; Guan, Z.; Cheng, X. Scalable decentralized privacy-preserving usage-based insurance for vehicles. *IEEE Internet Things J.* **2020**, *8*, 4472–4484. [CrossRef]
10. Khan, S.N.; Loukil, F.; Ghedira-Guegan, C.; Benkhelifa, E.; Bani-Hani, A. Blockchain smart contracts: Applications, challenges, and future trends. *Peer-To-Peer Netw. Appl.* **2021**, *14*, 2901–2925. [CrossRef] [PubMed]
11. Fiege, U.; Fiat, A.; Shamir, A. Zero knowledge proofs of identity. In Proceedings of the Nineteenth Annual ACM Symposium on Theory of Computing, New York, NY, USA, 25–27 May 1987; pp. 210–217.
12. Sharma, B.; Halder, R.; Singh, J. Blockchain-based interoperable healthcare using zero-knowledge proofs and proxy re-encryption. In Proceedings of the 2020 International Conference on COMmunication Systems & NETworkS (COMSNETS), Bangalore, India, 7–11 January 2020; pp. 1–6.
13. Wan, Z.; Guan, Z.; Zhou, Y.; Ren, K. Zk-AuthFeed: How to feed authenticated data into smart contract with zero knowledge. In Proceedings of the 2019 IEEE International Conference on Blockchain (Blockchain), Atlanta, GA, USA, 14–17 July 2019; pp. 83–90.
14. Wan, Z.; Zhou, Y.; Ren, K. Zk-AuthFeed: Protecting data feed to smart contracts with authenticated zero knowledge proof. *IEEE Trans. Dependable Secur. Comput.* **2022**, *20*, 1335–1347. [CrossRef]
15. Raikwar, M.; Mazumdar, S.; Ruj, S.; Gupta, S.S.; Chattopadhyay, A.; Lam, K.Y. A blockchain framework for insurance processes. In Proceedings of the 2018 9th IFIP International Conference on New Technologies, Mobility and Security (NTMS), Paris, France, 26–28 February 2018; pp. 1–4.
16. Brophy, R. Blockchain and insurance: A review for operations and regulation. *J. Financ. Regul. Compliance* **2020**, *28*, 215–234. [CrossRef]
17. Yadav, A.S.; Charles, V.; Pandey, D.K.; Gupta, S.; Gherman, T.; Kushwaha, D.S. Blockchain-based secure privacy-preserving vehicle accident and insurance registration. *Expert Syst. Appl.* **2023**, *230*, 120651. [CrossRef]
18. Nizamuddin, N.; Abugabah, A. Blockchain for automotive: An insight towards the IPFS blockchain-based auto insurance sector. *Int. J. Electr. Comput. Eng. (IJECE)* **2021**, *11*, 2443–2456. [CrossRef]

19. Lamberti, F.; Gatteschi, V.; Demartini, C.; Pelissier, M.; Gomez, A.; Santamaria, V. Blockchains can work for car insurance: Using smart contracts and sensors to provide on-demand coverage. *IEEE Consum. Electron. Mag.* **2018**, *7*, 72–81. [CrossRef]
20. Bader, L.; Bürger, J.C.; Matzutt, R.; Wehrle, K. Smart contract-based car insurance policies. In Proceedings of the 2018 IEEE Globecom Workshops (GC wkshps), Abu Dhabi, United Arab Emirates, 9–13 December 2018; pp. 1–7.
21. Chiu, W.Y.; Meng, W. Towards decentralized bicycle insurance system based on blockchain. In Proceedings of the 36th Annual ACM Symposium on Applied Computing, Virtual, 22–26 March 2021; pp. 249–256.
22. Nanda, S.K.; Panda, S.K.; Das, M.; Satapathy, S.C. Automating vehicle insurance process using smart contract and Ethereum. In Proceedings of the Advances in Micro-Electronics, Embedded Systems and IoT: Proceedings of Sixth International Conference on Microelectronics, Electromagnetics and Telecommunications (ICMEET 2021), Bhubaneswar, India, 27–28 August 2021; Springer: Berlin/Heidelberg, Germany, 2022; pp. 237–247.
23. Kumar, S.; Dohare, U.; Kaiwartya, O. FLAME: Trusted fire brigade service and insurance claim system using blockchain for enterprises. *IEEE Trans. Ind. Inform.* **2022**, *19*, 7517–7527.
24. Pawar, V.; Sachdeva, S. ParallelChain: A scalable healthcare framework with low-energy consumption using blockchain. *Int. Trans. Oper. Res.* **2023**. [CrossRef]
25. Iyer, V.; Shah, K.; Rane, S.; Shankarmani, R. Decentralised Peer-to-Peer Crop Insurance. In Proceedings of the 3rd ACM International Symposium on Blockchain and Secure Critical Infrastructure, Virtual, 7 June 2021; pp. 3–12.
26. Jha, N.; Prashar, D.; Khalaf, O.I.; Alotaibi, Y.; Alsufyani, A.; Alghamdi, S. Blockchain based crop insurance: A decentralized insurance system for modernization of Indian farmers. *Sustainability* **2021**, *13*, 8921. [CrossRef]
27. Nakamoto, S. Bitcoin: A peer-to-peer electronic cash system. *Decentralized Bus. Rev.* **2008**, *13*, 21260.
28. Nofer, M.; Gomber, P.; Hinz, O.; Schiereck, D. Blockchain. *Bus. Inf. Syst. Eng.* **2017**, *59*, 183–187. [CrossRef]
29. Zheng, Z.; Xie, S.; Dai, H.N.; Chen, X.; Wang, H. Blockchain challenges and opportunities: A survey. *Int. J. Web Grid Serv.* **2018**, *14*, 352–375. [CrossRef]
30. Buterin, V. A next-generation smart contract and decentralized application platform. *White Pap.* **2014**, *3*, 1–36.
31. Hewa, T.; Ylianttila, M.; Liyanage, M. Survey on blockchain based smart contracts: Applications, opportunities and challenges. *J. Netw. Comput. Appl.* **2021**, *177*, 102857. [CrossRef]
32. Goldwasser, S.; Micali, S.; Rackoff, C. The knowledge complexity of interactive proof-systems. In *Providing Sound Foundations for Cryptography: On the Work of Shafi Goldwasser and Silvio Micali*; ACM: New York, NY, USA, 2019; pp. 203–225.
33. Gennaro, R.; Gentry, C.; Parno, B.; Raykova, M. Quadratic span programs and succinct NIZKs without PCPs. In Proceedings of the Advances in Cryptology–EUROCRYPT 2013: 32nd Annual International Conference on the Theory and Applications of Cryptographic Techniques, Athens, Greece, 26–30 May 2013; Proceedings 32; Springer: Berlin/Heidelberg, Germany, 2013; pp. 626–645.
34. Canetti, R.; Chen, Y.; Holmgren, J.; Lombardi, A.; Rothblum, G.N.; Rothblum, R.D.; Wichs, D. Fiat-Shamir: From practice to theory. In Proceedings of the 51st Annual ACM SIGACT Symposium on Theory of Computing, Phoenix, AZ, USA, 23–26 June 2019; pp. 1082–1090.
35. Eberhardt, J.; Tai, S. Zokrates-scalable privacy-preserving off-chain computations. In Proceedings of the 2018 IEEE International Conference on Internet of Things (iThings) and IEEE Green Computing and Communications (GreenCom) and IEEE Cyber, Physical and Social Computing (CPSCom) and IEEE Smart Data (SmartData), Halifax, NS, Canada, 30 July–3 August 2018; pp. 1084–1091.
36. Baghery, K.; Pindado, Z.; Ràfols, C. Simulation extractable versions of Groth's zk-SNARK revisited. In Proceedings of the International Conference on Cryptology and Network Security, Vienna, Austria, 14–16 December 2020; Springer: Berlin/Heidelberg, Germany, 2020; pp. 453–461.
37. Partala, J.; Nguyen, T.H.; Pirttikangas, S. Non-interactive zero-knowledge for blockchain: A survey. *IEEE Access* **2020**, *8*, 227945–227961. [CrossRef]
38. Dannen, C. *Introducing Ethereum and Solidity*; Springer: Berlin/Heidelberg, Germany, 2017; Volume 1.
39. Bentov, I.; Lee, C.; Mizrahi, A.; Rosenfeld, M. Proof of activity: Extending bitcoin's proof of work via proof of stake [extended abstract] y. *Acm Sigmetrics Perform. Eval. Rev.* **2014**, *42*, 34–37. [CrossRef]
40. Lee, W.M.; Lee, W.M. Testing smart contracts using ganache. In *Beginning Ethereum Smart Contracts Programming: With Examples in Python, Solidity, and JavaScript*; Springer: Berlin/Heidelberg, Germany, 2019; pp. 147–167.
41. Mohanty, D.; Mohanty, D. Frameworks: Truffle and embark. In *Ethereum for Architects and Developers: With Case Studies and Code Samples in Solidity*; Springer: Berlin/Heidelberg, Germany, 2018; pp. 181–195.
42. Demir, M.; Turetken, O.; Ferworn, A. Blockchain based transparent vehicle insurance management. In Proceedings of the 2019 Sixth International Conference on Software Defined Systems (SDS), Rome, Italy, 10–13 June 2019; pp. 213–220.
43. Roriz, R.; Pereira, J.L. Avoiding insurance fraud: A blockchain-based solution for the vehicle sector. *Procedia Comput. Sci.* **2019**, *164*, 211–218. [CrossRef]
44. Liu, X.; Yang, H.; Li, G.; Dong, H.; Wang, Z. A blockchain-based auto insurance data sharing scheme. *Wirel. Commun. Mob. Comput.* **2021**, *2021*, 3707906. [CrossRef]

5. Bhadra, O.; Sahoo, S.; Kumar, C.M.; Halder, R. Decentralized Insurance Subrogation Using Blockchain. In Proceedings of the 2022 5th International Conference on Blockchain Technology and Applications, Xi'an, China, 16–18 December 2022; pp. 1–9.
6. Loukil, F.; Boukadi, K.; Hussain, R.; Abed, M. Ciosy: A collaborative blockchain-based insurance system. *Electronics* **2021**, *10*, 1343. [CrossRef]

electronics

MDPI

Article

A Blockchain-Based Method for Optimizing the Routing of High-Frequency Carbon-Trading Payment Channels

Yu Song [1,*], Ao Xiong [1], Xuesong Qiu [1], Shaoyong Guo [1], Dong Wang [2], Da Li [2], Xin Zhang [3] and Yue Kuang [3]

1 State Key Laboratory of Network and Switching Technology, Beijing University of Posts and Telecommunications, Beijing 100876, China; xiongao@bupt.edu.cn (A.X.); xsqiu@bupt.edu.cn (X.Q.); syguo@bupt.edu.cn (S.G.)

2 State Grid Digital Technology Holding Co., Ltd., Beijing 100053, China; wangdong@sgdt.sgcc.com.cn (D.W.); lida@sgdt.sgcc.com.cn (D.L.)

3 Chinese Research Academy of Environmental Sciences, Beijing 100012, China; zhangxin@craes.org.cn (X.Z.); jwjtzx@craes.org.cn (Y.K.)

* Correspondence: songyu@bupt.edu.cn

Abstract: Carbon trading is an effective way to achieve carbon neutrality. It is a market mechanism aimed at reducing global greenhouse gas emissions and carbon dioxide emissions. Blockchain technology can be applied to the carbon-trading scenario using characteristics that guarantee the security, decentralization, data immutability, and data traceability of the carbon-trading process. It would be difficult to implement carbon trading on blockchains for all enterprises and individuals, as the current performance of blockchains does not meet the requirements. There has been extensive research conducted on blockchain performance optimization, and the off-chain payment channel is one of the more mature solutions. This approach involves the transfer of transactions to off-chain transactions, thus avoiding high transaction fees. Existing research has addressed the problem of routing security and efficiency, with less emphasis on factors such as routing transaction costs, node reputation, and routing path considerations. This paper researches the optimization of payment routing in Payment Channel Networks (PCNs) and proposes the Multi-Factor Routing Payment Scheme (MFPS), which integrates factors such as the node reputation, transaction fee cost, and distance to select the route for payment transactions. In order to improve the success ratio of routing transactions, the transaction-splitting algorithm is proposed. To ensure the security and privacy of the transaction process, the Asymmetric Time-Lock Contract (ATLC) protocol is proposed. The results of extensive experimental simulations show that the MFPS proposed in this paper outperforms the ShortestPath, Cheapest, and SplitDistance algorithms. It achieves an approximately 13.8%∼49% improvement in the transaction success ratio and reduces the average transaction processing cost. The security and privacy measures can defend against wormhole and double-flower attacks and exhibit better performance in terms of computational verification and message overhead.

Keywords: carbon trading; off-chain payment channels; blockchain; route optimization; privacy and security

Citation: Song, Y.; Xiong, A.; Qiu, X.; Guo, S.; Wang, D.; Li, D.; Zhang, X.; Kuang, Y. A Blockchain-Based Method for Optimizing the Routing of High-Frequency Carbon-Trading Payment Channels. *Electronics* **2023**, *12*, 2586. https://doi.org/10.3390/electronics12122586

Academic Editor: Hamed Taherdoost

Received: 15 May 2023
Revised: 3 June 2023
Accepted: 4 June 2023
Published: 8 June 2023

1. Introduction

Carbon trading refers to the trading of greenhouse gas emission rights and is a market mechanism aimed at reducing global greenhouse gas emissions and carbon dioxide emissions [1]. In December 1997, the governments of the United Nations adopted the Kyoto Protocol, which introduced a market mechanism for carbon dioxide emission rights. These emission rights are treated as a commodity that can be traded, which is referred to as carbon trading [2]. When the Kyoto Protocol came into effect in 2005, the carbon emissions trading market reached $10.9 billion and grew at an annual rate of 108%, reaching approximately $669 billion in 2013 [3]. As the scale of carbon trading increased, so did the need for a platform that was stable, secure, and decentralized, with tamper-proof data and traceable transaction information, to support companies and individuals involved in carbon trading. The features of the blockchain platform can be applied to the carbon-trading scenario.

In 2008, Satoshi Nakamoto published the white paper "Bitcoin: A Peer-to-Peer Electronic Cash System" [4], Bitcoin was born, and blockchain technology entered the public domain and began to develop rapidly. Blockchains are decentralized through P2P networks, consensus mechanisms, and cryptography to ensure the security and consistency of user nodes [5]. At present, blockchains have a wide range of applications in government, finance, the Internet of Things, healthcare, energy, and electricity [6]. In addition, blockchains in carbon-trading applications have also been studied in recent years [7,8].

Due to the technical characteristics of blockchains [9], they can be applied in many industries but public blockchains have certain security problems [10]. Blockchains can be either public blockchains or permissioned blockchains. The biggest difference between them is that permissioned blockchains can only be accessed by users with the correct permissions [11], making them more secure than public blockchains. Permissioned blockchains are represented by Hyperledger Fabric [12]. Although permissioned blockchains have improved security, their poor scalability remains a challenge. When faced with a large number of users conducting transactions concurrently, current blockchains cannot support faster execution.

Bitcoin and Ethereum are digital cryptocurrencies represented using blockchain technology. Bitcoin's Transactions Per Second (TPS) is only 7, and Ether's TPS is in the range of 30 to 40 [13]. Clearly, Bitcoin and Ethereum are no longer sufficient for the current digital payment scenario, and it is now difficult to support the current scale of corporate and personal carbon trading. With the explosive growth in the number of blockchain users, the scalability issue has become a significant challenge, limiting the development of blockchains. The transaction process in blockchains involves broadcasting, verification, and a large number of calculations to achieve consensus confirmation, making it less scalable [14]. In addition, the fee for Bitcoin on-chain transactions is determined by the size of the transaction data in the Unspent Transaction Output (UTXO), the number of transactions, and other factors, independent of the transaction amount [15]. Users need to pay substantial fees for each transaction, making these systems unsuitable for high-frequency, small-scale personal carbon-trading applications. In order to solve the scalability problem of blockchains, many solutions have been proposed in the literature [16,17]. Solutions that have been proposed to improve the performance of blockchains include the sharding mechanism [18], directed acyclic graph (DAG)-based blockchains [19], off-chain payment channel networks [20,21], and cross-chain technology [22]. The off-chain payment channel network allows for the transfer of small, high-frequency, and time-consuming on-chain transactions to off-chain channels, enabling fast, low-fee transactions.

Payment channels refer to technologies such as the Bitcoin-based Lightning Network [20] and Ethereum-based Thunderbolt Network technologies [21]. Transferring on-chain transactions to off-chain transactions is the key idea of payment channels. In terms of applied carbon trading, this concept can provide a fast and low-cost off-chain payment channel for small-scale carbon transactions between individuals. In addition, larger transactions involving corporate carbon-trading quotas can be conducted on the blockchain, effectively reducing the burden on the blockchain and increasing transaction throughput, as well as reducing processing fees for small-scale transactions. Payment channels between users form a network of payment channels.

Current research on routing algorithms for payment channel networks focuses on improving privacy and security during transactions and optimizing routing and scheduling problems to improve transaction efficiency and success rates. However, there is a lack of analysis regarding transaction fees in transaction routing. In a small-scale transaction scenario, the more relay nodes that are involved, the higher the fees charged, which can be a significant expense for the user. At the same time, node transactions can cause an imbalance between the two ends of the channel among certain nodes. This imbalance can cause channel blocking and an imbalance of funds, impacting the path selection for certain transaction amounts, which can degrade performance and decrease transaction success rates.

In order to address the above problems, this paper focuses on the following:

1. The Multi-Factor Routing Payment Scheme (MFPS)-based algorithm is proposed. The path selection takes into account the transaction fees, node reputation, and distance to select the optimal path for the transaction. In order to reduce channel congestion and avoid the problem of channel capacity imbalances, the channel balance is adjusted by applying different charging percentages based on the channel balance difference. In order to improve the transaction success ratio, when there is no path in the payment channel network to fulfill the amount required for the transaction, the transaction amount is split according to the largest channel balance, employing a greedy transaction splitting approach to minimize the number of splits.
2. The Asymmetric Encryption-Based Time-Locked Contract payment protocol (ATLC) is proposed. This protocol uses a combination of asymmetric encryption algorithms and data signing algorithms for encryption, decryption, and verification. It ensures the atomicity of the transaction process, data consistency, balance security, and value privacy.
3. Extensive experimental simulations are conducted, demonstrating that the MFPS algorithm achieves higher transaction payment success ratios with lower average handling expenses. The proposed solutions also provide security and privacy protection against wormhole attacks, double-flower attacks, and false-transaction attacks, demonstrating improved performance in terms of computational verification and message overhead.

The remainder of this paper is as follows. In Section 2, the current research in this area is presented. In Section 3, the knowledge related to the study of this paper is introduced. In Section 4, the system model is described and the related equations are given. In Section 5, the design and specific implementation process of the routing algorithm, MFPS, is presented, and the ATLC protocol is introduced in detail. In Section 6, simulation experiments are carried out for algorithm verification. Section 7 summarizes and concludes this paper.

2. Related Works

Ref. [23] analyzes the challenges and latest research trends in off-chain payment channel networks. Current routing algorithms for payment channel networks can be divided into two categories: single-way routing and multi-way routing [24]. In cases where a single route does not allow a transaction to be split, an alternative approach is to find a route path that satisfies the requirements for the transfer of transaction assets. This method is mainly used in scenarios where the network size and the number of transaction transfers are small. Ref. [25] proposes an application of the ant-routing algorithm in lightning networks, inspired by the behavior of ants, using the characteristics of pheromones released on the paths taken by ants to achieve a de-neutralized routing mechanism.

Multi-way routing aims to find at least one path for a transaction, and if necessary, the transaction amount can be split across different paths, which can effectively reduce the problem of channel blocking. Ref. [26] was the first to propose the implementation of the SilentWhispers algorithm for route forwarding using a distributed credit network, which is based on the idea of Landmark routing. This algorithm enables the selection of nodes with the highest network connectivity to generate transaction paths and uses a combination of password-sharing multi-party computation and digital signature chains to enhance user security and privacy. By combining PCN concurrency and privacy protection protocols, a new protocol, MHTLC, was developed. Ref. [27] proposes the SpeedyMurmurs algorithm, a routing algorithm for decentralized path-based transactions, with privacy protection for transaction nodes and transaction amounts using the VOUTE algorithm within Friend-to-Friend networks. The architecture of SpeedyMurmurs is similar to that of SilentWhisper but it offers better efficiency and performance. Ref. [28] proposes the CoinExpress algorithm, which uses distributed dynamic routing to find a routing path for state transfer based on a breadth- or depth-first algorithm. The algorithm effectively addresses the problem of parallelizing transaction processing. Ref. [29] proposes the Flash algorithm, which categorizes transactions into elephant stream payments and mouse stream payments according to their size. Larger transaction amounts use elephant stream payments, which require multiple routing paths to collaborate to complete the transaction. Smaller, single transaction amounts use mouse stream payments, where the transaction path is directly selected from the routing table for transferring assets. Ref. [30] proposes the Spider algorithm, which uses a packet-forwarding-like mechanism to achieve balanced routing and address the issues of network congestion and channel capacity imbalances. Although multi-route splitting transactions can reduce channel blocking and improve network flow, the use of multi-route paths requires real-time network state information and imposes significant communication overhead on nodes, which introduces concerns about transaction atomicity and reduces the efficiency of routing.

Ref. [31] proposes a new mechanism called anonymous multi-hop locking (AMHL) to protect against wormhole attacks in off-chain payment channel networks, effectively reducing the computational and communication overhead in the network. The privacy security protocol of k-HTLC based on PCN transactions proposed in Ref. [32] uses symmetric encryption algorithms to protect against wormhole attacks on payment channel networks, but the security of key transmission cannot be guaranteed. Ref. [33] analyzes congestion attacks on off-chain payment channels and proposes mitigation techniques to increase the difficulty of carrying out such attacks. Ref. [34] proposes the PnP algorithm, which mainly addresses the estimated payment needs between nodes. The PnP algorithm effectively balances channel funds to meet these needs, without relying on any trusted third party, thereby providing strong protection against malicious attacks with minimal overhead.

The above-mentioned works on off-chain payment channels mainly focus on enhancing the security and efficiency of routing, with other factors, such as routing transaction costs, node reputation, and routing paths, often being overlooked. Ref. [35] proposes the CheaPays algorithm to minimize transaction costs while meeting the feasibility and timeliness constraints. However, it does not take into account the node reputation. Ref. [36] proposes a comprehensive routing scheme that takes into account factors such as fee offers, path distance, and historical reputation. However, it does not take into account channel congestion or channel capacity imbalances. In this paper, we design a routing scheme that takes into account factors such as the routing transaction cost, node reputation, and routing path to improve the transaction success ratio in off-chain payment channel networks and reduce the problem of congestion and channel capacity imbalances in the channel. In addition, we propose an ATLC to ensure the consistency and atomicity of transactions, improve the security and privacy of transactions, and protect against wormhole and double-flower attacks.

3. Related Knowledge

3.1. *Off-Chain Payment Channels*

Off-chain payment channels are used to carry out high-frequency and small-scale transactions off-chain, enabling the transfer of certain on-chain transactions to off-chain. This indirectly improves the transaction throughput of the blockchain while reducing the fees incurred by users for on-chain transactions. The complete payment channel process is shown in Figure 1 and consists of three phases: the establishment phase, the payment phase, and the closing phase. During the establishment phase, a fixed amount of money is deposited on the blockchain as the initial balance of the channel, and this money cannot be accessed on the blockchain until the channel is closed after opening. During the payment phase, both users make payments within the channel, transferring funds and updating the channel's status. The payment channel can be classified as either a one-way channel or a two-way channel. A one-way channel only allows transactions to be conducted with a specific party, whereas a two-way channel enables transactions between both parties. In this paper, we focus on the analysis of a two-way channel. The closing phase occurs once the payment is finalized and the channel's status is updated. During this phase, a transaction to close the channel is submitted to the blockchain, and the node transfers the remaining balance from the channel to the blockchain. If any party behaves maliciously during the payment phase, the other party can submit the channel status to the blockchain for broadcasting at any time to penalize the node that engaged in malicious behavior.

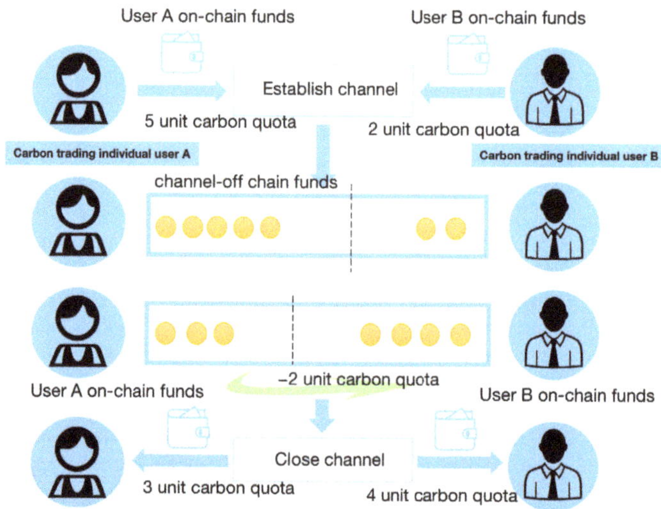

Figure 1. Flowchart of off-chain payment channel.

3.2. *Off-Chain Payment Channels*

Cross-channel payments occur when an intermediate user receives a payment from the previous node and refuses to perform the next transfer of funds, and the recipient refuses to receive the payment. In order to address this problem, a Hashed Time-Lock Contract (HTLC) is used, with a hash lock and a time lock at its core. As shown in Figure 2, the receiver first randomly generates the secret R, and after hashing, generates H to send to the sender, which is delivered through a secure messaging channel and not the payment channel. The sender and the intermediate node must include H in the transaction contract, and only after the receiver provides the secret R to the previous hop forwarder within the specified time can the verification process unlock the funds, confirming that Hash(R) = H. The payment process in the HTLC is bound by the time lock, and if the node does not receive the secret R within the specified time, the funds will be returned to the forwarder. The time locks in the transaction path are specified in decreasing order to prevent nodes from experiencing losses in terms of the transaction amount.

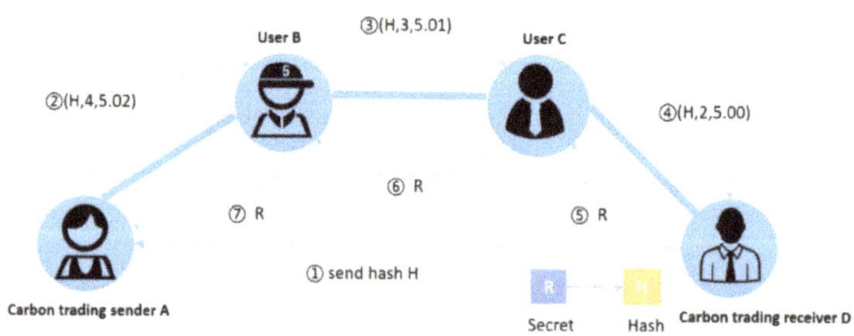

Figure 2. Hashed Time Lock Contract.

4. System Model

This section describes the overall model, which includes the network and payment models. The model architecture is shown in Figure 3.

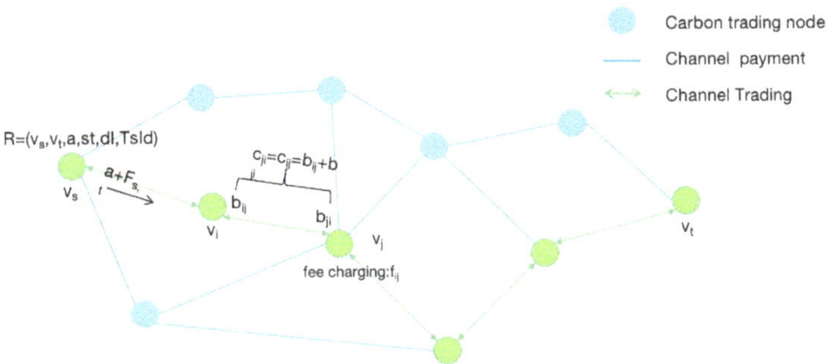

Figure 3. Flow chart of off-chain payment channel.

4.1. Network Model

The topology of a PCN can be represented as a directed acyclic graph of the form G = (V,E), where V denotes the set of nodes in the network and E is the set of payment channels. Each node $v_i \in V$ denotes a user who has a funding account and has established an off-chain payment channel with at least one other node. Each edge $e = (v_i, v_j) \in E$ denotes a payment channel, where v_i is a sender node and v_j is a receiver node, and each edge has the following properties:

- Set each edge $e = (v_i, v_j) \in E$ to have a transaction fee f_{ij}, which is the fee charged by node v_i when transferring the transaction amount from v_i to v_j.
- Set each edge $e = (v_i, v_j) \in E$ to have a channel balance b_{ij}, which represents the maximum transaction amount that can be transferred from v_i to v_j.
- Set each edge $e = (v_i, v_j) \in E$ to have an HTLC tolerance time τ_{ij}, which is the longest time for v_i to wait for v_j to provide the secret R.
- Set the total balance of each edge $e = (v_i, v_j) \in E$ channel as c_ij, which denotes the sum of the channel balances on both sides. Therefore, $b_{ij} + b_{ji} = c_{ij} = c_{ji}$ and the balance is always $b_{ij} \leq c_{ij}$.

For simplicity, this paper does not consider cases where the total channel balance c_{ij} is 0 or the channel balance $b_{ij} \geq 0$. Only non-negative channel balances are considered, and the payment channels are assumed to be two-way channels.

4.2. Payment Model

Define a payment request $R = (v_s, v_t, a, st, dl, TsId)$, where v_s is the initiator of the payment request; v_s is the sender; v_t is the receiver; a is the amount of the transaction, excluding the processing fee to be paid in the transfer path; *st* is the start time of the transaction; *dl* is the transaction cutoff time; and *TsId* is the unique identifier of the transaction. P is the set of all paths in the PCN, P_R is a set of paths that satisfy the $v_s - v_t$ payment request R, and $p \in P_R$ denotes a path that satisfies the $v_s - v_t$ payment request. Suppose the associated path p is serialized as $p = v_0 \rightarrow v_1 \rightarrow ...... \rightarrow v_L$, where $v_0 = v_s$, $v_L = v_t$, and $v_i \in p$ denotes a node in path p. Define the function $F_{l,m}$ to denote the sum of fees to be paid for a transaction on path p from v_l to v_m. Additionally, $f_{i,i+1}$ is used to denote the fees charged for a transaction path passing through node i in channel $(i, i+1)$, where $0 \leq l < m \leq L$. The function is shown below:

$$F_{l,m} = \begin{cases} \sum_{i=l+1}^{m-1} f_{i,i+1}, & v_l = v_s \\ \sum_{i=l}^{m-1} f_{i,i+1}, & v_l \neq v_s \end{cases} \tag{1}$$

To successfully complete the payment request, certain constraints need to be satisfied:

- Timing constraint: To complete the payment request R, the next route must be forwarded within the time-lock tolerance time $\tau_{i,i+1}$ of the HTLC in each payment channel, as shown in Equation (2). Timeliness is achieved through the HTLC's time lock, which guarantees the integrity of transactions in intermediate payment channels.

$$\tau_{i,i+1} \geq L - i, \forall i \in [0, L - l] \tag{2}$$

- Feasibility constraint: To successfully complete the payment request R by passing through each channel in the path p , it is necessary for the channel balance $b_{i,i+1}$ to be at least equal to the sum of the payment amount and all the handling fees incurred after passing through node v_i, as shown in Equation (3).

$$b_{i,i+1} \geq a + F_p(i+1, L), \forall i \in [0, L - l] \tag{3}$$

- Timeliness constraint: To successfully complete the payment request R, the payment must be finalized within the specified completion time of the payment request and reach the intended recipient. Therefore, it is necessary to ensure that the sum of the payment start time st and the time needed to complete the transaction transfer in each link along the path does not exceed the transaction deadline dl specified in the payment request. d_e indicates the transaction transfer time required in channel link e, as shown in Equation (4).

$$st + \sum_{e \in p} d_e \leq dl \tag{4}$$

5. PCN Routing Optimization Algorithm

Suppose an individual carbon transaction initiator initiates a transaction to a recipient using a payment channel link $p = v_0 \rightarrow v_1 \rightarrow \ldots\ldots \rightarrow v_L$, which goes through many node choices in between. In this paper, the selection of paths and nodes is mainly based on the channel handling fee, the historical reputation value of nodes, and the nearest distance. This section proposes an MFPS-based routing optimization algorithm and an ATLC payment protocol. The implementation algorithms include the Route-Init, PaySplit, PaymentRoute, and Recipient routing algorithms.

5.1. MFPS: Multi-Factor Routing Payment Scheme

In this subsection, the MFPS is proposed. The transaction initiator selects the optimal path for the transaction payment based on a combination of factors, including the total transaction fee charged by the path, the node reputation value, and the distance of the path.

5.1.1. Transaction Fees

An important factor to consider in path selection is the cost of the intermediate nodes in the transaction path so that the transaction costs are minimized while satisfying the timing, feasibility, and timeliness constraints. Three types of transaction fees are considered in this paper, namely, a proportional fee for the channel node based on the transaction amount, a fixed fee for the node, and a fee based on the difference in balance between the two ends of the channel. The fee to be charged for a transaction path through node i is shown in Equation (5), where t_{ij} is the amount of money transferred in the channel e_{ij}, including the transaction amount and the handling fee charged by the nodes in the chain. is the proportion parameter of the fee charged for the transaction amount, is the proportion parameter of the fee charged according to the balance difference between the two ends of the channel, and H is a constant that indicates the fixed fee charged for the transaction passing through a node. f_{ij} denotes the transaction passing through the channel e_{ij} and the handling fee charged by node v_i. The total fee F to be paid for the entire transaction path from the sender side v_s to the receiver side v_t is shown in Equation (6).

$$f_{i,j} = \begin{cases} 0, & v_i = v_j \quad or \quad v_i = v_s \\ t_{i,j} * \alpha_1 + \frac{|b_{ij} - b_{ji}|}{b_{ij}} * \alpha_2 + H, & otherwise \end{cases} \tag{5}$$

$$F_{s,t} = \sum_{i=s+1}^{i=t-1} f_{i,i+1} \tag{6}$$

In this paper, we consider the case of channel congestion, i.e., the scenario where there is an imbalance in the channel's balance at both ends. In the u, v node channel, when the channel is established, both users, that is, both ends of the channel, deposit a balanced amount. However, with the subsequent occurrence of transactions, if the transaction direction has been from node u to node v, the result is that one side of the channel (node v) has sufficient funds, whereas the other side (node u) has insufficient funds, resulting in an imbalance between the two ends of the balance. If a transaction later needs to be transferred from $u \rightarrow v$ but the funds available at node u are insufficient to meet the transaction amount, it will result in congestion of the transaction at the node. The transaction will remain congested until the time lock stipulates that the transaction will not be released until the funds are available. The imbalance in the channel balance at both ends of the channel causes a lower transaction success ratio, resulting in channel congestion. Before the transaction is canceled, the balance of the channel that was passed before the transaction will be locked, resulting in the channel-funds deadlock phenomenon. In this paper, we use different percentages of the difference between the two ends of the channel balance to determine the fee. Thus, choosing a path with a higher channel balance results in a lower fee compared to choosing the other end of the path. If there are two or more available paths, the fee becomes a consideration in selecting the path. Under the same conditions, the path with a relatively abundant channel balance at one end is selected first, which can reduce the occurrence of channel congestion due to balance imbalances.

5.1.2. Node Reputation Value

An important factor to consider in transaction path selection is the reputation value of the nodes. If the reputation value of a node is low, there is a higher probability of it being malicious, which can cause the transaction not to reach the intended recipient within the specified time lock, leading to transaction failure. Therefore, in order for the transaction to successfully reach the intended recipient within the specified time, the channels associated with nodes with a higher reputation value should be considered in the transaction path selection. In this paper, we consider that the reputation value of a node is related to the number of times it participates in a transaction and exhibits correct/malicious behavior. Correct/malicious behaviors are defined as follows:

- Correct behavior: The node successfully transfers the transaction amount and required secret within the specified time lock.
- Malicious behavior: The node fails to transfer the amount of the transaction within the prescribed time lock, fails to provide the required secret, or provides a forged secret.

The node reputation value is shown in Equation (8). In order to ensure that the reputation gap between nodes is within a certain range and avoid node centralization, the sigmoid function is used to limit the reputation value within the range of (0, 1). R_t is the reputation value at the iteration and is a sigmoid function that takes values within the range of (0, 1). The impact factor I_t represents the reward factor when node v_i exhibits correct behavior and the penalty factor when node v_i exhibits malicious behavior. g_i denotes the number of correct behaviors exhibited by node i, m_i denotes the number of malicious behaviors exhibited by node v_i, and n_i denotes the total number of transactions that node v_i has participated in. The penalty for a node engaging in malicious behavior is greater than the reward for engaging in correct behavior so the penalty factor is generally higher than the reward factor. In the case of the initial state, the I_t influence factor is 0 and the reputation value R_t is 0.5.

$$I_t = \begin{cases} I_{t-1} + \beta_1 * \frac{g_i}{n_i}, & node \quad v_i \quad behaves \quad correctly \\ I_{t-1} - \beta_2 * \frac{m_i}{n_i}, & node \quad v_i \quad behaves \quad maliciously \end{cases} \tag{7}$$

$$R_i = \frac{1}{1 + e^{-I_t}} \tag{8}$$

5.1.3. Distance

The time to perform transactions in the off-chain payment channel mainly considers the processing time at each node while ignoring the time required to transfer funds within the channel. In the case of multiple available paths, the path that passes through as few nodes as possible, i.e., the shortest path, is chosen if all other conditions are the same. A weighted directed graph is constructed by considering factors such as the node reputation, transaction cost, and distance length. The node reputation and transaction fee are used as the weight factors of the edges. The value of the node reputation is a sigmoid function with a range of (0, 1), and the value of the transaction fee is normalized using the max-min technique, as shown in Equation (9). In considering the distance problem, which is equivalent to the transaction time problem, each node has a specified time-lock tolerance time τ_i for transaction processing. This paper considers the minimum number of hops, without considering the comparison of processing times at each node. Therefore, when constructing the directed graph with weights, a constant $\gamma_3 \in (0,1)$ is added to the weight of each edge. The weight w_{ij} of each edge e_{ij} is shown in Equation (10), where is an argument of $\gamma_{i \in 1,2,3}(0,1)$. Weights are assigned to each edge and a non-negative weighted directed graph is constructed. The Dijkstra algorithm is used to find the shortest path, i.e., the optimal path, considering the node reputation, transaction fees, and distance.

$$f_{ij}^{'} = \frac{f_{ij} - min(f_{1,2}, f_{2,3}, \ldots, f_{L-1,L})}{max(f_{1,2}, f_{2,3}, \ldots, f_{L-1,L}) - min(f_{1,2}, f_{2,3}, \ldots, f_{L-1,L})} \tag{9}$$

$$w(i,j) = \gamma_1 * f_{ij}^{'} + \gamma_2 * \frac{1}{R_i} + \gamma_3 \tag{10}$$

5.2. ATLC: Time-Locked Contract Payment Protocol Based on Asymmetric Encryption

This subsection proposes an asymmetric encryption-based time-locked contract payment protocol that uses a combination of asymmetric encryption algorithms and digital signatures to secure the transaction process and ensure the atomicity, data consistency, balance security, and value privacy of the transaction process.

5.2.1. Introduction to ATLC Protocol

In this paper, we propose the ATLC protocol for the scenario of off-chain payment channel network transactions. The security of the ATLC protocol relies on the security features of the asymmetric encryption algorithm RSA and the digital signature algorithm DSA. The steps and operational details of the ATLC are described in detail below. In this paper, we use an example of an indirect payment from sender v_0 to receiver v_4 with a payment path $p = v_0 \rightarrow v_1 \rightarrow v_2 \rightarrow v_3 \rightarrow v_4$ and a transaction amount value a, as shown in Figure 4. The specific steps are as follows:

1. The sender v_0 generates a random numeric identifier $\{\varphi_i\}_{i\in[0,4]}$ and r_4 using the asymmetric encryption algorithm RSA $\{r_i\}_{i\in[0,3]}$. The sender node v_0 transmits $\{\varphi_i, r_i\}_{i\in[1,4]}$ through the anonymous secure communication channel to each intermediate node and the receiver.

2. Transaction lock phase: The sender v_0 transmits transaction-related information $< v_0, v_4, a + F_{0,4}, r_0, TsId >$ to the next hop node v_1. The intermediate node v_i receives the transaction information from the previous hop and verifies the correctness of the conditions of the incoming contract using the RSA algorithm for decryption and verification. It checks if $r_i.equals(RSA_{sk_i}(r_{i-1})$. If the match is successful, v_i provides funds $a + F_{i,4}$ to the next hop node v_{i+1} and locks them in the channel. If the match is unsuccessful, the transaction is terminated and the funds locked in the channel are unlocked and returned to the original section.
3. Transaction release phase: After receiving the transaction information, the receiver v_4 determines whether the current time is less than dl. If it is satisfied, it proceeds to the verification phase and enters the fund release phase upon successful verification of the match. v_4 uses the digital signature algorithm RSA for signature generation $\varphi_4^{'} \leftarrow DSA_{sk_4}(\varphi_4, r_3)$ and sends $\varphi_4, \varphi_4^{'}$ to the previous hop stage v_3. When the intermediate node v_i receives $\varphi_{i+1}, \varphi_{i+1}^{'}$ from the next hop node of the transaction, it uses the DSA algorithm to verify and determine if $DSA_{pk_{i+1}}(\varphi_{i+1}^{'}).equals(DSA_{pk_{i+1}}(\varphi_{i+1}, r_i))$. If the verification is successful, the funds locked in the channel are released to v_{i+1}. If the verification is unsuccessful, the transaction is terminated and the funds locked in the channel are unlocked and returned to the original node.
4. When the transaction release phase is successfully verified at sender v_0, the channel funds are released to node v_1, the transaction is closed, and sender v_0 successfully transfers the funds to receiver v_4.

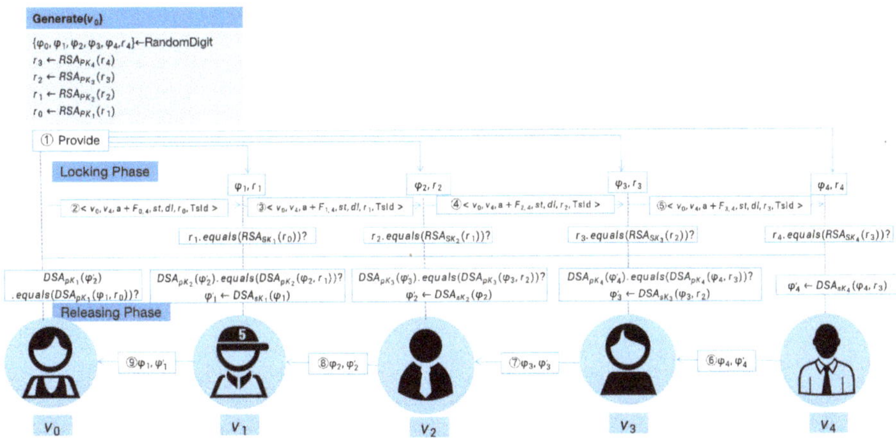

Figure 4. ATLC: Asymmetric encryption-based time-locked contract for off-chain payment channel networks.

5.2.2. Privacy Protection Analysis

In this paper, we use the asymmetric encryption algorithm RSA and digital signature algorithm DSA to ensure the security of off-chain payment channel transactions. The n htlc and ktlc protocols proposed in Ref. [32] use a hash algorithm and a symmetric encryption algorithm. However, the symmetric encryption algorithm requires the transmission of a key, whose secure distribution is a security issue, To address this issue, this paper uses an asymmetric encryption algorithm, which avoids the problems associated with key transmission. During the transaction release phase, this paper uses a digital signature for verifying that the channel amount is released gradually in the order of the transactions. This approach has the advantages of being more resistant to attacks, eliminating the need for key protection, and providing non-repudiation and privacy protection compared to hash verification and symmetric encryption algorithms. The following is an analysis of the nature and privacy protection of the ATLC protocol proposed in this paper:

- Atomicity: Transactions are executed or aborted. Atomicity is a necessary property in the proposed multi-hop payment mechanism. During the transaction process, if there is a node verification mismatch, the transaction will terminate, and the transaction amount in the channel will be released back to the original node, ensuring the atomicity of the transaction.
- Data consistency: During transactions, the balances in the payment phase and release phase are changed and transferred to the node channel. If the verification does not match at any stage of the transaction process, the transaction amount will be returned to the original node, and the channel balance of each node will remain unchanged. The consistency of the balance throughout the channel is ensured.
- Balance security: If there is a malicious node, any intermediate node participating in the transaction does not lose its balance or the transaction fee obtained by participating in the transaction. If an honest node v_i is in the transaction lock phase and encounters a situation where v_{i-1} is a malicious node and the decryption r'_{i-1} received by v_i using the private key does not match r_i, the transaction terminates and v_i does not lose any fees. In the transaction release phase, if node v_{i+2} directly sends $\varphi_{i+2}, \varphi'_{i+2}$ to v_i, v_i uses the public key φ'_{i+2} of node v_{i+1} to verify and sign. If it does not match $DSA_{pk_{i+1}}(\varphi_{i+2}, r_i)$, the channel transaction balance of user v_i will remain locked, and the amount will be returned to the original node after the transaction is terminated. In this case, the node will not lose any fees, which ensures the balance security of the node.
- Value privacy: In addition to the sender and receiver intermediate nodes not knowing the transaction amount information of the transaction, the intermediate nodes in the payment path can only see the payment value transmitted by the previous hop node and the payment value paid to the next hop node.

5.3. Design of Routing Algorithm

In this subsection, the implementation process of the MFPS routing algorithm is described in detail and its computations are shown in Algorithms 1–4.

The route initialization phase, Route-Init, is shown in Algorithm 1. The algorithm describes that when a payment request is made, a weighted directed graph is constructed specifically for that payment request. The nodes traverse the graph G (V,E) and build weights for the edges. Table $T^E(i,j)$ is the weight value of stored edge e_{ij}. If $v_i = v_j$, $T^E(i,j) = 0$, and if there is no channel between v_i and v_j, $T^E(i,j)$ is assigned an infinite value. If there is a channel between v_i and v_j, the balance of the v_i channel segment is $b_{ij}a + F_p(i+1, L)$. If the feasibility constraint of the transaction is not satisfied, the weight of the edge $T^E(i,j) \leftarrow \infty$ is assigned. In the remaining cases, $w(i,j)$ is calculated according to Equation (10) and assigned to the $T^E(i,j)$ transaction weight table. When the node has traversed all nodes of G, the loop is finished and $T^E(i,j)$ is output. After route initialization, the Breadth-First-Search (BFS) algorithm is used to traverse the entire graph to find the set of paths $P_R(s,t)$ between v_s and v_t.

Algorithm 1: Route-Init

Input: network G(V,E), payment request $R = (v_s, v_t, a, st, dl, TsId)$;
Output: edge weight table $\{T^E(i,j)\}_{v_i \in V}$;
for $v_i, v_j \in V$ **do**
  if $v_i = v_j$ **then** $T^E(i,j) = 0$;
  else if $e_{i,j} \notin E$ **then** $T^E(i,j) \leftarrow \infty$;
  else if $b_{i,j} < a + F_p(i+1, L)$ **then** $T^E(i,j) \leftarrow \infty$;
  else $T^E(i,j) = w(i,j)$;
end
return $T^E(i,j)$

Before routing transactions, it should be determined whether there is a path with a sufficient channel balance for the transfer of the funds needed for the transaction. If such a path does not exist, the transaction amount is split and the transaction is forwarded separately. The specific steps of Algorithm 2 (PaySplit algorithm) are shown below and are as follows. Find the channel paths that exist between v_s and v_t and sort the transferable funds from largest to smallest. Next, calculate the total transferable funds and determine if the funds needed for the transaction are reached. If not, return null to indicate that the transaction failed. If the funds needed for the transaction are reached, split transactions are performed from the path with the most transferable funds, thus minimizing the number of split transactions and re-initializing the routes. When the split amount reaches the funds required for the transaction, the cycle ends and the split transaction set is returned.

Algorithm 2: PaySplit

Input: payment request $R = (v_s, v_t, a, st, dl, TsId)$;
Output: splitting the transaction set partval;
if $P_R(s,t) == null$ **then**
  return noSplit
else
  funds=getFundsEdges(s,t).sort(largest to smallest);
  if *funds.sumVal()<a* $+ F_{s,t}$ **then**
    return null
  else
    sumFunds=0,partval=[] ;
    for $i \leftarrow 0$ **to** $funds.length - 1$ **do**
      Sumfunds=Sumfunds+funds[i].val;
      partval.add(funds[i]);
      Route-Init($R = (v_s, v_t, funds[i].val - F_{s,t}, st, dl, TsId)$) ;
      $P_R(s,t).add(funds[i].path)$;
      if *funds.sumVal()>a* $+ F_{s,t}$ **then** break;
    end
  end
end
return partval;

The routing path selects the optimal path for the transaction, as shown in Algorithm 3 (PaymentRoute). The aim of the algorithm is to find the optimal path for forwarding transactions. First, it is necessary to check if $P_R(s,t)$ is not null. Then, define $S = \phi$ and Q as the set V of all vertices in graph G. When $Q = \phi$, carry out a loop. Select the intermediate node with the minimum weight from the starting point v_s to v_t in $T^E(s,t)$, remove the node from the set of nodes Q, and add the node to the set S. p_i is an s-t path in $P_R(s,t)$ and is considered the optimal path if all nodes are in set $S \supseteq p_i$. The transaction request R will then choose this path for the transaction. When the transaction path p_i is determined, the sender randomly generates the digital identifier of the participating transaction nodes and the r_t digital identifier $\{\varphi_i\}_{i\in[0,p_i.length-1]}$ of the receiver. In addition the digital identifiers of the intermediate nodes and the sender are generated according to $\{r_i\} \leftarrow RSA_{pk_{i+1}(\varphi_{i+1})}$, and the sender v_s sends $\{\varphi_i, r_i\}_{i\in[0,p_i.length-1]}$ to node v_i through the anonymous communication channel.

Algorithm 3: PaymentRoute

```
Input: edge weight table TE(i,j), payment request R = (v_s, v_t, a, st, dl, TsId);
Output: optimal path p;
if P_R(s,t)! = null then
    S = ϕ, Q = G.V;
    while Q ≠ ϕ do
        v_i = argmin(T^E(s,t));
        Q = Q − v_i ;
        S = S ⋃ {v_i} ;
    end
end
for p_i ∈ P_R(s,t) do
    if p_i ⊆ s then
        p_i is optimal path;
        {r_t} ← RandomDigit;
        for i ← 0 to p_i.length − 1 do
            {φ_i} ← RandomDigit;
        end
        v_s through anonymous communication channels send {φ_t, r_t} to v_t;
        for i ← p_i.length − 2 to 0 do
            r_i ← RSA_{pk_{i+1}}(φ_{i+1});
            v_s through anonymous communication channels send {φ_i, r_i} to v_i
        end
    end
end
return p_i;
```

The transaction reaches the receiver, as shown in Algorithm 4 (Recipient). When the receiver receives the fund transfer, it first verifies whether the transaction timeliness constraint is satisfied. If the receiving time exceeds the deadline specified in the transaction, the receiver has the right to reject the transaction. Subsequently, the intermediate nodes involved in the transaction release the funds and return the fund transfer generated by the transaction to the original owner. If the timeliness constraint is satisfied, the receiver receives the transaction, generates a signature $\varphi_t^{'} \leftarrow DSA_{sk_t}(\varphi_t, r_{t-1})$, and sends$\{\varphi_t, \varphi_t^{'}\}$ to v_{t-1}. When the intermediate node v_i receives $\{\varphi_{i+1}, \varphi_{i+1}^{'}\}$, verify that $DSA_{pk_{i+1}}(\varphi_{i+1}^{'}) =?DSA_{pk_{i+1}}(\varphi_{i+1}, r_i)$. If verification is successful, generate $\varphi_{i+1}^{'} \leftarrow DSA_{sk_i}(\varphi_i, r_{i-1})$ and send $\{\varphi_i, \varphi_i^{'}\}$ to node v_{i-1}. Node v_i then releases the locked funds in the channel to node v_{i+1}. If the match is unsuccessful, the funds generated by the transaction are transferred back to node v_i.

Algorithm 4: Recipient

Input: Payment request $R = (v_s, v_t, a, st, dl, TsId)$;
if *current.time* $> dl$ **then**
  v_t refuse payment R;
  for $v_i \in p$ **do**
    release funds return to v_i;
  end
else
  v_t accept fund a;
  $\varphi_t^{'} \leftarrow DSA_{sk_t}(\varphi_t, r_{t-1})$ send $\{\varphi_t, \varphi_t^{'}\}$ to v_{t-1};
  for $v_i \in p$ **do**
    if $DSA_{pk_{i+1}}(\varphi_{i+1}^{'}) == DSA_{pk_{i+1}}(\varphi_{i+1}, r_i)$ **then**
      $\varphi_i^{'} \leftarrow DSA_{sk_i}(\varphi_i, r_{i-1})$ send $\{\varphi_i, \varphi_i^{'}\}$ to v_{i-1};
      release funds return to v_{i+1};
    else
      release funds return to v_i
    end
  end
end

6. Simulation Experiments

6.1. Environments and Parameter Setting

Experimental data and environment: In this paper, the experimental data were obtained from Ref. [37]. The off-chain payment channel network used was the Lightning Network, where the experimental topology is a real snapshot of lightning. The experimental hardware environment was as follows. CPU, Intel(R) Core(TM) i7-8565U 1.80 GHz; memory, 8.00 GB; and operating system, Windows 10. IntelliJ IDEA 2021.1 x64 software was used for the simulation experiments.

Parameter setting: The parameters for the transaction commission were set as follows. The function f_{ij} parameters were α_1= 0.01, α_2 = 0.03, and H = 0.005. In the credibility function, the parameters were β_1 = 0.2 and β_2 = 0.5. In the weighting function, the parameters were $\gamma_1 = \gamma_2 = \gamma_3 = \frac{1}{3}$.

6.2. Security Analysis

The ATLC security analysis proposed in this paper is described below:

- Anti-wormhole attack: In the path $p = v_0 \rightarrow v_1 \rightarrow ...... \rightarrow v_L$, if v_i is an honest node, v_{i-1} and v_{i+1} are malicious nodes. If v_{i-1} and v_{i+1} collude to obtain the transaction fee of v_i, the malicious attack is a wormhole attack. In this paper, we propose an ATLC that can effectively prevent wormhole attacks. If node v_{i+1} directly sends the transaction information to node v_{i-1} for the digital signature verification phase, and node v_{i-1} does not have the r_i identifier of node v_i, the verification result cannot be matched. As a result, the channel balance will not be released or unlocked and the honest node will not lose any funds.
- Anti-double-flower attack: A double-flower attack is when an attacker repeatedly uses the same funds to make payments. In the asymmetric encryption and digital signature used by the ATLC, each transaction node has a unique digital signature and public key for verification, and an attacker cannot forge the digital signature. Therefore, the attacker cannot perform a double-flower attack.
- Anti-fraudulent trading attack: A fake transaction attack is when an attacker sends fake transactions to deceive the trading node and obtain the fee. In the ATLC protocol, each transaction has a unique identification called TsId, and nodes need to sign and

verify the transaction. Only through the verification of the node can the transaction be completed, thereby preventing attackers from sending false transactions.

- Anti-replay attack: This is when an attacker retransmits the transmitted data during the data transmission between two communication parties. The ATLC uses the digital signature algorithm (DSA) and encryption algorithm to protect against replay attacks.
- Denial-of-service attack: This is when an attacker sends a large number of invalid requests by forging identities. As a result, communication between two parties cannot proceed as normal. The ATLC uses the DSA to protect against denial-of-service attacks.

6.3. Experimental Results

The comparison algorithms used in the experimental evaluation were as follows:

- ShortestPath: The algorithm chooses the shortest path for trading under the trade constraints;
- Cheapest: In the transaction path selection, this algorithm selects the path with the lowest transaction fee;
- SplitDistance: If the channel balance does not meet the amount needed for the transaction, this algorithm splits the transaction amount along the shortest path.

The performance metrics used in the experimental evaluation were as follows:

- Success ratio: This refers to the percentage of total transactions successfully received by the recipient;
- Average transaction fee: This refers to the average transaction fee paid in successfully completed transactions;
- Average message spending: This refers to the overhead in terms of the average number of messages involved in a successful transaction request;
- Delay: This refers to the time required to complete a transaction (in mesc).

The independent variables used in the experimental evaluation were as follows:

- Number of transactions: This refers to the total number of transactions performed in the off-chain payment channel network;
- Initial channel balance: This refers to the initial balance in the channel of the node that establishes the off-chain payment channel network;
- Transaction value: This refers to setting the transaction amount for each transaction in the off-chain payment channel network;
- TNumber of nodes: This refers to the total number of nodes present in the off-chain payment channel network.

The impact of the number of transactions and the initial channel balance on the transaction success ratio in the MFPS algorithm is illustrated in Figure 5. In Figure 5a, the effect of the number of transactions on the transaction success ratio is shown. The initial channel balance was set to 200 and the transaction amount was set to 25. With the increase in the number of transactions, the transaction success ratio remained consistent at around 84%. In addition, it was observed that the change in the number of transactions had little effect on the success ratio of the MFPS routing transactions. In Figure 5b, the effect of the initial channel balance on the transaction success ratio is shown. The number of transactions was set to 10,000 and the transaction amount was set to 250. As the channel balance increased, the transaction success ratio also increased. With an initial channel balance of 500, the transaction success ratio improved by 17% compared to a balance of 50.

We compared the performance of the MFPS algorithm with the ShortestPath, Cheapest, and SplitDistance algorithms. The initial channel balance was set to 25 and the number of transactions was set to 10,000. The independent variable was the transaction amount. The results of the performance comparison are shown in Figures 6 and 7.

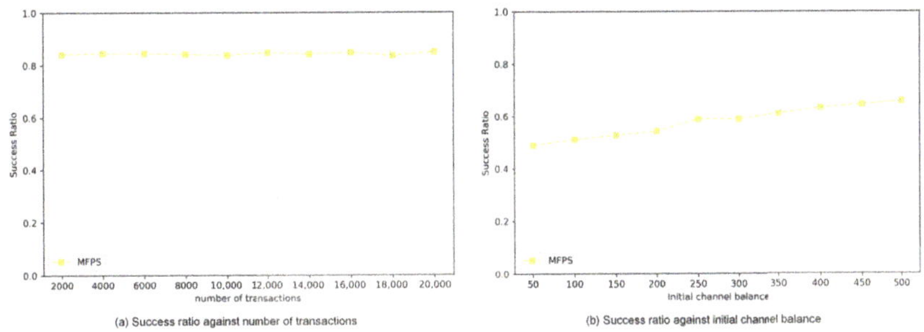

Figure 5. Factors affecting the success ratio of trading. (**a**) Effect of the number of transactions on the transaction success rate. (**b**) Effect of initial channel balance on the transaction success rate.

Figure 6 shows a comparison of the transaction success ratios when using the ShortestPath, Cheapest, SplitDistance, and MFPS algorithms. The MFPS algorithm improved the transaction success ratio by about 13.84%∼49% compared to the other algorithms for a transaction amount of 50. We can see that the Cheapest algorithm had the fastest decrease in the success ratio as the transaction amount increased. This is because the Cheapest algorithm only chooses the path with the lowest transaction fee, without considering the feasibility constraint of the transaction. This led to a significant decrease in the transaction success ratio. The success ratio of the SplitDistance algorithm was higher than that of the ShortestPath algorithm. This is because the SplitDistance algorithm splits the transaction amount according to the shortest path. This improved the transaction success ratio. When the transaction amount was greater than the initial channel capacity, the transaction success ratio of the SplitDistance and MFPS algorithms tended to be stable. In contrast, the transaction success ratio of the ShortestPath and Cheapest algorithms decreased faster. This is because the SplitDistance and MFPS algorithms used the split amount for transactions, whereas the ShortestPath and Cheapest algorithms did not split the transaction amount. As a result, the number of paths capable of meeting the transaction amount gradually decreased, and the transaction success ratio also gradually decreased. As the amount per transaction increased, the success ratio of all the algorithms decreased because the percentage of paths that could satisfy the feasibility constraint of the transaction also decreased.

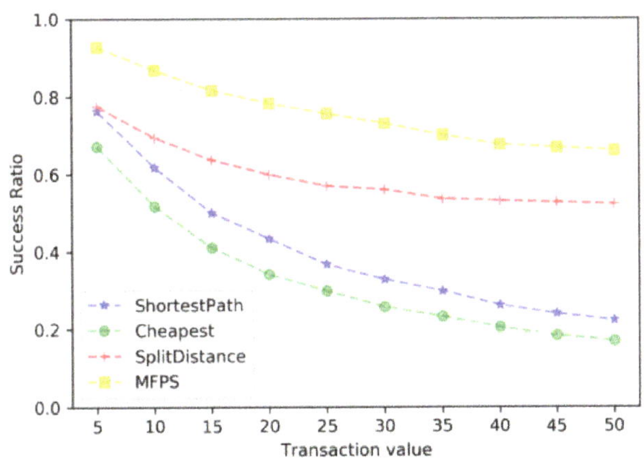

Figure 6. Success ratio against transaction value.

Figure 7 shows a comparison of the average transaction fees for the ShortestPath, Cheapest, SplitDistance, and MFPS algorithms. As the transaction amount increased, the transaction fee became more significant due to its correlation with the transaction amount.

Subsequently, the average fee for all the algorithms exhibited an upward trend. The Cheapest algorithm's average transaction fees were lower than those of the MFPS because the Cheapest algorithm chose the path with the lowest transaction fees so its average transaction fees were the lowest. The ShortestPath algorithm had the highest average transaction fees because the ShortestPath algorithm only considered the shortest path and did not consider the transaction fee. The MFPS algorithm, when compared with the ShortestPath and SplitDistance algorithms, achieved significant reductions in the average transaction fee of 41.72% and 28.25%, respectively, for a transaction amount of 50.

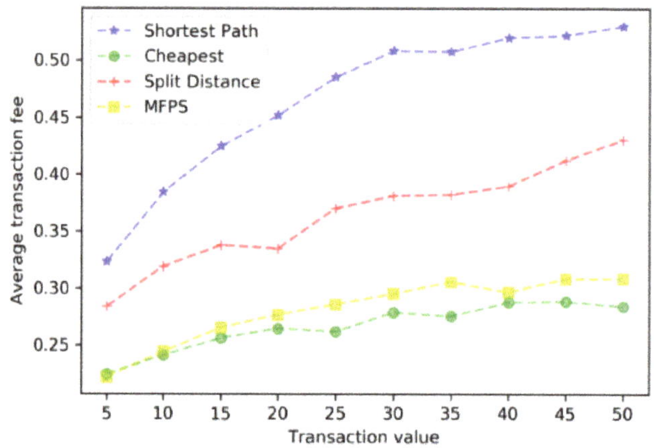

Figure 7. Average transaction fee against transaction value.

A latency comparison of the MFPS and ktlc algorithms is shown in Figure 8. The independent variable was the change in the number of nodes, the response variable was the latency of the transaction (in msec), and the number of transactions was set to 10. Since the asymmetric encryption algorithm is slower than the symmetric encryption algorithm, the experimental comparison in this paper excluded the time required for key generation in both algorithms. The aim was to compare the computational verification time delay between the two encryption algorithms. In the figure, it can be seen that the MFPS algorithm computed the verification with lower latency than the ktlc algorithm.

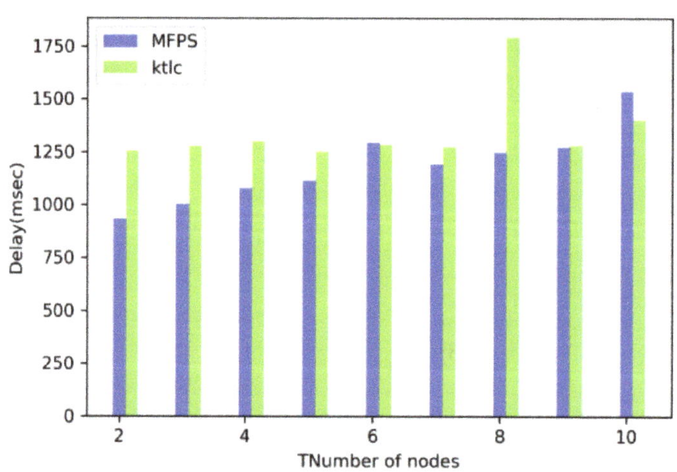

Figure 8. Time delay against Tnumber of nodes.

The average message spending values of the MFPS and ktlc algorithms are shown in Figure 9. The independent variable was the change in the number of nodes, the response variable was the average message overhead, and the number of transactions was set to 10. As the number of network nodes increased, the number of transaction hops increased, and the average message overhead also tended to increase. In this figure, it is clear that the MFPS algorithm had a lower overall average message overhead compared to the ktlc protocol.

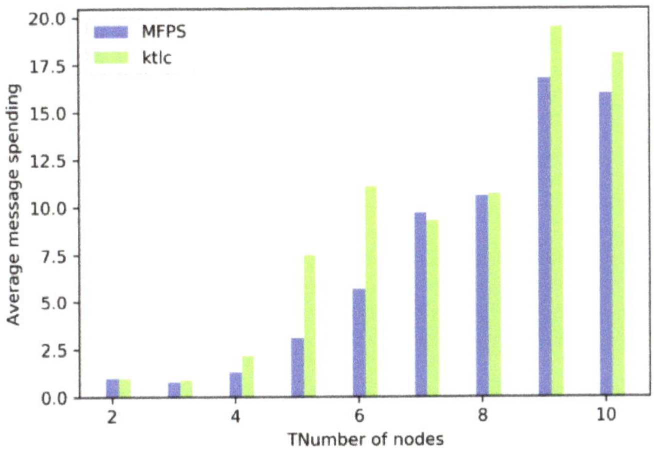

Figure 9. Average message spending against Tnumber of nodes.

In summary, the MFPS algorithm had a higher transaction success rate, lower transaction communication overhead, and reduced transaction processing costs compared to the other algorithms. The MFPS achieved better performance and effectively improved the transaction success rate of off-chain payment channels. In addition, it effectively ensured the privacy and security of transactions with minimal communication overhead.

7. Conclusions

Blockchain technology can be successfully applied to the carbon-trading scenario using the characteristics of blockchains to guarantee the security, decentralization, data immutability, and data traceability of the carbon-trading process. This paper uses an off-chain payment channel to improve the performance of carbon trading on the blockchain, moving the small-scale and high-frequency transactions of individuals to the off-chain for trading while keeping the carbon transactions of companies on the blockchain. This approach reduces the burden of trading on the blockchain and also allows individuals engaging in small-scale and high-frequency carbon transactions to avoid expensive processing fees on the blockchain. In this paper, the optimization of PCN routing based on off-chain payment channels is conducted, and a multi-factor-based routing payment scheme is proposed, which takes into account factors such as the transaction fee cost, node reputation, and distance for routing. In order to reduce channel congestion, we propose implementing transaction fees based on the difference between the balances at each end of the channel. In order to improve the transaction success rate, a greedy splitting algorithm is proposed for the transaction amount. In order to ensure the security and privacy of node transactions, an ATLC protocol is proposed in the off-chain payment channel, which is capable of preventing wormhole and double-flower attacks. The experimental results show that the MFPS scheme proposed in this paper achieves a high transaction success rate with low communication overhead and handling costs while guaranteeing the privacy and security of transactions.

The MFPS scheme aims to optimize the performance of the off-chain payment channel by transferring the high-frequency and small-scale carbon trading of individuals to the off-chain, This approach reduces the pressure of on-chain blockchain carbon trading and enables quick processing of large-scale carbon trading business between enterprises on the chain, thus improving the scalability of the blockchain. The MFPS scheme proposed in this paper can improve the success rate of individual carbon trading in the off-chain payment channel, improve the performance of the blockchain in processing individual carbon trading, and reduce the cost burden for individual users by reducing transaction costs. In terms of security and privacy, the security of individual users' carbon assets can be guaranteed. The MFPS solution addresses the problem of insufficient blockchain scalability, improves blockchain performance, and enables the blockchain to support high-concurrency transactions in large-scale carbon-trading scenarios.

In the future, the influence of a single factor, such as node reputation, transaction cost, or distance, on routing selection will be further analyzed. Additionally, the influence of other factors on routing selection that were not considered in this paper will also be investigated.

Author Contributions: Conceptualization, X.Z.; Methodology, Y.S. and S.G.; Validation, Y.S.; Formal analysis, A.X. and Y.K.; Investigation, D.L.; Resources, X.Q.; Data curation, D.W.; Writing—original draft, Y.S.; Writing—review & editing, Y.S. and A.X.; Supervision, X.Q.; Project administration, D.W. All authors have read and agreed to the published version of the manuscript.

Funding: This work was supported by the National Key R&D Program of China (2022YFB2703400).

Data Availability Statement: No new data were created or analyzed in this study. Data sharing is not applicable to this article.

Conflicts of Interest: The authors declare no conflict of interest.

References

1. Fang, G.; Tian, L.; Liu, M. How to optimize the development of carbon trading in China—Enlightenment from evolution rules of the EU carbon price. *Appl. Energy* **2018**, *211*, 1039–1049. [CrossRef]
2. Almer, C.; Winkler, R. Analyzing the effectiveness of international environmental policies: The case of the Kyoto Protocol. *J. Environ. Econ. Manag.* **2017**, *82*, 125–151. [CrossRef]
3. Kim, S.K.; Huh, J.H. Blockchain of carbon trading for UN sustainable development goals. *Sustainability* **2020**, *12*, 4021. [CrossRef]
4. Nakamoto, S. Bitcoin: A peer-to-peer electronic cash system. *Decentralized Bus. Rev.* **2008**, 21260.
5. Zheng, Z.; Xie, S.; Dai, H.N.; Chen, X.; Wang, H. Blockchain challenges and opportunities: A survey. *Int. J. Web Grid Serv.* **2018**, *14*, 352–375. [CrossRef]
6. Abou Jaoude, J.; Saade, R.G. Blockchain applications–usage in different domains. *IEEE Access* **2019**, *7*, 45360–45381. [CrossRef]
7. Richardson, A.; Xu, J. Carbon Trading with Blockchain. In *Mathematical Research for Blockchain Economy*; Pardalos, P., Kotsireas, I., Guo, Y., Knottenbelt, W., Eds.; Springer: Cham, Switzerland, 2020; pp. 105–124.
8. Pan, Y.; Zhang, X.; Wang, Y.; Yan, J.; Zhou, S.; Li, G.; Bao, J. Application of blockchain in carbon trading. *Energy Procedia* **2019**, *158*, 4286–4291. [CrossRef]
9. Vaig, I.K.K.; Karne, R.; Siluveru, M.; Kesoju, M. Review on Blockchain Technology: Architecture, Characteristics, Benefits, Algorithms, Challenges and Applications. *Mesopotamian J. Cybersecur.* **2023**, *2023*, 73–85.
10. Yaseen, M.; Bahari, M.; Hammood, O.A. Blockchain technology applications, concerns and recommendations for public sector. *Mesopotamian J. Comput. Sci.* **2021**, *2021*, 1–6. [CrossRef] [PubMed]
11. Helliar, C.V.; Crawford, L.; Rocca, L.; Teodori, C.; Veneziani, M. Permissionless and permissioned blockchain diffusion. *Int. J. Inf. Manag.* **2020**, *54*, 102136. [CrossRef]
12. Nasir, Q.; Qasse, I.A.; Talib, M.A.; Nassif, A.B. Performance analysis of hyperledger fabric platforms. *Secur. Commun. Netw.* **2018**, *2018*, 3976093. [CrossRef]
13. Li, W.; He, M. Comparative analysis of bitcoin, Ethereum, and libra. In Proceedings of the 2020 IEEE 11th International Conference on Software Engineering and Service Science, Beijing, China, 21 September 2020; pp. 545–550.
14. Pan, C.; Liu, Z.; Liu, Z.; Long, Y. Research on Scalability of blockchain: Issues and methods. *J. Comput. Res. Dev.* **2018**, *55*, 2099–2110.
15. Lee, J.; Long, A.; McRae, M.; Steiner, J.; Handler, S.G. Bitcoin basics: A primer on virtual currencies. *Bus. Law Int.* **2015**, *16*, 21.
16. Zhou, Q. Research on Blockchain Scalability Based on State Channel and Cross-Chain Protocol. Master's Thesis, Shanghai Jiaotong University, Shanghai, China, 2020.

17. Yang, D.; Long, C.; Xu, H. A review on scalability of blockchain. In Proceedings of the 2020 the 2nd International Conference on Blockchain Technology, Hilo, HI, USA, 12 March 2020; pp. 1–6.
18. Yang, D.; Narayanan, V.; Zheng, C.D. A Secure Sharding Protocol for Open Blockchains. In Proceedings of the 23rd ACM Conference on Computer and Communications Security, Vienna, Austria, 24–28 October 2016; pp. 17–30.
19. Huang, J.; Kong, L.; Chen, G.; Wu, M.-Y.; Liu, X.; Zeng, P. Towards Secure Industrial IoT: Blockchain System with Credit-Based Consensus Mechanism. *IEEE Trans. Ind. Inform.* **2019**, *15*, 3680–3689. [CrossRef]
20. Poon, J.; Dryja, T. The Bitcoin Lightning Network: Scalable Off-Chain Instant Payments. 2016. Available online: https://coinrivet.com/research/papers/the-bitcoin-lightning-network-scalable-off-chain-instant-payments/ (accessed on 14 January 2016).
21. Raiden Network. Available online: https://raiden.network/ (accessed on 1 September 2019).
22. Herlihy, M. Atomic cross-chain swaps. In Proceedings of the 2018 ACM Symposium on Principles of Distributed Computing, Egham, UK, 23–27 July 2018; pp. 245–254.
23. Papadis, N.; Tassiulas, L. Blockchain-based payment channel networks: Challenges and recent advances. *IEEE Access* **2020**, *8*, 227596–227609. [CrossRef]
24. Jia, L.; Pei, Q.; Wang, X.; Zhang, H.; Yu, L.; Zhang, J.; Sun, Y. Overview of off-chain Channel Routing Algorithms. *J. Softw.* **2022**, *22*, 233–253.
25. Grunspan, C.; Pérez-Marco, R. Ant routing algorithm for the lightning network. *arXiv* **2018**, arXiv:1807.00151.
26. Malavolta, G.; Moreno-Sanchez, P.; Kate, A.; Maffei, M. Silentwhispers: Enforcing Security and Privacy in Decentralized Credit Networks. Cryptology ePrint Archive. 2016. Available online: https://eprint.iacr.org/2016/1054 (accessed on 15 November 2016).
27. Roos, S.; Moreno-Sanchez, P.; Kate, A.; Goldberg, I. Settling payments fast and private: Efficient decentralized routing for path-based transactions. *arXiv* **2017**, arXiv:1709.05748.
28. Yu, R.; Xue, G.; Kilari, V.T.; Yang, D.; Tang, T. Coinexpress: A fast payment routing mechanism in blockchain-based payment channel networks. In Proceedings of the 27th International Conference on Computer Communication and Networks, Hangzhou, China, 30 July–2 August 2018; pp. 1–9.
29. Wang, P.; Xu, H.; Jin, X.; Wang, T. Flash: Efficient dynamic routing for offchain networks. In Proceedings of the 15th International Conference on Emerging Networking Experiments and Technologies, Orlando, FL, USA, 9–12 December 2019; pp. 370–381.
30. Sivaraman, V.; Venkatakrishnan, S.B.; Ruan, K.; Negi, P.; Yang, L.; Mittal, R.; Fanti, G.; Alizadeh, M. High throughput cryptocurrency routing in payment channel networks. In Proceedings of the 17th USENIX Symposium on Networked Systems Design and Implementation (NSDI 20), Santa Clara, CA, USA, 25–27 February 2020; pp. 777–796.
31. Malavolta, G.; Moreno-Sanchez, P.; Kate, A.; Maffei, M.; Ravi, S. Concurrency and privacy with payment-channel networks. In Proceedings of the 2017 ACM SIGSAC Conference on Computer and Communications Security, Dallas, TX, USA, 30 October–3 November 2017; pp. 455–471.
32. Mohanty, S.K. Tripathy S.n-htlc: Neo hashed time-lock commitment to defend against wormhole attack in payment channel networks. *Comput. Secur.* **2021**, *106*, 102291. [CrossRef]
33. Mizrahi, A.; Zohar, A. Congestion Attacks in Payment Channel Networks. In Proceedings of the Financial Cryptography and Data Security, Virtual Event, 1–5 March 2021; pp. 170–188.
34. Li, P.; Miyazaki, T.; Zhou, W. Secure Balance Planning of Off-blockchain Payment Channel Networks. In Proceedings of the IEEE INFOCOM 2020—IEEE Conference on Computer Communications, Toronto, ON, Canada, 6–9 July 2020; pp. 1728–1737.
35. Zhang, Y.; Yang, D.; Xue, G. Cheapay: An optimal algorithm for fee minimization in blockchain-based payment channel networks. In Proceedings of the 2019 IEEE International Conference on Communications (ICC), Shanghai, China, 20–24 May 2019; pp. 1–6.
36. Zhang, Q.; Cao, S.; Zhang, X. Cross-chain payment routing scheme based on multi-factor reverse auction. *J. Comput. Res. Dev.* **2022**, *59*, 2233–2246.
37. Eckey, L.; Faust, S.; Hostáková, K.; Roos, S. Splitting Payments Locally While Routing Interdimensionally. Cryptology ePrint Archive. Paper 2020/555. 2020. Available online: https://eprint.iacr.org/2020/555 (accessed on 15 May 2020).

electronics

MDPI

Article

Blockchain-Based Authentication Protocol Design from a Cloud Computing Perspective

Zhiqiang Du [1], Wenlong Jiang [1], Chenguang Tian [1], Xiaofeng Rong [1,2,*] and Yuchao She [2]

[1] School of Computer Science and Engineering, Xi'an Technological University, Xi'an 710021, China; duzhiqiang@xatu.edu.cn (Z.D.); jiangwenlong@st.xatu.edu.cn (W.J.); tianchenguang@st.xatu.edu.cn (C.T.)

[2] Center of Information Technology, Xi'an Technological University, Xi'an 710021, China; sheyuchao@xatu.edu.cn

* Correspondence: rongxiaofeng@xatu.edu.cn

Abstract: Cloud computing is a disruptive technology that has transformed the way people access and utilize computing resources. Due to the diversity of services and complexity of environments, there is widespread interest in how to securely and efficiently authenticate users under the same domain. However, many traditional authentication methods involve untrusted third parties or overly centralized central authorities, which can compromise the security of the system. Therefore, it is crucial to establish secure authentication channels within trusted domains. In this context, we propose a secure and efficient authentication protocol, HIDA (Hyperledger Fabric Identity Authentication), for the cloud computing environment. Specifically, by introducing federated chain technology to securely isolate entities in the trust domain, and combining it with zero-knowledge proof technology, users' data are further secured. In addition, Subsequent Access Management allows users to prove their identity by revealing only brief credentials, greatly improving the efficiency of access. To ensure the security of the protocol, we performed a formal semantic analysis and proved that it can effectively protect against various attacks. At the same time, we conducted ten simulations to prove that the protocol is efficient and reliable in practical applications. The research results in this paper can provide new ideas and technical support for identity authentication in a cloud environment and provide a useful reference for realizing the authentication problem in cloud computing application scenarios.

Keywords: cloud computing; federated chains; zero-knowledge proofs; formal formalized semantic analysis

Citation: Du, Z.; Jiang, W.; Tian, C.; Rong, X.; She, Y. Blockchain-Based Authentication Protocol Design from a Cloud Computing Perspective. *Electronics* **2023**, *12*, 2140. https://doi.org/10.3390/electronics12092140

Academic Editor: Hamed Taherdoost

Received: 7 April 2023
Revised: 3 May 2023
Accepted: 4 May 2023
Published: 7 May 2023

1. Introduction

Cloud computing [1], as an Internet-based computing model, provides users with flexible, convenient, and economical computing and storage capabilities that greatly facilitate our lives. It quickly provides computing resources according to user needs to meet various business requirements and workload changes without requiring users to manage hardware and software. In daily use, cloud computing ensures a high availability of services through multiple data centers and corresponding backup mechanisms, reducing the risk of service interruption and providing users with more reliable services.

There is a growing trend of users preferring convenient and affordable cloud services as a platform for software development and server building. This trend has reduced labor and energy costs to some extent. However, traditional identity management models usually have a domain central node that manages the entire domain's business, which means that all data in the domain may be stored in the central node [2]. In this case, the security of the entire system depends entirely on the central node. Once a single point of failure occurs (such as a central node attack or central server downtime), the security of the system will be difficult to guarantee.

Although an increasing number of solutions are adopting distributed and decentralized approaches for deployment, designing a comprehensive identity management system and a reasonable access control scheme in cloud computing environments remains a critical task, as many security and performance-related issues are yet to be resolved. Traditional identity management models are typically managed by central authorities or organizations [3], which can be easy targets for attackers and may suffer from poor management or abuse of power issues [4]. In addition, during identity authentication, users are required to provide a significant amount of personal identity information, which may result in the leakage of personal privacy. Simple data encryption alone is insufficient to ensure the security of communication between both parties, as eavesdropping by an attacker on sensitive data within the conversation can occur, ultimately leading to the desired attack effect. Examples of such attacks include the classic IP spoofing attack and SSL/TLS man-in-the-middle attacks. In fact, most existing identity management models suffer from single-point-of-failure and low efficiency issues, with the central server exposed to potential attacks that can result in the entire system being paralyzed [5].

Zero-knowledge proof (ZKVP) is used as a secure cryptographic technique. It is often used to prove the authenticity of a fact or information without revealing any specifics about the fact or piece of information, which means that the proving party does not need to reveal any additional information to the verifying party [6]. During the communication between two parties, a protocol based on zero-knowledge proof can encrypt the communication, thus effectively preventing man-in-the-middle attacks and other malicious behaviors.

A consortium blockchain network, also known as a private or permissioned blockchain network, is a type of blockchain network that is jointly managed and operated by multiple organizations or entities, and in which only authorized members are allowed to participate in transactions and verification [7]. These members typically have common interests or goals, and compared to public blockchain networks, consortium blockchain networks have stricter access permissions, making them more suitable for cooperation and transactions between enterprises and organizations [8]. Consortium blockchain has been used as an auxiliary support technology for identity management. By utilizing the consortium blockchain, sensitive nodes or organizations in the system can be securely isolated and protected, thus improving the security and reliability of the system. In addition, consortium blockchains can provide more efficient solutions for identity management. For example, VeChain is a supply chain management platform that utilizes a federated chain network to enhance the quality and security of products for companies. The platform also features a reliable identity management system that ensures traceability at every point in the supply chain, effectively protecting both businesses and consumers. Another example is Corda, an enterprise-grade blockchain platform focused on the financial services sector, which provides a secure and efficient transaction and identity management experience, ensuring that participants' identities are protected during transactions and identity verification. These federated chain applications are built on common interests and goals, leveraging blockchain technology to provide organizations and businesses with a more secure and efficient transaction and identity management solution.

Combining zero-knowledge proof technology with blockchain technology can bring the following benefits to identity management solutions:

- Better privacy protection: The audit access mechanism of consortium blockchain technology ensures that all nodes in the chain can only access and view the information they need, while zero-knowledge proof technology can achieve verification without disclosing any personal identity information, thereby better protecting personal privacy.
- More efficient identity verification: Consortium blockchain technology can provide a more efficient identity verification mechanism because it does not rely on traditional centralized identity verification institutions but instead implements decentralized identity verification based on blockchain technology. At the same time, zero-knowledge proof technology can help verifiers complete verification without the need to disclose identity information.

- More reliable identity management: Consortium blockchain technology can establish a more reliable identity management system because all identity information is stored on a distributed ledger to ensure that it is not tampered with. Using zero-knowledge proof technology for identity verification can enhance the reliability of identity verification, thereby reducing the risk of identity fraud and theft.

To better solve various problems arising from the traditional identity management model, an identity authentication scheme (HIDA) based on the combination of blockchain technology and zero-knowledge proof technology is proposed in this paper. It aims to achieve more secure, efficient, and reliable identity authentication. Our specific contributions are as follows:

- In this paper, we propose an efficient and secure authentication scheme (HIDA) based on the combination of blockchain technology and zero-knowledge proof technology, which can support efficient authentication of users to service providers in cloud computing scenarios.
- BAN logic was chosen to perform a formal security analysis of our solution to demonstrate the security and privacy protection that HIDA can provide.
- Finally, the performance of the HIDA scheme was evaluated by conducting simulation experiments and comparing it with the previous scheme. The experimental results show that HIDA not only has advantages in terms of security but also performs well in terms of efficiency. Specifically, our scheme can provide efficient authentication services in a cloud computing environment with a shorter response time and lower computational resource consumption compared to traditional schemes.

The following sections of this article are organized as follows: Section 2 reviews related work in the field of identity management; Section 3 provides a detailed problem statement and presents a threat model, upon which our design goals are based; Section 4 introduces relevant theoretical knowledge; Section 5 elaborates on the system model and specific solution; Section 6 provides a proof and analysis of the security of our solution; Section 7 conducts experiments in a simulated environment and compares our solution with previous works; and finally, Section 8 concludes the article.

2. Related Work

Identity authentication is a security mechanism that confirms a user's identity through certain technical means to ensure that only legitimate users can access the corresponding services [9]. Currently, blockchain-based identity authentication can be divided into three methods: anonymous authentication, real-name authentication, and controllable anonymous authentication [3].

Anonymous authentication means that users do not need to reveal their true identity during the registration and authentication process. However, due to the openness and multi-party confirmation of the ledger, privacy protection of transaction identities cannot be guaranteed. Real-name authentication is similar to the traditional CA-based authentication scheme, where a third party issues an authentication certificate to prove the user's legitimacy. Currently, controllable anonymous authentication is more popular, and most schemes use ring signatures or blind signatures to anonymously operate user identities. However, the association between user identity information and account addresses is still stored in the third-party authentication institution. If the third party cannot guarantee its own security, the user's anonymous identity may still be obtained. Keltoum Bendiab and Nicholas Kolokotronis combined blockchain technology with a cloud environment to design a blockchain-based cloud identity management scheme. The proposed trust model allows CSPs to autonomously manage their trust relationships in a dynamic and distributed manner. Subsequently, domestic scholars designed an Ethereum-based Identity Management (EIDM) [10] scheme using the CIDM (Consolidated Identity Managcment) [11] protocol, smart contracts and reputation systems, and EIDM. The EIDM does away with the traditional Identity Management (IDM) proxy and uses blockchain technology to establish a trust relationship between the CSP and the cloud subscriber, thus solving the problems of the single point of failure and over-reliance on third parties that exist in traditional authentication. However, this solution suffers from user privacy and man-in-the-middle attacks during initialization.

OAuth [12] is a well-known identity management protocol designed to help users manage their identities. In this protocol, users obtain a token from an authentication server and use it to access resources on a resource server. However, the protocol lacks trust in data and servers, and the authentication process depends entirely on the authentication server. The UPort [13], ShoCard [14], and Sovrin [15] solutions are all Distributed Ledger Technology (DLT) projects [15] currently implemented on blockchain platforms. These solutions utilize the idea of decentralization to varying degrees, but mainly aim to reshape the role of centralization and intermediaries. For the UPort solution, if an attacker can compromise the Uport application and replace the trusted party with a controller without being noticed, the Uport ID will be permanently compromised. For the ShoCard solution, the intermediary role does introduce uncertainty to the vertical existence of the ShoCard ID; if the company no longer exists, ShoCard users will not be able to use their authenticated results to access the system [16]. As for the Sovrin solution, users must rely on the institutions that represent them and maintain the distributed ledger in the Sovrin network. Depending on the selection and implementation of the intermediary agency, a lot of information may be in its hands [16].

In addition, there are also some blockchain-based IoT device identity authentication schemes. For example, the scheme proposed by Liangqin Gong et al. [17] considers recording device identity information in a distributed ledger to ensure identity verification transactions are recorded in the blockchain network. However, the threshold and feature weight settings in this scheme are static and require regular training updates. The scheme proposed by Gan et al. [18] uses a private chain to store node public key information, but this scheme is based on a completely trusted central CA node, which has a single point of failure. The scheme proposed by Li Wenjie [19] uses an improved model based on the claim identity authentication model and declares user attribute ownership using digital signatures and hash values. However, the authentication authority in this scheme is centrally managed, and if the manager of the authentication center behaves improperly or is attacked, attackers may steal or tamper with user identity information, leading to further fraud or infringement of user privacy.

Therefore, these schemes all need further improvement to enhance security and stability. According to the existing solutions for identity authentication on the blockchain, most of them heavily rely on authentication nodes or third-party platforms. Exposing the core nodes of the system on the public network itself poses a risk of attack. Once these nodes are compromised, there is a possibility of all user identity information being leaked. Our scheme attempts to isolate the authentication nodes from user access, logically avoiding the possibility of the authentication nodes being attacked, thereby improving the robustness and security of the system model.

3. Problem Statement

In this section, we describe two application scenarios with authentication requirements in a cloud-based environment. Different organizations complete the corresponding tasks. We discuss the possible security threats during the authentication process and summarize the objectives of our design.

3.1. Scenario Description

With the rapid development of cloud computing, more and more services are being deployed on lightweight and convenient cloud platforms. Figures 1 and 2 depict scenarios for identity authentication based on CA certificates and user credentials. Figure 1 includes three parts: a center authentication (CA), a cloud user, and a cloud service provider (CSP). The user requests centralized authentication from the CA, which verifies and stores the user's information and finally issues a certificate for future access. Figure 2 includes three parts: a cloud service provider, a cloud data center (CDC), and a cloud user. Users within the same security domain no longer need to be authenticated by the CA before requesting services from the cloud service provider. The specific authentication process is completed by the cloud service provider. Users submit registration requests to the cloud service provider, which then transfers the authentication business to the data processing center for review and registration of user identities. Multiple service providers may form a single alliance organization and share the same group of data centers, but more schemes choose to combine CSP and CDC designs.

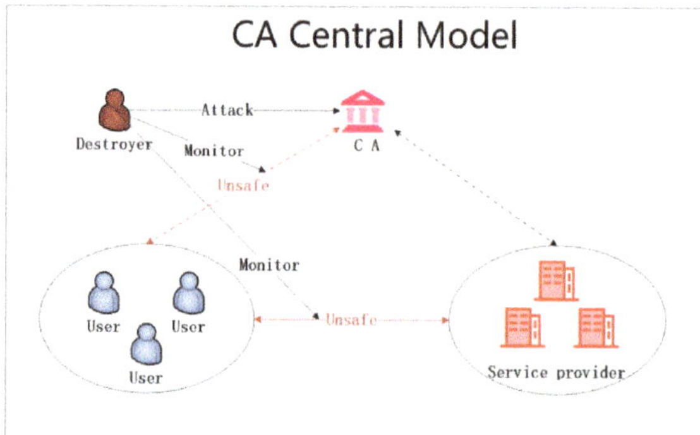

Figure 1. CA Centre based authentication model.

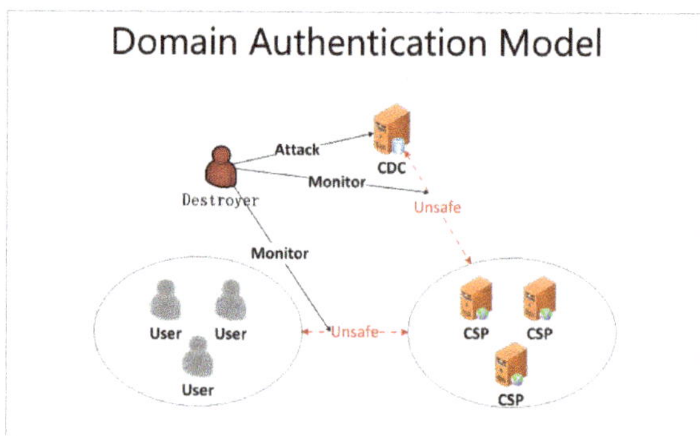

Figure 2. Intra-domain-based business separated authentication model.

3.2. Security Threats

Different models are suitable for different business needs. Centralized authentication based on authoritative organization authentication is more suitable for government and educational organizations, while credential-based identity authentication based on service provider registration seems to be more suitable for flexible small enterprises or companies. When deploying user data into public clouds, these resources naturally become the target of cybercriminals. Therefore, we should consider their security and privacy issues more carefully.

For the model in Figure 1, all system business will flow to the authentication center for data filtering and storage. First, we should consider the performance issue with CA. With the continuous growth in the number of users, the lightweight central service may not be able to support simultaneous multi-user access. In addition, when attackers attempt to block and destroy CA using worm attacks and other methods, it may cause the entire system to be paralyzed. Secondly, CA has the highest authority in the system, and all user data will flow into the central organization for review and storage. If CA is tempted by other larger interests and causes corruption, all information will be directly disclosed, leading to privacy leaks.

For the model in Figure 2, users can undergo identity authentication in different security domains based on different service providers. Although private CDCs in each security domain can to some extent avoid the single point of failure problem brought on by over-centralization of CA, from another perspective, not all domain users are completely trustworthy. This model only separates service businesses from authentication businesses, improving the overall operating efficiency of the system model, but does not further perform security maintenance on the CDC. When there are channel listeners in the system, attackers can intercept and listen to the communication between the two parties to obtain credential information. In addition, after the CDC is separated, attackers are more likely to launch attacks on specific business servers to block the normal operation of the system.

3.3. Design Goals

Identity security: All user requests should be transmitted through reliable channels, the information used for authentication should only belong to the user, and no adversary can impersonate a legitimate user to access services.

Business separation: When users request services in different domains, they should register with a specific service provider. The specific identity authentication is then forwarded to the corresponding data center for processing. To ensure high performance of the system as a whole, the service provider should split the data center into another independent server, focusing only on user access to business.

Entity isolation: To prevent independent data centers from being easily destroyed by attackers, corresponding measures should be taken to ensure their security isolation. When the data center is hidden in a specific security domain and does not accept any unfamiliar requests, the overall security and robustness of the system can be greatly improved.

Secure transmission: To protect user data from eavesdroppers, more security measures are needed to prevent information leakage. Therefore, we can implement permitted access within the domain and encryption using the zero-knowledge proof method to improve the security of the system.

Mutual authentication: During the authentication process, users and service providers can authenticate each other. Service providers only provide services to authenticated users, while users only trust services provided by authenticated service providers.

Scalability: The system has no single point of failure and, therefore, can support large-scale identity authentication business processing.

4. Preliminaries

This section focuses on the technical knowledge covered in the article. We first review some basic concepts and definitions of cryptography [20] and then normalize the Fait–Shamir protocol in zero-knowledge proofs. Finally, we refer to blockchain technology and describe the possible benefits of choosing this technology.

4.1. Elliptic Curve [21]

Definition 1. (the elliptic curve discrete logarithm problem). *Let P be a base point arbitrarily chosen from the elliptic curve EP(a,b). For any probabilistic polynomial time (PPT) adversary $\mathcal{A}$, the probability of finding $k \in \mathbb{Z}_n^*$ satisfying equation $Q = kP$ is negligible when the parameters are given.*

4.2. RSA [22] Secure Public Key Encryption

RSA is a public-key encryption algorithm based on the large prime number decomposition puzzle, proposed by Ron Rivest, Adi Shamir, and Leonard Adleman in 1977. The

RSA algorithm uses a pair of keys, a public key and a private key, to encrypt and decrypt data [23].

Definition 2 (RSA). *Let ℓ be a function such that for all n, $\ell(n) \leq 2^{(n-2)}$. The public key encryption scheme is defined as follows, where Algorithm 1 specifies the GenRSA function's specific process:*

1. *Gen: Given input 1^n, run GenRSA (1^n) to obtain (N,e,d). Output the public key pk = <N,e> and private key sk = <N,d>.*
2. *Enc: Given a public key pk = <N,e> and a message m∈ {0, 1}^ℓ(n), choose a random string r ← {0, 1}^{| |N| | −ℓ(n) − 1}, and interpret r| |m as an element of $\mathbb{Z}_n$. The output ciphertext is c := [(r| |m)e mod N].*
3. *Dec: Given the private key sk = <N,d> and the ciphertext c∈$\mathbb{Z}_n^*$, compute the message m := [cd mod N], and output the low l(n) bits of m.*

Algorithm 1. GenRSA

Input: security parameter 1^n
Output: N,e,d
1. (N,e,d) ← GenModulus (1^n);
2. ϕ(N) = (p − 1) (q − 1);
3. Choose e satisfied with gcd (e,ϕ(N)) = 1;
4. Computer d:= [e^{-1} mod ϕ(N)];
5. Return N,e,d;

4.3. Zero-Knowledge Proof (Fait–Shamir Protocol [24])

Zero-knowledge proofs were proposed by S. Goldwasser, S. Micali, and C. Rackoff in the early 1980s. This proof method allows the prover to prove to the verifier that it possesses a particular piece of information without revealing its secret information. In communication between a communicating entity B (the prover) and A (the verifier), if B is able to successfully prove to A that it possesses the secret but A is unable to infer the secret information, the proof is shown to have zero knowledge. In addition, a zero-knowledge proof needs to satisfy correctness, i.e., the inability of A to master a proof method that makes it highly probable that B possesses the secret. Finally, a zero-knowledge proof also needs to satisfy completeness, i.e., B possesses a theorem-proving method that makes A believe that B can complete the proof. Next, we give specific definitions in Algorithms 2 and 3, and describe the proof process in detail. The flow of the scheme is shown in Figure 3.

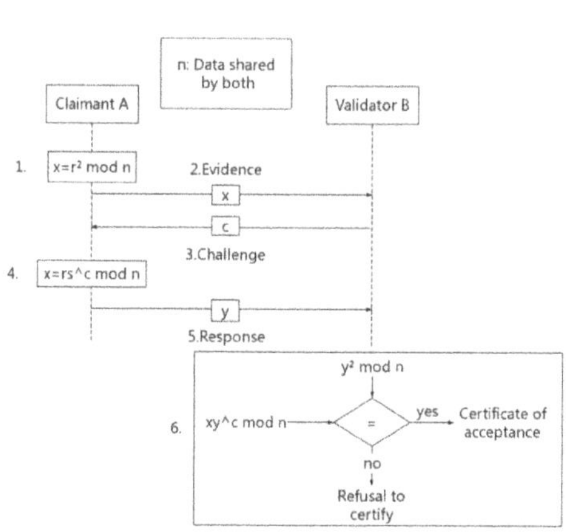

Figure 3. The Fait–Shamir protocol

Definition 3. *The zero-knowledge proof based on the Fait–Shamir protocol consists of four algorithms (Gen,Prf,Chg,Vrfy) for the prover A and the verifier B. The details of the algorithms are as follows:*

- *(pk,sk) $\leftarrow$ FS.Gen(1^λ): The key generation algorithm takes a security parameter $\lambda \in \mathbb{N}$ as input and outputs the public prover key pk and the signing key sk.*
- *(x) $\leftarrow$ FS.Prf(r): The evidence generation algorithm selects a random number r, where $(0 \leq r < n) \cap r \leftarrow \mathbb{Z}_n$, and computes the evidence x to be used in the subsequent proof based on r.*
- *(y) $\leftarrow$ FS.Chg(c): The challenge–response algorithm computes the corresponding response y by A after receiving the random number $c \in \{0, 1\}^*$ generated by B for the challenge.*
- *$\{0, 1\}^* \leftarrow$ FS.Vrfy(): The key generation algorithm outputs the public prover key pk and the secret key sk for signing when given a security parameter $\lambda \in \mathbb{N}$ as input.*

Algorithm 2. Proof/Challenge information generation

Input: Prover public key Pk and secret private key Sk, r (commitment random number)/c (challenge random number).
Output: Output corresponding evidence x and response y
1. Compute n = (pk,sk) $\in \mathbb{Z}_n^*$;
2. Compute x = r^2 mod n;
3. Or Compute y = rs c mod n;
Return x or y;

Algorithm 3. Verify secret

Input: corresponding evidence x and response y
Output: Verification result: succeed or fail.
1. Compute x == y;
2. if true return 1;
3. else return 0;

4.4. Blockchain Technology

Blockchain technology is a decentralized, public, and distributed ledger technology used to record and verify transactions and data transfers. It consists of a series of blocks, each containing information about specific transactions such as transaction amount, timestamp, and participant addresses. Each block is linked to the previous block, forming an immutable chain that makes it impossible for anyone to alter previous transaction records. This means that blockchain technology has a high level of security and transparency, as all participants can view and verify transactions, and no centralized institution or single entity can manipulate it. Additionally, due to its decentralized nature, blockchain technology can address some of the problems that exist in traditional centralized systems, such as single points of failure, data leaks, and security concerns.

A federated chain is a private chain based on blockchain technology that consists of a group of pre-authorized participants, which are usually businesses, government agencies, or organizations. Unlike public blockchains, federated chains allow participants to selectively disclose or protect their data, thus allowing for trustworthiness and security while maintaining data privacy. In a federated chain, participants need to be authenticated and authorized to join the network and participate in transactions. Each participant has a local copy of all transactions and associated data that are verified and authorized by a consensus algorithm. Unlike public blockchains, federated chains typically use more efficient consensus algorithms because of the relatively small number of participants and the higher speed and throughput required for transactions. In addition, federated chains are also more flexible and customizable, as participants can configure and deploy them to meet specific needs.

Hyperledger Fabric is a permissioned blockchain platform that serves as the foundation for many federated chains. It was selected as the basis for this experiment because of its features that support the creation and operation of permissioned networks. Hyperledger Fabric allows for fine-grained control of permissions, providing greater flexibility in the management of the network. Additionally, it supports a modular architecture that allows for easy customization and integration with existing systems. A federated chain based on Hyperledger Fabric consists of a group of pre-authorized participants who are authenticated and authorized to join the network and participate in transactions. Each participant has a local copy of all transactions and associated data that are verified and authorized by a consensus algorithm. Hyperledger Fabric's pluggable consensus algorithm allows for a more efficient consensus mechanism that can be customized based on the requirements of the network. In summary, Hyperledger Fabric was selected as the foundation for this experiment because of its features that support permissioned networks, fine-grained control of permissions, modular architecture, and pluggable consensus algorithm.

5. Design of HIDA

Starting from this section, we provide in Table 1 the symbols used in the protocol and their related meanings to facilitate a better understanding in the subsequent description.

Table 1. The symbol description in the identity authentication protocol.

Symbol	Meaning
PK_n	Public key for entity n
SK_n	Private key of entity n
K_n	Key for entity n
E (K_n, M)	Encryption of M using the key of entity n
D (K_n, M)	Decryption of M using the key of entity n
A→B:m	Entity A sends a message to Entity B m
ID_n	Unique ID of entity n
T_n	Entity n generated timestamp and signature
Hash (M)	Hash for M using MD5

In this section, we first introduce the system model and then provide a detailed description of a secure authentication scheme based on blockchain technology and zero-knowledge proofs in a cloud computing environment.

5.1. System Model

To achieve a secure and efficient identity authentication scheme, we have designed a system model combining blockchain technology and zero-knowledge proof, as shown in Figure 4. The model consists of several parts: cloud users, cloud service organizations, cloud data centers, and blockchain networks.

In the CDC organization, a node can communicate with a designated node in the CSP, and they share a blockchain in the same channel. Separating business and identity management has the benefit of improving CSP service efficiency because identity authentication consumes computing resources, while CSP provides cloud resource usage to users. Therefore, separating identity authentication functionality allows CSP to focus on business processing. Another advantage is scalability. If additional CSP organizations want to join the blockchain network, the organization need only apply for a node in the CDC group to achieve subsequent user access management. In the identity management and authentication model based on the Fabric network, CSP provides interfaces for users to register and log in with their identity information. In Hyperledger Fabric, there is more than one node that provides services to users on the same channel, and they share a ledger. As the administrator of the blockchain network, they monitor the state of the entire network and can view the information in the corresponding blocks. In a network environment, the most trusted party is oneself, so the user's identity credentials must be stored in ciphertext in the block and transmitted in ciphertext form during transmission to prevent man-in-the-middle attacks.

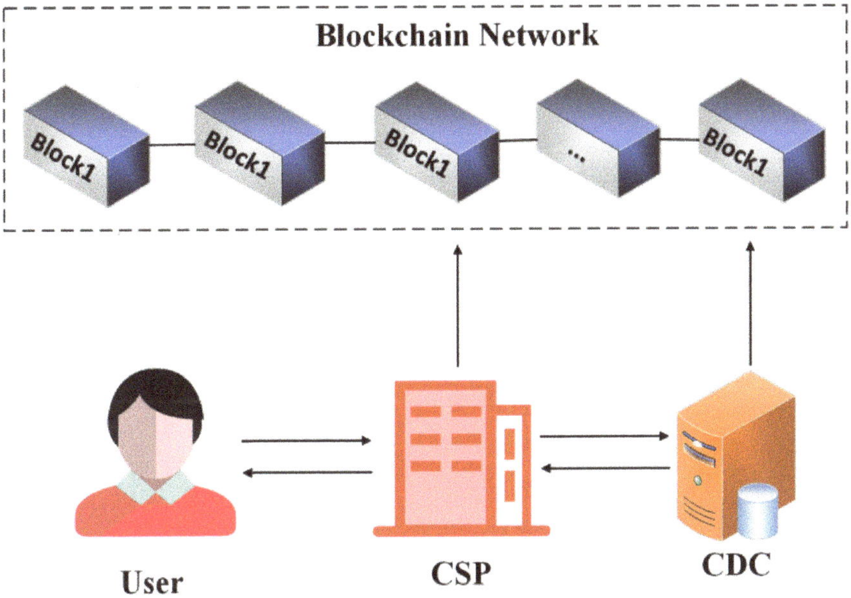

Figure 4. Secure and efficient identity authentication system model.

5.2. Specific Steps

As shown in Figure 5, the proposed scheme consists of seven steps. In these steps, we can observe that users only need to interact with CSP, and all related authentication tasks will be performed by CDC, thereby reducing the computational burden on CSP. The specific authentication process of the scheme is as follows:

(1) Initializing the user environment means that all system parameters have been instantiated, including <N,e,d> for encryption, which is computed by the user running GenRSA(1^n) and inputting the security parameter 1^n. The public key is represented by <N,e>, and the private key is represented by <N,d>. In addition, system parameters for a specific elliptic curve are pre-defined, and mathematical operations and calculations necessary for encryption and decryption are performed. These system parameters include the elliptic curve identifier cid, parameters for the base field Fq of the elliptic curve, parameters a and b for the elliptic curve equation, the order N and cofactor cf of the curve, as well as the embedding degree k and f of the curve E(Fq). Using these system parameters, a bilinear mapping from G1 to G2 can be defined, where P1 and P2 are generators of the cyclic subgroups G1 and G2 that generate the curve E(Fq). In addition, a bilinear pairing identifier eid and a homomorphic mapping Ψ from G2 to G1 are required. The user sets the password as the user s secret s and hash s to get x. It then finds the two points G and H on the elliptic curve, and multiply points G and H by x to get xG and xH.

(2) After the user has initialized the environment, they send the following registration request information to the CSP (where ID_u is the user's ID, Username is the user's registered name, and T_u is the user's timestamp to prevent replay attacks):

$$\text{User} \rightarrow \text{CSP: Request\_Enroll}(ID_u, \text{Username}, xG, xH, T_u)$$

(3) After receiving the registration request from the user, CSP checks its local ledger to see if there is any corresponding block information with ID_u. If the information exists, CSP responds to the user with a message m_1 indicating that the user already exists with Tcsp. Otherwise, CSP sends the user's registration identity information, including IDu and EGH, to CDC.

$$\text{CSP} \rightarrow \text{CDC: Message}(ID_u, xG, xH, T_s)$$

(4) After receiving the message from CSP, CDC generates a random value of c. CDC encrypts c using the secret key K_{cdc}, resulting in $E_c = E(K_{cdc}, c)$, and stores it along with xG and xH in the local ledger corresponding to ID_u.

(5) The registration result m_2 returned to CSP includes E_c, the ciphertext $EK_{cdc} = E(PK_u, K_{cdc})$ obtained by encrypting Kcdc with the user's public key, and the user ID.

(6) After receiving the message from CDC, CSP stores more detailed user information, including IDu, username, and user registration time-related information in a block for later user information querying operations. CSP then returns the registration result message to the user as follows (where m_3 represents the message containing E_c, EK_{cdc}, and the registration success information):

$$\text{CSP} \rightarrow \text{User: Respond}(m_3, T_s)$$

(7) After receiving the registration result message from CSP, the user decrypts $E(PK_u, K_{cdc})$ using the private key SK_n to obtain K_{cdc}. Then, the user decrypts Ec using Kcdc to obtain the plaintext c and stores it locally for the next identity verification. When the user needs to verify their identity to access cloud resources, The user only needs to provide a random value v which is calculated by (v − cx) to obtain an r (r is called a promise) to complete the operation of verifying the user's identity. The user sends a login request to CSP, and the request message includes the following content (where vG and vH are obtained by doubling G and H points by v, and Tu is the user-generated timestamp):

$$\text{User} \rightarrow \text{CSP: Request\_Verify}(ID_u, \text{Username}, r, vG, vH, T_u)$$

(8) After receiving the login request from the user, the CSP checks if the user with IDu is registered. If the user is not registered, the CSP returns a message m1 indicating that the user does not exist. Otherwise, the CSP sends IDu, r, vG, and vH to the CDC.

$$\text{CSP} \rightarrow \text{CDC: Message }(ID_u, r, vG, vH)$$

(9) After receiving the message from CSP, CDC uses the doubling method to multiply the G and H points by r to obtain rG and rH. Then, CDC queries the corresponding information of the user in the local ledger using IDu to obtain (xG, xH) and $E(K_{cdc}, c)$, and decrypts E(Kcdc, c) using Kcdc to obtain c. Next, CDC multiplies the xG and xH points by c using the doubling method to obtain cxG and cxH, and sets a = rG + c(xG) and b = rH + c(xH). Finally, CDC verifies if a equals vG and b equals vH as sent by the user and returns the verification result m2 to CSP. The reason why the equality of the two sides can prove that the user owns s is as follows:

$$vG = rG + c(xG) = (v - cx)G + cxG = vG$$

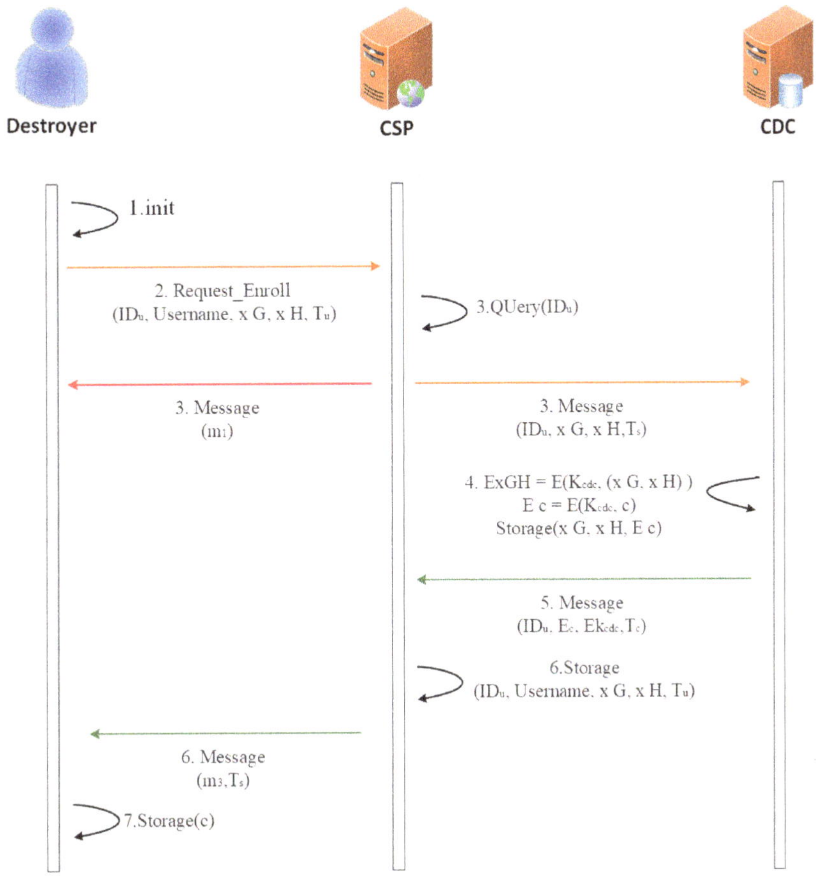

Figure 5. Specific protocol of identity authentication.

6. Security Analysis

The proposed HIDA authentication model in this paper involves three entities $U \in \mathbb{P} = (\mathbb{U} \cup \mathbb{S} \cup \mathbb{C})$, which are cloud users (to be authenticated), cloud service providers, and cloud data centers. We assume that $U \in \mathbb{P}$ has long-term asymmetric keys (sku, pku). During the operation of an entire system, let us assume the existence of adversary $\mathcal{A}$ who tries to disrupt the system in polynomial time. The possible ways of disruption mainly include attacks on the three-party entities and eavesdropping on the communication between the two parties to obtain sensitive data transmitted over the channel, thereby causing user identity leakage. We will analyze and prove the infeasibility of various types of attacks below.

6.1. Security Analysis

Claim 1: The CDC in this model has a high level of security protection and can effectively prevent system crashes caused by single-point attacks.

Analysis: The system model constructed in this article is based on a secure consortium chain using Fabric as a permission-granted blockchain. This means that only authorized entities can participate in the chain and access and process sensitive data and business logic. At the same time, permission-granted consortium chains have higher transaction processing speeds and scalability because only a small number of authorized nodes participate in verifying and adding new blocks, rather than all nodes needing to participate in the process. As shown in Figure 6, when an attacker is in a complex environment and wants to perform access attacks on the CDC, the likelihood of a successful attack is negligible due to being unauthorized.

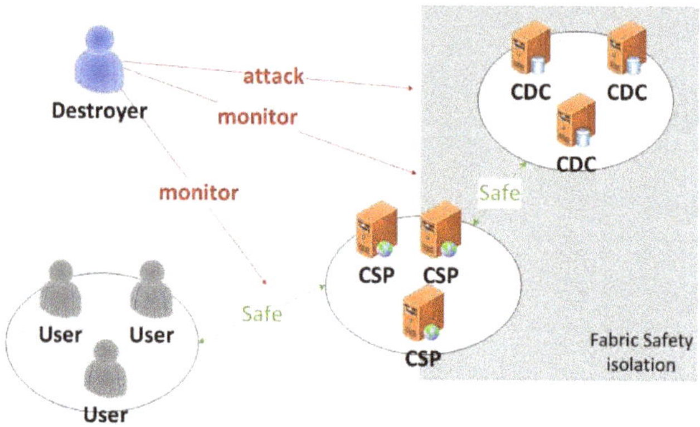

Figure 6. Safety model under HIDA.

6.2. Privacy Analysis

Claim 2: User data can be securely transmitted in the proposed model.

Analysis: This paper cleverly combines asymmetric encryption and symmetric encryption techniques to protect sensitive user data and uses zero-knowledge proof technology and blockchain technology to achieve secure authentication and storage of user identities. Therefore, unauthorized attackers find it difficult to eavesdrop on the channel. Secondly, assuming the difficulty of the RSA problem and the relatedness of GenRSA, selecting H as a random oracle [25] and using Π to represent the construction method in Algorithm 1, it can be proved that Π has indistinguishable encryption in the presence of eavesdropping by an adversary $\mathcal{A}$. Define $\varepsilon(n) = \Pr[\mathrm{PubK}^{\mathrm{eav}}_{A,\Pi}(n) = 1]$ and use the random oracle model to prove its security.

6.3. Formal Analysis

BAN logic is a logical system for analyzing the security of communication protocols, proposed by computer scientists Michael Burrows, Martín Abadi, and Roger Needham in 1989 [26]. Its main purpose is to analyze and prove the security of communication protocols, which are sets of rules used to exchange information between two parties. BAN logic defines a set of logical formulas and rules to analyze the security of these protocols. These formulas and rules describe the processes of sending, receiving, and processing information, and allow for the derivation of properties such as non-attackability and confidentiality of the protocol.

- BAN logical reasoning rule:
  (1) Rules of message meaning:
    (1.1) $\frac{\mathrm{P}|\equiv \mathrm{Q}\overset{K}{\leftrightarrow}P, P\lhd\{X\}_K}{P|\equiv Q\sim X}$
    (1.2) $\frac{\mathrm{P}|\equiv \overset{K}{\rightarrow}\mathrm{Q}, P\lhd\{X\}_{K-1}}{P|\equiv Q\sim X}$
    (1.3) $\frac{\mathrm{P}|\equiv Q\underset{Y}{\leftrightarrow}\mathrm{P}, P\lhd\{X\}_Y}{P|\equiv Q\sim X}$
  (2) Temporary value verification rules:

$$\frac{\mathrm{P}|\equiv \#(X), \mathrm{P}|\equiv \mathrm{Q} \mid \sim \mathrm{X}}{P|\equiv Q|\equiv X}$$

(3) Arbitration rules:

$$\frac{P|\equiv Q|\Rightarrow X, P|\equiv Q|\equiv X}{P|\equiv X}$$

(4) Faith rules:

$$\frac{P|\equiv X, P|\equiv Y}{P|\equiv (X,Y)} \quad \frac{P|\equiv (X,Y)}{P|\equiv X} \quad \frac{P|\equiv Q|\equiv (X,Y)}{P|\equiv Q|\equiv X}$$

(5) Send rules:

$$S\frac{P|\equiv Q|\sim (X,Y)}{P|\equiv Q|\sim X}$$

(6) Receive rules:

$$\frac{P\lhd (X,Y)}{P\lhd X} \quad \frac{P\lhd \langle X\rangle_Y}{P\lhd X} \quad \frac{P|\equiv Q\overset{K}{\leftrightarrow}P, P\lhd \{X\}_K}{P\lhd X}$$

$$\frac{P|\equiv \overset{K}{\rightarrow}P, P\lhd \{X\}_K}{P\lhd X} \quad \frac{P|\equiv Q|\sim (X,Y)}{P|\equiv Q|\sim X}$$

(7) Fresh rules:

$$\frac{P|\equiv \#(X)}{P|\equiv \#(X,Y)}$$

(8) Share key rules:

$$\frac{P|\equiv R\overset{X}{\leftrightarrow}R'}{P|\equiv R'\overset{X}{\leftrightarrow}R} \quad \frac{P|\equiv Q|\equiv R\overset{X}{\leftrightarrow}R'}{P|\equiv Q|\equiv R'\overset{X}{\leftrightarrow}R}$$

(9) Sharing of secret rules:

$$\frac{P|\equiv R\underset{x}{\leftrightarrow}R'}{P|\equiv R'\underset{x}{\leftrightarrow}R} \quad \frac{P|\equiv Q|\equiv R\underset{X}{\leftrightarrow}R'}{P|\equiv Q|\equiv R'\underset{x}{\leftrightarrow}R}$$

- The agreement is idealized:

  Message 1: U→S:{IDu, Username, xG, xH, Tu}Ks
  Message 2: S→C:{IDu,xG, xH, Ts}Kc
  Message 3: C→S:{IDu, {Ec, EKcdc}Kuc, Tc}Ks
  Message 4: S→U:{ {Ec, EKcdc}Kuc, Ts}Ku

- Initialize the hypothesis:

  (1) $U|\equiv \overset{K_s}{\rightarrow} S$
  (2) $U|\equiv \overset{K_c}{\rightarrow} C$
  (3) $S|\equiv \overset{K_u}{\rightarrow} U$
  (4) $S|\equiv \overset{K_c}{\rightarrow} C$
  (5) $C|\equiv \overset{K_u}{\rightarrow} U$
  (6) $C|\equiv \overset{K_s}{\rightarrow} S$
  (7) $S|\equiv U(|\Rightarrow xG, xH)$
  (8) $C|\equiv S(|\Rightarrow xG, xH)$
  (9) $S|\equiv C\left(|\Rightarrow \{E_c, EK_{cdc}\}_{K_{uc}}\right)$
  (10) $U|\equiv S\left(|\Rightarrow \{E_c, EK_{cdc}\}_{K_{uc}}\right)$
  (11) $U|\equiv \sharp(T_s)$
  (12) $U|\equiv \sharp(T_c)$
  (13) $S|\equiv \sharp(T_u)$

(14) $S \mid\equiv \sharp(T_c)$
(15) $C \mid\equiv \sharp(T_u)$
(16) $C \mid\equiv \sharp(T_s)$

- The purpose of certification:

(1) $S \mid\equiv \{xG, xH\}$
(2) $C \mid\equiv \{xG, xH\}$
(3) $S\left\{\{E_c, EK_{cdc}\}_{K_{uc}}\right\}$
(4) $U\left\{\{E_c, EK_{cdc}\}_{K_{uc}}\right\}$

- Logical inference:

(i)

message 1 $\Rightarrow S \triangleleft \{ID_u, Username, xG, xH, T_u\}_{K_s}$	(1a)
(1a) $\wedge$ assumption (3) $\Rightarrow S \mid\equiv U \sim \{ID_u, Username, xG, xH, T_u\}$	(1b)
rule (7) $\wedge$ assumption (12) $\Rightarrow S\sharp\{ID_u, Username, xG, xH, T_u\}$	(1c)
(1b), (1c) $\wedge$ rule (2) $\Rightarrow S \mid\equiv U \mid\equiv \{ID_u, Username, xG, xH, T_u\}$	(1d)
(1d) $\wedge$ rule (4) $\Rightarrow S \mid\equiv U \mid\equiv \{xG, xH\}$	(1e)
(1e) $\wedge$ assumption (7) $\Rightarrow S \mid\equiv \{xG, xH\}$	(a)

(ii) Similarly, the above message and assumptions under the reasoning of the BAN logic rules lead to proof (b): $C \mid\equiv \{xG, xH\}$ (b)

(iii)

message 3 $\Rightarrow S \triangleleft \left\{ID_u, \{Ec, EK_{cdc}\}_{K_{uc}}, Tc\right\}_{K_s}$	(2a)
(2a) $\wedge$ assumption (4) $\Rightarrow S \mid\equiv C \sim \left\{ID_u, \{E_c, EK_{cdc}\}_{K_{uc}}, T_c\right\}$	(2b)
rule (7) $\wedge$ assumption (14) $\Rightarrow S\sharp\left\{ID_u, \{E_c, EK_{cdc}\}_{K_{uc}}, T_c\right\}$	(2c)
(2b), (2c) $\wedge$ rule (2) $\Rightarrow S \mid\equiv C \mid\equiv \left\{ID_u, \{E_c, EK_{cdc}\}_{K_{uc}}, T_c\right\}$	(2d)
(2d) $\wedge$ rule (4) $\Rightarrow S \mid\equiv C \mid\equiv \{E_c, EK_{cdc}\}_{K_{uc}}$	(2e)
(2e) $\wedge$ assumption (7) $\Rightarrow S \mid\equiv \{E_c, EK_{cdc}\}_{K_{uc}}$	(c)

(iv) Similarly, the above message and assumptions under the reasoning of the BAN logic rules lead to proof (d): $U \mid\equiv \{E_c, EK_{cdc}\}_{K_{uc}}$ (d)

In summary, the steps of the HIDA protocol have been proven to be secure by means of a formal language, and to satisfy the security properties of confidentiality, integrity, and authentication.

7. Efficiency Analysis

We conducted simulation testing on the overall model of the system in the Fabric network and implemented the chaincode related to HIDA user identity information registration. At the same time, we compared and displayed the performance indicators and key elements of each link. The relevant configuration of the experimental environment is detailed in Table 2.

Table 2. The experimental environment.

Name	Configure
Processor	Intel(R) Core(TM) i5-2430M
Run memory	4 GB
Operating System	Ubuntu 20.04
Docker	20.10.1
Docker-compose	1.25.0-rc1
Go	go1.14.6 linux/amd64
Fabric	2.3.0

7.1. Functional Testing

The HIDA functional test implemented a simple client. Firstly, xG, xH, G, and H were generated for user registration. The user's password was used as a parameter to call the user proof generation interface to produce the corresponding user proof string. Secondly, r, vG, and vH were generated for verifying the user's identity. After registration, the user would receive user challenge data, which, along with the password and user proof string, were used as parameters to call the user verification generation interface, generating the user verification string. The user proof string and the user verification string were used in the user identity access request and verify user identity access request, respectively.

During the registration process, the user provided their ID, username, and user-proof string to the CSP and called the user registration interface with the appropriate parameters to complete registration. The registration result contained the user's basic information (ID, username, and registration status), as well as the encrypted challenge encoded in Base64. The functionality was successfully executed and met expectations based on testing.

7.2. Performance Testing

The SIDM (Secure Identity Management) protocol is a privacy-preserving protocol designed with a zero-knowledge proof on the basis of the CIDM protocol, which solves the problems of the man-in-the-middle attack and user privacy leakage in traditional identity authentication. We have conducted corresponding comparative statistics on the data transmission of the three-party entities, and the results show that our scheme is significantly better than the above scheme in terms of data volume on both the user and authentication sides. Table 3 shows the relevant comparison data.

Table 3. The relevant comparison data.

	SIDM			HIDA		
	User	**Idms**	**Csp**	**User**	**Cdc**	**Csp**
Send (bytes)	106	50	110	100	8	113
Receive (bytes)	80	116	70	16	97	108
Total (bytes)	186	166	180	116	105	221

The authentication time of the designed HIDA scheme for different numbers of users was calculated using a Shell script. The authentication for each user count was statistically measured ten times, and the longest and shortest times were excluded before calculating the average. The results are shown in Figure 7. It can be observed from the results that the HIDA scheme performs significantly faster than the EIDM scheme for different numbers of users.

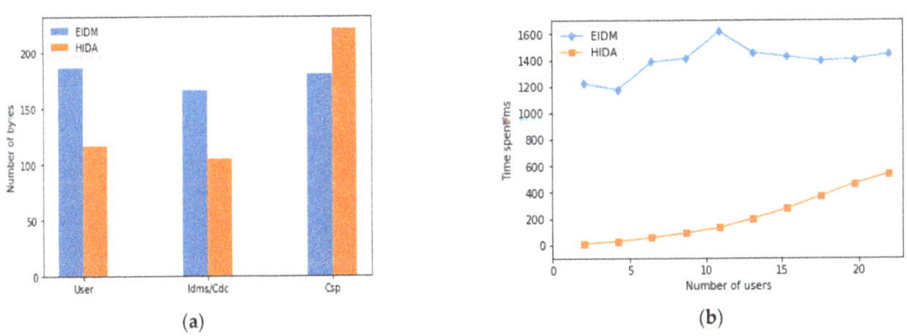

Figure 7. (**a**,**b**) performance comparison.

7.3. Experimental Summary

The experimental results show that the HIDA protocol has the smallest byte-wise overhead in each step of the verification process. In addition, we compared the HIDA protocol with CIDM and EIDM protocols, and found that the HIDA protocol has a higher value in terms of privacy protection for users. Furthermore, we studied the time overhead of user verification between HIDA and EIDM. The experimental results demonstrate that the time overhead of the HIDA protocol for user verification is much lower than that of EIDM. This finding further confirms the innovation and value of the HIDA protocol in the field of privacy protection. Therefore, we believe that the HIDA protocol has broad application prospects, and future research can further explore its application in other fields.

8. Conclusions

The popularization of cloud computing has made our lives more convenient, but it has also brought many challenges to traditional identity authentication. To solve the problems of single point of failure, privacy security, efficiency, and transparency in traditional identity authentication in a cloud environment, this paper designs and implements the HIDA identity authentication scheme based on the Hyperledger Fabric platform, with the main work as follows:

This paper mainly introduces the HIDA identity authentication scheme designed based on the Hyperledger Fabric platform. First, we introduce the challenges faced by traditional identity authentication in cloud computing environments, including single points of failure, privacy security, efficiency, and transparency issues. Then, we elaborate on the design ideas and implementation of the HIDA scheme, including key technologies such as user identity information registration, user identity verification, and user access control. Finally, we conduct experimental simulations and performance tests to verify the security and efficiency advantages of the scheme.

Through the research in this paper, we have drawn the following conclusions: First, the HIDA scheme can effectively solve the privacy and security problems of traditional identity authentication in a cloud environment, ensuring the security and controllability of user data. Second, the HIDA scheme adopts modern cryptography technologies such as zero-knowledge proof, which can effectively avoid security threats such as man-in-the-middle attacks, ensuring the security of the system. Third, through experimental simulations and performance tests, we have verified the efficiency and superiority of the HIDA scheme under different numbers of users, demonstrating its practicality and feasibility. Overall, the HIDA identity authentication scheme designed and implemented in this paper is a feasible, secure, and efficient solution that can provide strong support and guarantees for identity authentication in the field of cloud computing. In the future, we will continue to optimize the scheme to improve its security and efficiency, better meeting user and market needs.

Since the birth of blockchain technology, its special underlying architecture and security model have been widely sought after. It has been widely used not only in the field of cryptocurrency but also in various work scenarios. In recent years, research combining blockchain technology with cloud computing has also increased. This combination is expected to bring better data security, higher efficiency, and lower costs. In the future, we hope to see more excellent solutions being borrowed and cited to better promote the development of this field.

Author Contributions: Methodology, W.J.; Formal analysis, W.J.; Investigation, Z.D. and Y.S.; Resources, X.R.; Data organization, C.T. and Y.S.; Writing-original draft, W.J.; Writing-review and editing, W.J., X.R. and Z.D.; Visualization, X.R.; Supervision, Z.D.; Funding acquisition, X.R. All authors have read and agreed to the published version of the manuscript.

Funding: This research work was partially supported by the Shaanxi Natural Science Basic Research Project (Grant No. 2020JM-565), the Shaanxi International Science and Technology Cooperation Program Project (Grant No. 2021KW-07), and Xi'an Weiyang District Science and Technology Plan Project, No. 202025.

Informed Consent Statement: Not applicable.

Data Availability Statement: All the data are included in the article.

Conflicts of Interest: The authors declare no conflict of interest.

References

1. Leavitt, N. Is cloud computing really ready for prime time. *Growth* **2009**, *27*, 15–20. [CrossRef]
2. Li, W.; Wu, J.; Cao, J.; Chen, N.; Zhang, Q.; Buyya, R. Blockchain-based trust management in cloud computing systems: A taxonomy, review and future directions. *J. Cloud Comput.* **2021**, *10*, 35. [CrossRef]
3. Yao, Q.; Zhang, D. Survey on identity management in blockchain. *J. Softw.* **2021**, *32*, 2260–2286.
4. Carlin, S.; Curran, K. Cloud computing security. In *Pervasive and Ubiquitous Technology Innovations for Ambient Intelligence Environments*; IGI Global: Hershey, PA, USA, 2013; pp. 12–17.
5. Shukla, S.; Patel, S.J. A novel ECC-based provably secure and privacy-preserving multi-factor authentication protocol for cloud computing. *Computing* **2022**, *104*, 1173–1202. [CrossRef]
6. Goldwasser, S.; Micali, S.; Rackoff, C. The knowledge complexity of interactive proof-systems. In *Providing Sound Foundations for Cryptography: On the Work of Shafi Goldwasser and Silvio Micali*; Association for Computing Machinery and Morgan & Claypool Publishers: Manhattan, NY, USA, 2019; pp. 203–225.
7. Kamboj, P.; Khare, S.; Pal, S. User authentication using Blockchain based smart contract in role-based access control. *Peer-to-Peer Netw. Appl.* **2021**, *14*, 2961–2976. [CrossRef]
8. Androulaki, E.; Barger, A.; Bortnikov, V.; Cachin, C.; Christidis, K.; De Caro, A.; Enyeart, D.; Ferris, C.; Laventman, G.; Manevich, Y.; et al. Hyperledger fabric: A distributed operating system for permissioned blockchains. In Proceedings of the Thirteenth EuroSys Conference, Porto, Portugal, 23–26 April 2018; pp. 1–15.
9. Hammi, M.T.; Bellot, P.; Serhrouchni, A. BCTrust: A Decentralized Authentication Blockchain-Based Mechanism. In Proceedings of the 2018 IEEE Wireless Communications and Networking Conference (WCNC), Barcelona, Spain, 15–18 April 2018; pp. 1–6.
10. Suguna, M.; Anusia, R.; Shalinie, S.M.; Deepti, S. Secure Identity Management in Mobile Cloud Computing. In Proceedings of the 2017 International Conference on Nextgen Electronic Technologies: Silicon to Software (ICNETS2), Chennai, India, 23–25 March 2017; pp. 42–45.
11. Khalil, I.; Khreishah, A.; Azeem, M. Consolidated Identity Management System for secure mobile cloud computing. *Comput. Netw.* **2014**, *65*, 99–110. [CrossRef]
12. Jones, M.; Hardt, D. *No. RFC6750*; The Oauth 2.0 Authorization Framework: Bearer Token Usage; Internet Engineering Task Force (IETF): Fremont, CA, USA, 2012.
13. Lundkvist, C.; Heck, R.; Torstensson, J.; Mitton, Z.; Sena, M. Uport: A Platform for Self-Sovereign Identity. 2017. Available online: https://whitepaper.uport.me/uPort_whitepaper_DRAFT20170221.pdf (accessed on 3 May 2023).
14. Shrier, D.; Wu, W.; Pentland, A. Blockchain & infrastructure (identity, data security). *Mass. Inst. Technol.-Connect. Sci.* **2016**, *1*, 1–19.
15. Tobin, A.; Reed, D. The inevitable rise of self-sovereign identity. *Sovrin Found.* **2016**, *29*, 18.
16. Liang, X.; Zhao, J.; Shetty, S.; Liu, J.; Li, D. Integrating Blockchain for Data Sharing and Collaboration in Mobile Healthcare Applications. In Proceedings of the 2017 IEEE 28th Annual International Symposium on Personal, Indoor, and Mobile Radio Communications (PIMRC), Montreal, QC, Canada, 8–13 October 2017; pp. 1–5.
17. Gong, L.; Alghazzawi, D.M.; Cheng, L. BCoT sentry: A blockchain-based identity authentication framework for IoT devices. *Information* **2021**, *12*, 203. [CrossRef]
18. Gan, S. An IoT Simulator in NS3 and a Key-Based Authentication Architecture for IoT Devices Using Blockchain. Master's Thesis, Indian Institute of Technology Kanpur, Kanpur, India, 2017.
19. Alsayed Kassem, J.; Sayeed, S.; Marco-Gisbert, H.; Pervez, Z.; Dahal, K. DNS-IdM: A blockchain identity management system to secure personal data sharing in a network. *Appl. Sci.* **2019**, *9*, 2953. [CrossRef]
20. Cheng, Y.; Jia, Z.; Gong, B.; Wang, L.P.; Lei, Y. F. Threshold Signature Scheme with Strong Forward Security Based on Chinese Remainder Theorem. In *Security and Privacy in New Computing Environments, Proceedings of the Second EAI International Conference, SPNCE 2019, Tianjin, China, 13–14 April 2019*; Springer International Publishing: Manhattan, NY, USA, 2019; pp. 15–28.
21. Koblitz, N. Elliptic curve cryptosystems. *Math. Comput.* **1987**, *48*, 203–209. [CrossRef]
22. Rivest, R.L.; Shamir, A.; Adleman, L. A method for obtaining digital signatures and public-key cryptosystems. *Commun. ACM* **1978**, *21*, 120–126. [CrossRef]
23. Kaltz, J.; Lindell, Y. *Introduction to Modern Cryptography: Principles and Protocols*; CRC Press: Boca Raton, FL, USA, 2008.
24. Fiege, U.; Fiat, A.; Shamir, A. Zero knowledge proofs of identity. In Proceedings of the Nineteenth Annual ACM Symposium on Theory of Computing, New York, NY, USA, 1 January 1987; pp. 210–217.
25. Canetti, R.; Goldreich, O.; Halevi, S. The random oracle methodology, revisited. *J. ACM* **2004**, *51*, 557–594. [CrossRef]
26. Wessels, J. Application of BAN-logic. *CMG Finance BV* **2001**, *19*, 23.

electronics

MDPI

Article

Blockchain-Based Security Configuration Management for ICT Systems

Dimitrios Chatziamanetoglou * and Konstantinos Rantos

Department of Computer Science, International Hellenic University, 65404 Kavala, Greece; krantos@cs.ihu.gr
* Correspondence: diehatz@cs.ihu.gr

Abstract: The world has become increasingly dependent on large-scale and distributed information and communication technology (ICT) infrastructures and systems in sectors such as energy, transport, banking, healthcare, water supply, and digital services, while their protection is considered of paramount importance and has already drawn remarkable attention from governments and key industry players. Establishing common approaches by leveraging existing frameworks and cyber security practices for improving the security postures of those systems is one of the major objectives for ensuring an adequate level of protection and avoiding the detrimental effects of disruptions on society and citizens. Configuration management (CM) is one of those common practices for establishing and maintaining the integrity and consistency of a system and its elements with regard to the function, performance, and status of technical and physical attributes, and it contributes to a desirable security posture throughout the lifecycle of a system. This study addresses the importance of CM, and while considering the corresponding frameworks, standards, and best practices, it proposes a permissioned blockchain-based approach, that inherits the benefits of the blockchain technology and ensures the integrity of the systems' configuration across the complete lifecycle management of its products and services as an underlying model for mapping and integrating CM functions. Furthermore, this study briefly presents the benefits and challenges of the application of permissioned blockchain models and proposes a smart-contract-based role-based access control mechanism, in addition to presenting an operating concept based on brief but real-life lifecycle requirements of organizational configuration management.

Keywords: configuration management; change management; blockchain; critical infrastructure; distributed and large-scale ICT

Citation: Chatziamanetoglou, D.; Rantos, K. Blockchain-Based Security Configuration Management for ICT Systems. *Electronics* **2023**, *12*, 1879. https://doi.org/10.3390/electronics12081879

Academic Editor: Hamed Taherdoost

Received: 21 March 2023
Revised: 10 April 2023
Accepted: 13 April 2023
Published: 16 April 2023

1. Introduction

The cyber landscape is under constant changes as malicious actors exploit known cyber security gaps and discover new ones. Over recent years, cyber attacks on critical ICT systems have increased [1], and they have become more sophisticated and effective than before. The consequences of cyber attacks on a large-scale and distributed ICT infrastructure for financial, political, or military gain could include service degradation or disruption, environmental damage, financial loss and/or human injuries, or threats to human lives on a large scale, thus causing serious problems.

A fundamental understanding of how malware actors target critical systems and infrastructures can help organizations and key stakeholders comprehend how to conduct cyber security defense operations, respond to incidents, embed security in systems' design, understand risks and business impacts, implement strategic, operational, and tactical changes, and protect themselves from possible harm.

Collaboration between states and public or private entities facilitates the development of policies, frameworks, and guidelines for raising awareness, establishing best practices, leveraging and combining strengths, developing skills, identifying gaps, and increasing preparedness. These interactions also help teams work together more efficiently to deter, detect, and apply effective responses to cyber attacks, especially across distributed ICT infrastructure components, which, in many cases, deal with geographically dispersed legacy systems.

As cyber threats continue to evolve, organizations still have limited resources for minimizing their attack surfaces and improving their security postures. Security has become a risk-based activity, where the operational and economic costs define the appetite and tolerance thresholds of risk, and they are fully integrated to balance the needs of an organization's mission and business processes against cyber threats. In today's digital world, resources are scoped and tailored to fulfill this balance, and the practice of a risk management approach is fundamental to cyber security programs.

Information technology (IT) and operational technology (OT) systems are under a constant state of change, which spans technological, physical, business, and even human elements. Such changes could be hardware or software changes, incoming personnel with different skills, changes in system architectures, changes in the supply chain, changes in physical and technical access control measures, and many more. A disciplined and structured approach to monitoring, controlling, and documenting such changes is essential for the support of the security of IT and OT systems [2]. Such activities fall under configuration management (CM), the absence of which can have a significant impact on the security and privacy postures of these systems.

According to the National Institute of Standards and Technology (NIST) [3], Configuration Management is an umbrella definition that covers a collection of activities that are focused on establishing and maintaining the integrity of information technology products and information systems through the control of processes for initializing, changing, and monitoring the configurations of those products and systems throughout the system development lifecycle. Change management—despite being frequently shown as a standalone function—is a specific functional element of configuration management that is responsible for managing changes during the lifespan of a system. Its interactions with the rest of the configuration management functions will be shown and described below.

The concepts, principles, and processes described in existing publications set the overarching high-level framework of security configuration management (SCM) [3–5] with the objective of managing and monitoring the configurations of information and operational systems to achieve adequate security and minimize organizational risks while supporting the desired business outcomes and services.

Furthermore, during the rapid development and wide application of distributed ICT systems, the interest in blockchain technology has dramatically increased and has made it a widely accepted solution due to the efficiency, applicability, and essentials of its features [6]. Blockchain technology can be used to cryptographically sign the "who", "what", "where", "when", and "why" for the status of and changes in all critical cyber assets throughout the chain of custody. Traceability, auditability, decentralization, immutability, transparency, and peer communications are features of distributed ledger technology that are enablers in supporting the management and security of information technology products and information systems more effectively [7].

Focusing on the principles of security configuration management, this study proposes a blockchain-based model as the foundation of configuration management and change control applications; this model inherits all of the benefits and advantages of the integrity, tamper-resistance, trust, and scalability of distributed ledgers. Placing an emphasis on the protection of distributed ICT infrastructure and its complex nature, this study addresses the requirements of CM by proposing a permissioned ledger as a closed ecosystem with a defined governance structure, private transactions, and strict authentication access in order to maintain the security requirements of the critical components of IT/OT infrastructures.

The aim of this study is two-fold. The first objective is to present a high-level overview of the existing frameworks, publications, handbooks, and guidelines that underpin the CM process in IT/OT systems, especially in the domain of large-scale and distributed systems. The second objective, which is of equal importance, is to map the fundamental processes, practices, and phases of CM and demonstrate how those can be applied in a blockchain model to support real-life technical requirements by benefiting from the advantages and features inherited from distributed ledger technology.

The rest of this paper is structured as follows. Section 2 presents the existing set of publications, frameworks, handbooks, and guidelines that define CM, as well as the existing background research related to the scope. Section 3 presents the motivations and challenges that triggered the proposal of a blockchain-based CM model, since the existing research was found to be limited. Section 4 briefly presents the CM process and the phases that it comprises, as well as related definitions. Section 5 presents the proposed blockchain-based model and a related mapping of the functions of CM. Section 6 describes the operating concept using a practical application of a complete lifecycle of the CM process, while Section 7 summarizes the paper's proposals and depicts our objectives for future work.

2. Background and Research Review

Configuration management is a strong requirement in many cyber security frameworks and standards. It is used to enable the functional and physical attributes of IT/OT platforms, products, and their environments to determine the appropriate security features and assurances that are used to measure a system configuration state. It is also used to control modifications to hardware, firmware, software, and documentation in order to ensure that ICT systems are protected against unauthorized modifications prior to, during, and after system implementation by establishing baseline controls via policies, practices, and procedures.

The following publications are used to define the set of requirements, starting from risk management, ending with the implementation of configuration management, and spanning across various governmental and institutional entities:

1. ANSI/EIA-649C, Configuration Management Standard
2. Cloud Security Alliance (CSA) Cloud Controls Matrix (CCM)
3. EIA-836B, Standard for Configuration Management Data Exchange and Interoperability
4. ENISA Risk Management/Risk Assessment Framework
5. European Council Directive 2008/114/EC
6. IEC62443-Security for industrial automation and control systems-Parts 2-4, 4-1 and 4-2
7. ISO/IEC 27001, Information Security Management
8. ISO/IEC 20000:2018, IT Service Management System Requirements [8]
9. NIST Special Publication (SP) 800-37 Revision 2: Risk Management Framework for Information Systems and Organizations [2]
10. NIST SP 800-53 Revision 5: Security and Privacy Controls for Information Systems and Organizations [4]
11. NIST SP 800-82 Revision 2: Guide to Industrial Control Systems (ICS) Security [5]
12. NIST SP 800-128: Guide for Security-Focused Configuration Management of Information Systems [3]
13. NIST SP 800-160: Systems Security Engineering: Considerations for a Multidisciplinary Approach in the Engineering of Trustworthy Secure Systems [9]
14. NIST Cybersecurity Framework: Framework for Improving Critical Infrastructure Cybersecurity [10]
15. US DoD Protected Critical Infrastructure Information (PCII) Program

Alongside the above-mentioned frameworks, there are several other libraries of frameworks, standards, and best practices [11] that are focused on IT service management activities for service providers, enterprises, and military organizations. These publications play a significant role in the overall definition of CM and the set of requirements, processes, procedures, and business outcomes. These can be listed as follows:

1. ENISA Guidelines on Security Measures under the European Electronic Communications Code (EECC)

2. COBIT 2019, Management and Governance of Enterprise IT [12]
3. ISO 10007:2017, Quality management—Guidelines for configuration management
4. ITIL v4, IT Infrastructure Library
5. Mil-HDBK-61B, Configuration Management guidance [13]
6. TM Forum Business Process Framework (eTOM)

In the scope of the aforementioned publications that set the high-level frameworks and procedural guidelines, there are plenty of works found in the literature that address implementation proposals that follow the requirements of configuration management, but the research activity on blockchain-based solutions for supporting CM implementations is very limited.

Kinkelin et al. [14] proposed an abstract Byzantine fault-tolerant (BFT) configuration management system (CMS) based on the Hyperledger Fabric environment with the objective of managing configuration requests in an operational environment. Their proposal acts as an intermediate authority between administrators and managed devices, and it is able to conduct multi-party authorization for critical configurations to prevent individual malicious administrators from performing undesired actions; changes are applied only after a configuration has been validated and authorized by multiple experts. One of the drawbacks of the proposed system was found to be the potentially low number of validators, which would weaken the protection of this CMS. Kostal et al. [15] proposed a private blockchain approach for storing and loading configurations of Internet of Things (IoT) devices to manage and monitor network devices, and they highlighted the tamper-proof functionalities of the proposed method, which was also supported by off-chain databases. Furthermore, authorized network administrators used digital certificates to authenticate themselves, while they could modify the configuration of devices if they were authorized to do so for a given device or group/domain of devices. Alvarenga et al. [16] proposed a blockchain-based architecture for secure management, configuration, and migration of virtualized network functions, which ensured the immutability, non-repudiation, and auditability of the configuration update history and the anonymity of tenants and configuration information while guaranteeing the secure update and migration of configurations at the core of the network and being resilient to collusion attacks from up to one-third of the blockchain modules. Mylrea et al. [17] examined how blockchain technology can enable critical infrastructure protection compliance and aid in the security of software supply chains, patch management, and configuration management through an immutable cryptographically signed distributed ledger that enabled improved data security, provenance, and auditability while describing the challenges in applying blockchain technology due to the lack of existing policies and governance. Han [18] proposed a very abstract blockchain configuration management system that could protect the copyrights of software development projects and systems' configurations. Samaniego et al. [19] proposed a limited-scope virtualization of IoT components with the objective of supporting configuration management and provision across an IoT network, and they achieved efficiency in terms of latency and bandwidth. The solution was based on a permissioned blockchain with encrypted blocks for additional security.

The aforementioned research touched some elements of the overarching CM process, but did so in isolation—without a contextualized approach—and, most importantly, without identifying which function of CM they were addressing and how the proposed work might interact with the rest of the functions of CM. The present study covers this gap by addressing the challenges and opportunities of an efficient implementation of a CM process and presents an end-to-end blockchain-based model that enables all of the different functions of CM. Furthermore, it demonstrates how the benefits of blockchain technology can be leveraged in the CM lifecycle in alignment with existing frameworks, policies, recommendations, guidelines, and best practices.

3. Motivation and Challenges

Industries, enterprises, and governmental organizations use a plethora of tools that cover areas of the requirements of configuration management [20]. These tools have a certain overlapping coverage area. This is the reason for why more than one tool is required to cover the whole scope. Furthermore, there is no indication that these tools currently utilize blockchain technology in their production environments, especially for configuration management purposes.

We identified that even though there is a very mature framework addressing the requirements of that stem from the frameworks, policies, guidelines, and best practices of CM that were presented before, there is not a mature implementation showing how these requirements can be applied in a blockchain infrastructure by utilizing the value and benefits of this technology. In addition, the majority of the existing research covered CM topics in an abstract manner and did not address a holistic proposal or solution or how this may interact with the rest of the functions of CM.

It was found that the need for the development and application of blockchain technology to underpin the requirements of configuration management is evolving, and it is expected to expand and grow; when used, this technology will bring additional value and boost the business outcomes and overall security of large-scale and distributed ICT systems [17].

4. Configuration Management

4.1. Definitions and Value of Configuration Management

Information systems are discrete sets of information resources that are organized for the collection, processing, maintenance, use, sharing, dissemination, or disposition of information, and they are composed of several components, the configuration of which has a direct impact on the security posture and the operational functionality of the system [4]. A configuration item (CI) is a single component or set of components of an information system that is subject to configuration management and is considered as a single entity throughout the practice of configuration management [3,13]. Each CI should be properly identified, labeled, and tracked during its lifecycle, its interactions, and its contributions to an overall system's function. A CI could be a network element, a server, an application, a documentation, a security compliance checklist, a service, or even the information system as a whole, which when defined correctly, provides an organization with the means to apply the desired lifecycle management for security and operational requirements. Each CI has a baseline configuration.

A baseline configuration is a set of specifications for a CI or a set of CIs within a system that have been formally presented, reviewed, and approved at a given point in time. The baseline configuration can only be changed through a change control procedure triggered by a change request. The baseline configuration is used as a basis for future builds, releases, or changes and evolves over time depending on the stage and progress of the system development lifecycle (SDLC) requirements, such as development, testing, production and retirement. Early in the SDLC, when a system is being initiated and acquired, the baseline may be a set of functional requirements. As the system is developed and implemented, the baseline may expand to include additional configuration elements to fulfill the end-state objectives of production. When a new baseline configuration is established, all of the changes from the last baseline are approved. Older versions of approved baseline configurations are maintained and made available for review or rollback as required. There are also different types of baselines, such as functional baselines, product baselines, service baselines, etc., the differences of which, will be left out of this study.

CI records contain all information related to each CI, including the baseline configuration, unique identifiers, description, version, location, type, relationships with other CIs, status, etc. Practically, this information may include as many attributes/properties that can uniquely and unambiguously describe a CI, depending on the data model used by the system.

The way the configuration of the CIs is implemented, maintained, and managed requires a disciplined approach to providing adequate security and functionality and fulfilling the organizational outcomes and objectives. Changes in the CIs are often required to fulfill business requirements, adopt IT architectural changes, react to incidents and known IT problems, and be up to date on security needs. These changes can heavily and negatively impact the previously established security posture and/or operational requirements. This is the reason why a controlled, documented, and effective configuration management process is critical for the maintenance and improvement of the security posture, functional outcomes, and business objectives of an organization.

Security-focused configuration management (SecCM) is the management and control of secure configurations for an information system to enable security and facilitate the management of risk [3,21], which is still a subset of the overall configuration management process. The security-focused configuration management process is vital in maintaining a secure state during organizational operational management, ensuring that risks are properly assessed and changes are authorized to proceed, and managing a reliable and updated change record that is always available for auditing and accounting purposes.

A configuration management database (CMDB) is used to store configuration records throughout their lifecycle and maintain the relationships among them. It also helps an organization understand the relationships between the components of a system and track their configurations.

A configuration management system (CMS) comprises a minimum of one CMDB, while multiple CMDBs can be used by an organization to store and manage CI records from different domains. In addition, a CMS contains information related to incidents, problems, and known errors, which are functions that fall outside the scope of this study. The immutability of a CMS is considered vital, as it constitutes the overall IT baseline of the whole organization.

In practical terms, a CMS is part of a larger system called a service knowledge management system (SKMS). The SKMS consolidates and analyzes configuration item (CI) records to facilitate the design, development, delivery, operation, and improvement of services. The SKMS is based on a number of best practices and industry standards [8,12], and it fulfills organizational service management, security, and business intelligence objectives and requirements.

Finally, the SKMS also includes a definitive media library (DML), which is a secure repository in which an organization stores definitive and authorized versions of software, media, and data. When there is a requirement for the deployment of a new release, only available releases existing in the definitive media library can be used to build the new release. The entries in the DML can be considered as separate CIs, the records of which can be stored in the aforementioned CMDBs.

4.2. Configuration Management Functions

In general terms, the aforementioned configuration management (CM) standards, handbooks, and publications define the configuration management process with five key functions/disciplines [3,9,13] (shown in Figure 1):

- Management and planning;
- Configuration identification;
- Configuration control or change management (ChM);
- Configuration status accounting;
- Configuration verification, evaluation, and auditing.

Management and planning deal with defining the strategic aspects that are required to be in place in order to underpin the next stages. These aspects include the definition of roles, responsibilities, tools, interfacing processes, organizational criteria, and priorities, the definition of configuration items (CIs) and baselines, security considerations, conditions, and constraints, the development of documents such as directives, operational procedures, and guidelines, and the establishment of organizational boards that will enforce and oversee the configuration management process.

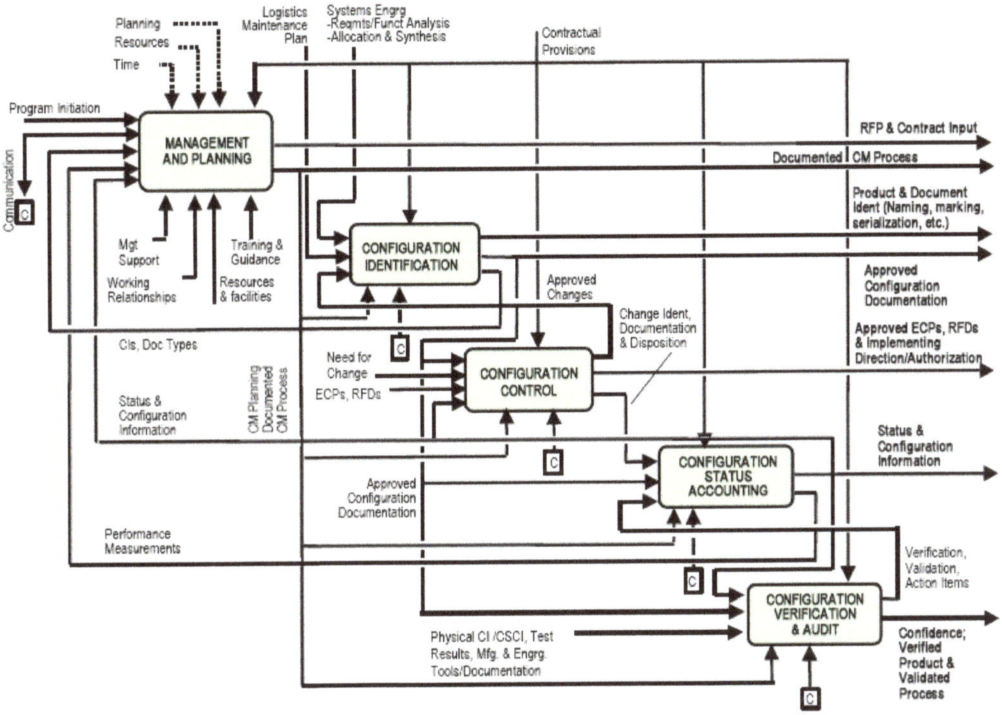

Figure 1. Configuration management functions (Mil-HDBK-61B [13]).

Configuration identification addresses the aspects of the proper identification, baseline, labeling, hierarchy, structure, and dependencies of CIs, which are subject to formal reviews and configuration audits. Via this function, all CIs should be uniquely and unambiguously identified, while the status of the configuration should be traceable for every past, present, or future (planned) configuration state.

Configuration control or change management (ChM) enforces the required security and control measures to maintain the secure, approved baseline of the system, minimize unauthorized and/or undocumented changes, further manage, document, and coordinate among all stakeholders, evaluate the quality, benefits, and costs, and assess the risk of all requested changes during the lifecycle management of all CIs by applying a broad range of procedures and criteria in order to finally deny or approve all change requests via the organizational regulatory boards, which are normally called configuration control boards (CCBs) or change advisory boards (CABs).

The configuration status accounting function captures, stores, maintains, and processes configuration management status information for CIs with respect to their baselines, approved changes, and releases in order to preserve, protect, and ensure the integrity, confidentiality, and availability of the CIs by supporting the planning and decision making for certification, accreditation, authorization, and reporting activities. By preserving current, accurate, and retrievable information on the status and configuration of CIs, an organization can ensure the effective and efficient lifecycle management of its systems, starting from the design phase, testing, and production and ending with full retirement.

The configuration verification, evaluation, and auditing function includes all necessary technical and non-technical capabilities for addressing security and operational concerns. These capabilities include security compliance checks, configuration gap analysis, and the difference between the "as-certified" and "as-maintained" status of CIs. Via these audit control mechanisms, an organization is able to assess whether a system conforms to the security aspects of the operational requirements against the defined baselines and is capable of assessing security-relevant discrepancies, variances, and deficiencies.

5. Proposed Model

This study consolidates the principles of configuration management and proposes a blockchain-based design that inherits all of the benefits and advantages of the integrity and immunity of distributed ledgers.

As mentioned before, the configuration management system is considered the overall baseline of a whole organization; thus, its confidentiality, integrity, and availability (CIA triad) are meant to be vital for the security posture and fulfillment of the operational objectives, especially when the organization underpins distributed ICT systems. The high-level components of the functions of CM and their proposed interrelationships with other functions and processes are shown in Figure 2, where the elements in green are enabled by blockchain technology.

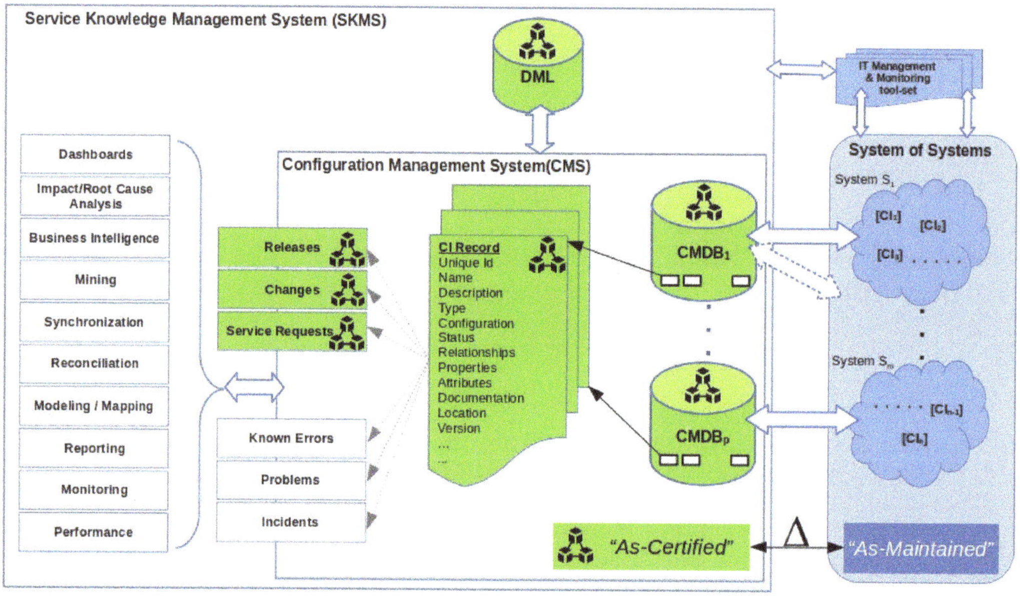

Figure 2. High-level components of CM.

A blockchain solution can be classified as public/permissionless or private/permissioned. Each one has differences in the nature of stakeholders' membership and authorization, which is required for participating in a blockchain network. In a public blockchain, anyone can register and interact without permission, while a permissioned network requires additional levels of access restriction policies, which narrow down the access, privileges, and rights of the participants.

Considering the increased security requirements and the nature of large-scale and distributed ICT systems, we propose the implementation of a permissioned distributed ledger with a well-defined governance structure, private transactions, and strict authentication for access in order to preserve and ensure the security requirements of critical IT/OT components. A permissioned ledger is proposed in order to allow only selected and verified participants of the organization to interact with the system by applying a role-based access control mechanism for giving specific privileges and access rights according to the roles and responsibilities in the CM process. Furthermore, the proposed blockchain-based system acts as a secure repository to ensure that all of the outputs of the configuration management process have not been tampered with.

This section will present the high-level architecture of the model by depicting all of the interfacing components and elements of the CM process, as well as the dependencies of the CIs of an IT system, while following an agnostic data model approach.

5.1. "As-Certified" vs. "As-Maintained"

Figure 2 depicts a helicopter view of the elements underpinning the concept of configuration management. The left part depicts the SKMS, which was defined earlier and represents the holistic configuration baseline of an organization, while the right part depicts the real-life status of the deployed IT/OT elements that serve the operational objectives of the enterprise.

An organization should maintain the status of its CIs, including the information on the stages of the whole lifecycle of those CIs, such as their design, planning, and status of being delivered/implemented, by utilizing various tools. These tools provide not only the capability of maintaining the baseline (golden data) of the organization, but also provide various levels of functionalities in order to model, map, reconcile, synchronize, analyze, report, and present the status of those CI records in a combined and controlled manner via either detailed or even executive-level views/dashboards. In our proposal, the core elements of the SKMS that hold the critical information of the organizational baseline are stored on a blockchain (green color) and benefit from its inherited advantages, while the rest of the depicted functions can be supported by conventional and existing methodologies/toolsets.

As mentioned before, the CIs of an organization are considered the building blocks of that organization's ICT infrastructure, and they consist of hardware, software, documentation, and many other components, which are either tangible or not. Proper management of CIs is critical for ensuring the availability, reliability, and security of ICT systems and services. Product lifecycle management (PLM) is a comprehensive approach that enables organizations to manage CIs throughout their lifecycle. Figure 3 depicts the lifecycle of a product from "cradle to grave", starting from the requirements and development of operational capabilities and ending with retirement and decommission [22]. Among all of the intermediate stages of a product lifecycle, in this study, we chose the "As-Certified" and the "As-Maintained" stages, which represent the "latest approved" baseline and the "As-Is" status of a CI, respectively, while in [13], these statuses were reflected as the "as-designed configuration" and "as-built configuration", respectively. These concepts can also be conceptually borrowed from product information modeling, building information modeling, and product lifecycle management [23–25], in conjunction with the digital thread and digital twin approaches [26–28].

On one hand, the set of information that constitutes the "As-Certified" status of the CIs of an ICT system of systems (SoS) includes the configuration status of the final approved design (latest approved baseline), including all changes authorized so far during the lifecycle of each CI. On the other hand, the real-life or actual configuration status of the deployed CIs constitutes the "As-Maintained" status of the SoS.

There are many ways to obtain the "As-Maintained" (or As-Is) status of the CIs, either by technical means (SNMP polling, network telemetry subscription, compliance checks, etc.) by utilizing specific IT management system tools, by physical means (inventory checks, site surveys, etc.), or even a combination of the above.

In theory, the "As-Certified" and "As-Maintained" status of each CI must be exactly the same, which means that everything that is implemented under the operational SoS should strictly follow the designed, planned, and authorized requirements. In reality, the situation is slightly different, as there is a mismatch between statuses, and organizations strive to keep this gap as short as possible. There are many reasons for having differences between the As-Certified and As-Maintained statuses, and some examples are provided below:

- Unauthorized changes;

- Emergency change implementations (e.g., responding to a major incident) with no post-documentation/CMS integration;
- Lack of change control/change management processes;
- Deviation from processes and guidelines;
- Malicious activities by insiders/outsiders;
- Malware that impacts the CI configuration;
- Wrongly or partially implemented changes;
- Failed changes without roll-back;
- Deficiencies during hand-over or take-over activities from project deliveries.

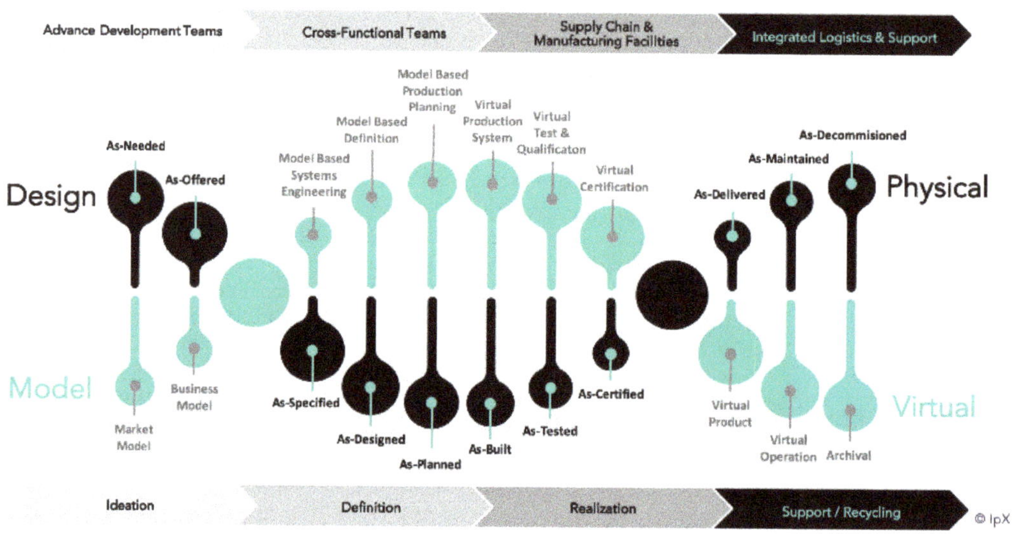

Figure 3. Product lifecycle [22].

5.2. Description, Benefits, and Limitations of the Blockchain CM Model

The type of blockchain proposed in this study is a permissioned ledger in order to provide increased security, transparency, and efficiency and allow for a more controlled and regulated environment for participants, thus enabling higher levels of trust and collaboration. Permissioned blockchains are increasingly used in the industry and their applicability is constantly benchmarked and evaluated. Depending on the application domain, blockchain technology has benefits and drawbacks compared to traditional databases and approaches [29–31], but there is continuous effort being made to identify gaps and challenges in terms of performance (execution time, latency, and throughput), scalability, and applicability [32–34] in order to meet the evolving technological requirements by indicating suitable strategies that can be deployed in blockchain systems [35–40].

The present model takes advantage of the existing features and benefits of distributed ledger technology and uses them as a vehicle for a new application area in the domain of configuration management. In essence, the main benefits of using blockchain technology—and, in particular, a permissioned blockchain—in support of security configuration management are as follows:

1. Tamper-proof CI records that enable integrity and constitute an immutable inventory baseline of an entire organization.
2. Historical records that enable traceability, verification, validation, auditability, restoration, and data recovery.
3. Transparency, which underpins the configuration status and accountability of actions and changes.

4. A shared ledger that enables the secure distribution of baseline configurations in the community.
5. Increased privacy and confidentiality via access control (AC) mechanisms by using smart contracts.

All of the above can further underpin real-time, near-real-time, or ad hoc/planned security compliance checks in order to detect malware or unauthorized changes/activities.

In addition, access control is a critical aspect of permissioned blockchain systems, which require the robust management of privileges and rights to ensure the integrity of the ledger and the security of the information stored on it. The application of CM requires a clear definition of specific roles and responsibilities for their functions to be performed inside an organization, such as in planning, submitting, and approving changes in CIs, as well as releasing and deploying changes and accessing data in the ledger. Hence, role-based access control (RBAC) is deemed necessary as a control mechanism to enable the implementation of a more scalable and manageable access control system by grouping users with similar roles together and then assigning them the appropriate permissions. Smart contracts can be used to enforce these access control policies [41] to provide an embedded, robust, flexible, and transparent mechanism for managing such permissions [42–46]. On the other hand, the effective implementation of RBAC requires careful planning, documentation, and continuous monitoring to ensure that access privileges remain appropriate and up to date.

The main stages of the configuration management process are shown in Figure 4, along with the interacting elements of the proposed blockchain-based model. The CM process is mainly a feedback-based cyclic model and conceptually starts with configuration management planning, where the governance, strategy, roles, responsibilities, and modeling are given a place on the organizational level, followed by the identification of the configuration items and their structure, relationships, and dependencies inside the system. These first two stages of the CM process do not interact with the blockchain model that we propose but are critical enablers for the following actions.

As mentioned before, the configuration control stage deals with the control of the changes in all configuration items of a system. In our proposal, it is clearly shown that the configuration management system (CMS) is completely based on the blockchain (highlighted in green), which leads to the conclusion that the overall baseline of all CIs of an organization benefits from the features of the immutable ledger. This design further leads to the outcome that the "As-Certified" status of the organization's critical elements or assets is stored in such a way that it grants full control and complete tracking of all lifecycle changes; it is also leveraged by the blockchain features that underpin the configuration status accounting stage of the CM process. In parallel, the "As-Maintained" status of an organization's ICT system is retrieved and made available via the applied technical monitoring and management solutions to show the current status of the CIs.

The combination of the above can lead to the detailed identification of the difference (delta) between the two statuses ("As-Certified" vs "As-Maintained") at a level of granularity that fulfills the expectations of an organization's auditing requirements. This delta is one of the fundamental benefits of the overall configuration management process; it provides a clear picture of the "As-Should-Be" vs. "As-Is" status of the CIs in the configuration, verification, and auditing stage of the CM process.

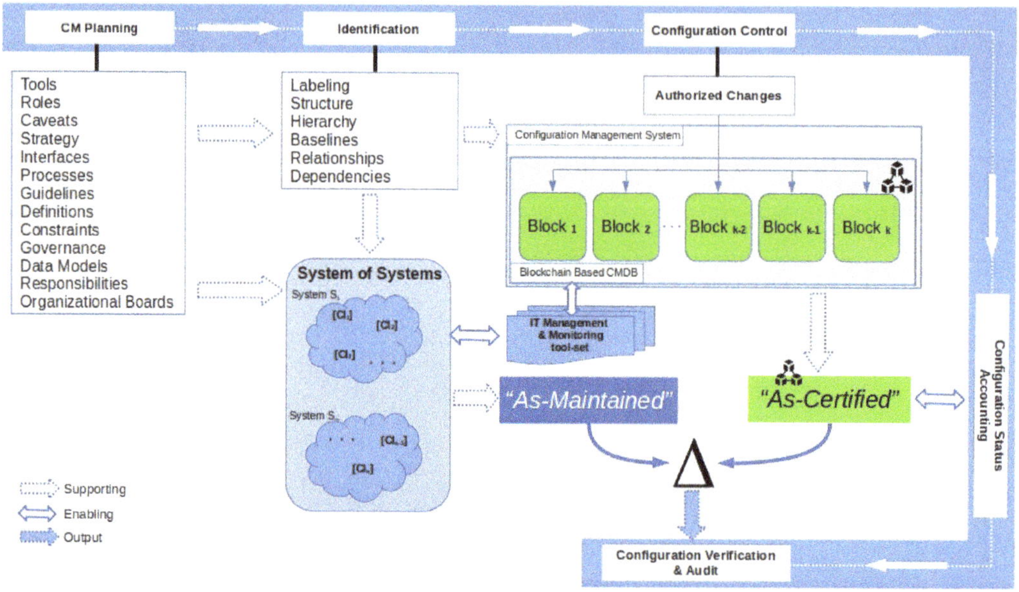

Figure 4. Configuration management process.

The identification of such differences can be applied in quantitative and qualitative analyses and is considered critical because an organization can assess the level of deviation of its configuration baselines, which can lead to more effective performance and more efficient compliance checks.

These compliance checks are not only applicable from the perspective of security, but also from many other perspectives that are linked together by the overall organizational requirements, such as asset management, release and deployment management, vulnerability assessments, financial management, and overall risk management activities.

6. Operating Concept

This section presents an operating concept of the proposed model, and a use case in a configuration management scenario is presented. For this use case, four entities with high-level roles and responsibilities are represented as follows:

- The change requestor's role is to plan, find the resources for, develop, test, and support the proposed changes in one or more configuration items of the system, as well as to present them to the configuration control board;
- The configuration control board (CCB) has the role of establishing and chartering a group of qualified people with responsibility for the process of assessing, controlling, approving, and recording changes throughout the development and operational lifecycle of the system;
- The implementation team has the role of releasing and deploying the approved changes;
- The security auditor has the duties of performing security compliance checks, configuration gap analyses, and post-implementation checks and comparing differences between the "As-Certified" and "As-Maintained" statuses of the system's CIs.

Figure 5 shows a flowchart of the actions performed by the defined actors when executing the roles mentioned above. More specifically, the change requestor defines the scope of the change, justifies its necessity, identifies and assesses the impacts on other configuration items, develops the test and coordination plan, and includes all of the other necessary documentation, such as the deployment, support, roll-back plan, etc., as required by the internal procedures of the organization. The change proposal is submitted as a technical package to the configuration control board (CCB) for the change evaluation by using existing IT service management tools.

The CCB reviews the change proposal, assesses the risk to the organization, and performs compliance checks as set by the organization, and it approves the change, rejects it, or escalates it to a higher-level CCB. If the change is escalated to a higher-level CCB, the next-level CCB performs the review and assessment of the change against a wider audience and additional criteria with two output options: change approval and change rejection.

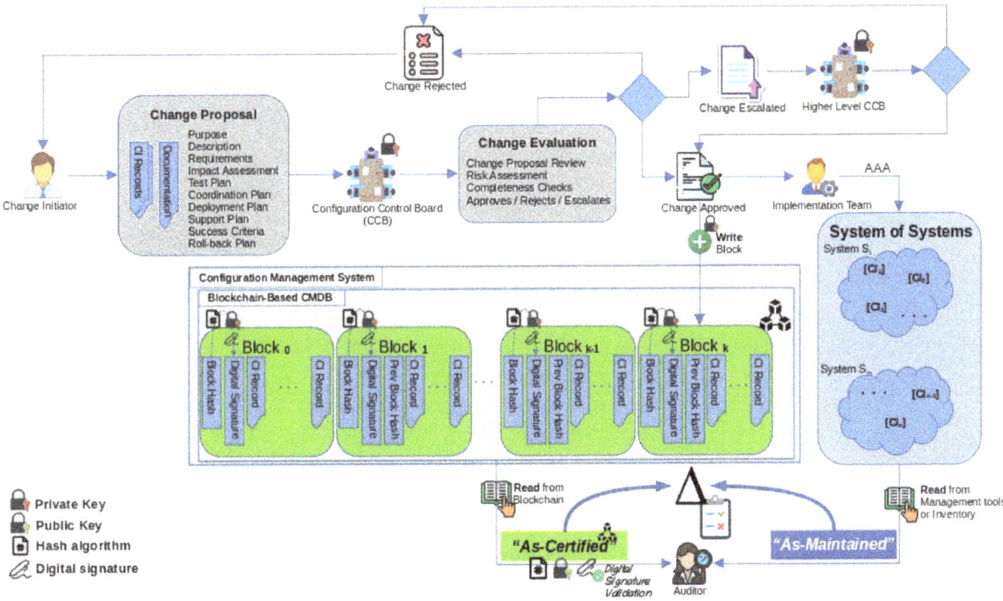

Figure 5. Flowchart of the operating concept.

If the change proposal is rejected for any reason, it is returned to the change requestor with full justification for further adjustments and further re-submission if deemed necessary. If the change proposal is approved, then the full details of the change are stored in the blockchain as a separate block, while, in the meantime, it is pipelined to the implementation team to streamline the change's implementation.

It is clear that any changes that are approved to be applied in the operational system can take place only after a change is approved by the configuration control board. Similarly, the same applies for adding blocks to the blockchain-based configuration management system (CMS). This implies that the "As-Certified" status of the system is under full control by using the features of the blockchain; moreover, a digital signature mechanism is applied for additional non-repudiation and accountability purposes [47]. On the other hand, the accountability and the non-repudiation in the operational ICT system are expected to be based on equivalent technical functions built on the system's management tools, such as AAA (authentication–authorization–accounting) mechanisms, which fall outside the scope of this study.

A digital signature is applied to the hash of the block that is planned to be added to the blockchain and further stored as separate data on the block itself. The digital signature ensures the identification of the change approval entity by utilizing the private keys of the involved CCBs that are authorized to approve changes. The use of a digital signature functionality requires an external public key infrastructure (PKI) to be in place in order to support the certificate's chain of trust. The establishment of a PKI is not covered in this study.

Block-0 represents the genesis block of the blockchain and contains the initial status and information of all configuration items in the system, thus representing the initial baseline; it is digitally signed by the highest-level CCB of the organization. The following blocks hold the information of the configuration items, which are the subjects of each change and are digitally signed by the authority that approved each change, thus progressively building the final/current approved ("As-Certified") baseline of the system.

As explained above, the output of the configuration, verification, and auditing process is based on the comparison between the "As-Certified" and "As-Maintained" statuses of the system being audited. The auditor verifies the digital signature of the blockchain blocks whose configuration items' records are within the scope of the audit, by using the public keys of the CCBs.

In essence, the proposed model underpins and enables the functions of the configuration management process, takes advantage of the existing benefits of blockchain technology, and ensures that all changes are securely recorded and added only after their approval and only by the authorized organizational elements. Furthermore, the immutable ledger can be used for configuration auditing purposes and for technical comparisons of the "As-Certified" and "As-Maintained" statuses of the system in order to identify unauthorized changes, security configuration gaps, lacking security compliance, wrong post-implementation changes, or potentially malware-impacted configuration items. All of the above are supported by a digital signature function that adds an additional security element of a non-repudiation feature to the ledger.

Finally, in support of the RBAC mechanism that was mentioned before, we present a brief code example of a smart contract in Listing 1 that implements just one simple feature. In this case, the smart contract owner can specify the authorized entities that can make changes in the ledger and add a configuration item into the ledger with the required information.

Listing 1. Smart Contract Code Example.

```
pragma solidity ^0.8.19;

contract ConfigurationItemRegistry {
    address public owner;
    mapping(address => bool) authorized;
    uint public authorizedCount;

    struct ConfigurationItem {
        uint uniqueId;
        uint configurationItemId;
        string description;
        string itemType;
        string configuration;
        uint timestamp;
        uint userId;
    }

    mapping(uint => ConfigurationItem) public configurationItems;

    constructor() {
        owner = msg.sender;
        authorized[msg.sender] = true;
        authorizedCount++;
    }

    modifier onlyOwner() {
        require(msg.sender == owner, "Only owners can perform this action");
        _;
    }

    modifier onlyAuthorized() {
        require(authorized[msg.sender], "Only authorized parties can perform this action");
```

```
        _;
    }

    function addAuthorized(address account) public onlyOwner {
        authorized[account] = true;
        authorizedCount++;
    }

    function removeAuthorized(address account) public onlyOwner {
        authorized[account] = false;
        authorizedCount--;
    }

    function addConfigurationItem(uint uniqueId, uint configurationItemId
        , string description, string itemType, string configuration, uint
         timestamp, uint userId) public onlyAuthorized {
        ConfigurationItem memory newItem = ConfigurationItem(uniqueId,
            configurationItemId, description, itemType, configuration,
            timestamp, userId);
        configurationItems[uniqueId] = newItem;
    }
}
```

7. Conclusions and Future Work

This study proposed a blockchain-based system for security configuration management, which is an enabler of the complete lifecycle management of IT/OT systems, especially in the domain of distributed and large-scale ICT systems. The existing standards, policies, handbooks, guidelines, and best practices were presented, which set the baseline for a robust framework for keeping such systems secure, especially in the context of tracking changes or baseline mismatches in operational environments.

However, although security configuration management is technically well embedded and exercised in current ICT operational environments, according to our research, current IT service management tools are not based on blockchain implementations; moreover, the existing academic research on this topic is very limited. Our approach intends to cover this identified gap, which could potentially underpin the efficiency of configuration management by using the proposed permissioned blockchain model. The proposal of a permissioned model was based on the fact that security and asset management information in such critical systems can only be shared inside a restrictive environment, where controlled access, confidentiality, and need-to-know principles play a critical role. By using such a blockchain model, configuration management can be exercised more effectively and efficiently by inheriting the advantages of distributed ledger technology, such as data integrity, confidentiality, fault tolerance, traceability, transparency, auditability, consistency, and controlled access management, thus supporting the security objectives of organizations.

On the other hand, while the application of blockchain-based models—and especially permissioned ones—grows, it is essential to evaluate the key performance properties of each platform before applying it in real use cases. This gap creates a challenging area of research that requires conducting more technical analyses of blockchain platforms regarding their performance, scalability, and applicability while taking blockchains' inherent limitations into consideration.

The focus of this study is on presenting a theoretical blockchain-based model of a service knowledge management system (SKMS) with a configuration management system (CMS) as a sub-element and on showing how the functions of those elements can benefit from the advantages of a permissioned ledger while interacting with the rest of the system management tools to underpin the objectives of the configuration management process. The end goal is to accurately identify gaps between the "As-Certified" ("As-Should-Be") status and the "As-Maintained" ("As-Is") status of the configuration items of the system (or system of systems) and further facilitate change control, accounting, verification, and auditing, which constitute the main functions of the CM process. This gap analysis is very critical for the security posture of an organization, since via this process, security risks can be easily identified, which could span across many activities, such as malicious activities, unauthorized changes, misconfigurations, incorrect baselining, etc. In addition, the proposed model is enabled by a role-based access control mechanism for smart contracts, as well as by a digital signature function that enables the non-repudiation of system changes.

Our proposal takes place on a theoretical basis, while the future intention is to build the proposed model in a proof-of-concept environment for performance testing, analysis, benchmarking, and evaluation of the operational utility.

Further research can be conducted to integrate deep learning and forecasting algorithms into the blockchain-based system to more accurately identify and prioritize the most important security gaps, as well as to assess how these findings can be integrated into existing risk management and analysis tools for informed decision-making processes.

Author Contributions: Conceptualization, D.C.; Investigation, D.C.; Resources, D.C.; Writing—original draft, D.C.; Writing—review and editing, K.R.; Supervision, K.R. All authors have read and agreed to the published version of the manuscript.

Funding: This research received no external funding.

Data Availability Statement: Not applicable.

Conflicts of Interest: The authors declare no conflict of interest.

References

1. ENISA. *Threat Landscape 2021*; Technical report; ENISA: Athens, Greece, 2021.
2. NIST. *SP 800-37 Revision 2: Risk Management Framework for Information Systems and Organizations*; 2018. Available online: https://nvlpubs.nist.gov/nistpubs/SpecialPublications/NIST.SP.800-37r2.pdf (accessed on 2 March 2023).
3. NIST. SP 800-128: Guide for Security-Focused Configuration Management of Information Systems. 2011. Available online: https://csrc.nist.gov/publications/detail/sp/800-128/final (accessed on 2 March 2023).
4. Joint Task Force Transformation Initiative Interagency Working Group. *Security and Privacy Controls for Federal Information Systems and Organizations*; Technical Report NIST Special Publication (SP) 800-53, Rev. 5, Includes updates as of 12 October 2020; National Institute of Standards and Technology: Gaithersburg, MD, USA, 2020. [CrossRef]
5. NIST. SP 800-82 Revision 2: Guide to Industrial Control Systems (ICS) Security. 2015. Available online: https://csrc.nist.gov/publications/detail/sp/800-82/rev-2/final (accessed on 2 March 2023).
6. Rajasekaran, A.S.; Azees, M.; Al-Turjman, F. A comprehensive survey on blockchain technology. *Sustain. Energy Technol. Assess.* **2022**, *52*, 102039. [CrossRef]
7. Berdik, D.; Otoum, S.; Schmidt, N.; Porter, D.; Jararweh, Y. A survey on blockchain for information systems management and security. *Inform. Process. Manag.* **2021**, *58*, 102397. [CrossRef]
8. *ISO/IEC 20000-1:2018*; ISO/IEC 20000—Information Technology—Service Management. International Organization for Standardization: Geneva, Switzerland, 2018.
9. NIST. SP 800-160v1 Systems Security Engineering: Considerations for a Multidisciplinary Approach in the Engineering of Trustworthy Secure Systems. 2016. Available online: https://csrc.nist.gov/publications/detail/sp/800-160/vol-1/archive/2018-03-21 (accessed on 2 March 2023).
10. NIST. *Cybersecurity Framework Version 1.1.*; Technical Report Cybersecurity Framework Version 1.1; National Institute of Standards and Technology: Gaithersburg, MD, USA, 2018. [CrossRef]
11. Iashchenko, V.V.; Orlova, E.D. A Model for Evaluating the Quality of the Configuration Management Process in the Energy Sector. In Proceedings of the 2021 IEEE Conference of Russian Young Researchers in Electrical and Electronic Engineering (ElConRus), St. Petersburg and Moscow, Russia, 26–29 January 2021; pp. 1895–1897. [CrossRef]
12. ISACA. *COBIT 2019 Framework: Introduction and Methodology*, 2nd ed.; ISACA: Rolling Meadows, IL, USA, 2019.
13. *Department of Defense Handbook: Configuration Management Guidance*; Springer International Publishing: New York City, NY, USA, 2020.
14. Kinkelin, H.; Hauner, V.; Niedermayer, H.; Carle, G. Trustworthy configuration management for networked devices using distributed ledgers. In Proceedings of the NOMS 2018—2018 IEEE/IFIP Network Operations and Management Symposium, Taipei, Taiwan, 23–27 April 2018; pp. 1–5.
15. Košťál, K.; Helebrandt, P.; Belluš, M.; Ries, M.; Kotuliak, I. Management and Monitoring of IoT Devices Using Blockchain. *Sensors* **2019**, *19*, 856. [CrossRef] [PubMed]
16. Alvarenga, I.; Rebello, G.; Duarte, O. Securing configuration management and migration of virtual network functions using blockchain. In Proceedings of the NOMS 2018—2018 IEEE/IFIP Network Operations and Management Symposium, Taipei, Taiwan, 23–27 April 2018; pp. 1–9. [CrossRef]

17. Mylrea, M.; Gourisetti, S.N.G. Blockchain for Supply Chain Cybersecurity, Optimization and Compliance. In Proceedings of the 2018 Resilience Week (RWS), Dencer, CO, USA, 20–23 August 2018; pp. 70–76. [CrossRef]
18. Han, S.H. Blockchain-based Configuration Management System. In Proceedings of the 2022 IEEE/ACIS 7th International Conference on Big Data, Cloud Computing, and Data Science (BCD), Danang, Vietnam, 4–6 August 2022; pp. 224–228. [CrossRef]
19. Samaniego, M.; Deters, R. Virtual Resources & Blockchain for Configuration Management in IoT. *J. Ubiquit. Syst. Pervas. Netw.* **2018**, *9*, 1–13.
20. Sarwar, M.I.; Abbas, Q.; Alyas, T.; Alzahrani, A.; Alghamdi, T.; Alsaawy, Y. Digital Transformation of Public Sector Governance With IT Service Management–A Pilot Study. *IEEE Access* **2023**, *11*, 6490–6512. [CrossRef]
21. Riascos Castaneda, R.; Ostrosi, E.; Majić, T.; Stjepandić, J.; Sagot, J.C. A Method to Explore Product Risk in Product Lifecycle Management of Configured Products. *Proc. Des. Soc. Des. Conf.* **2020**, *1*, 687–696. [CrossRef]
22. Wertel, S. *IpX*; Institute for Process Excellence: Denver, CO, USA, 2020.
23. Barrios, P.; Danjou, C.; Eynard, B. Literature review and methodological framework for integration of IoT and PLM in manufacturing industry. *Comput. Ind.* **2022**, *140*, 103688. [CrossRef]
24. Lim, K.Y.H.; Zheng, P.; Chen, C.H. A state-of-the-art survey of Digital Twin: Techniques, engineering product lifecycle management and business innovation perspectives. *J. Intell. Manuf.* **2020**, *31*, 1313–1337. [CrossRef]
25. Bagozi, A.; Bianchini, D.; Rula, A. Multi-perspective Data Modelling in Cyber Physical Production Networks: Data, Services and Actors. *Data Sci. Eng.* **2022**, *7*, 193–212. [CrossRef]
26. Xiong, M.; Wang, H. Digital twin applications in aviation industry: A review. *Int. J. Adv. Manuf. Technol.* **2022**, *121*, 1–16. [CrossRef]
27. Bolshakov, N.; Badenko, V.; Yadykin, V.; Celani, A.; Fedotov, A. Digital twins of complex technical systems for management of built environment. *IOP Conf. Ser. Mater. Sci. Eng.* **2020**, *869*, 6. [CrossRef]
28. Zhang, Q.; Zheng, S.; Yu, C.; Wang, Q.; Ke, Y. Digital thread-based modeling of digital twin framework for the aircraft assembly system. *J. Manuf. Syst.* **2022**, *65*, 406–420. [CrossRef]
29. Le, T.V.; Hsu, C.L. A systematic literature review of blockchain technology: Security properties, applications and challenges. *J. Internet Technol.* **2021**, *22*, 789–802.
30. Sargent, C.S.; Breese, J.L. Blockchain Barriers in Supply Chain: A Literature Review. *J. Comput. Inform. Syst.* **2023**, 1–12. [CrossRef]
31. Karumba, S.; Sethuvenkatraman, S.; Dedeoglu, V.; Jurdak, R.; Kanhere, S.S. Barriers to blockchain-based decentralised energy trading: A systematic review. *Int. J. Sustain. Energy* **2023**, *42*, 41–71. [CrossRef]
32. Pongnumkul, S.; Siripanpornchana, C.; Thajchayapong, S. Performance analysis of private blockchain platforms in varying workloads. In Proceedings of the 2017 26th International Conference on Computer Communication and Networks (ICCCN), Vancouver, BC, Canada, 31 July–3 August 2017; pp. 1–6.
33. Nasir, Q.; Qasse, I.A.; Abu Talib, M.; Nassif, A.B. Performance analysis of hyperledger fabric platforms. *Secur. Commun. Netw.* **2018**, *2018*, 3976093. [CrossRef]
34. Dabbagh, M.; Choo, K.K.R.; Beheshti, A.; Tahir, M.; Safa, N.S. A survey of empirical performance evaluation of permissioned blockchain platforms: Challenges and opportunities. *Comput. Secur.* **2021**, *100*, 102078. [CrossRef]
35. Monrat, A.A.; Schelén, O.; Andersson, K. Performance evaluation of permissioned blockchain platforms. In Proceedings of the 2020 IEEE Asia-Pacific Conference on Computer Science and Data Engineering (CSDE), Gold Coast, Australia, 16–18 December 2020; pp. 1–8.
36. Melo, C.; Oliveira, F.; Dantas, J.; Araujo, J.; Pereira, P.; Maciel, R.; Maciel, P. Performance and availability evaluation of the blockchain platform hyperledger fabric. *J. Supercomput.* **2022**, *78*, 12505–12527. [CrossRef]
37. Honar Pajooh, H.; Rashid, M.A.; Alam, F.; Demidenko, S. Experimental Performance Analysis of a Scalable Distributed Hyperledger Fabric for a Large-Scale IoT Testbed. *Sensors* **2022**, *22*, 4868. [CrossRef]
38. Wen, Y.F.; Hsu, C.M. A performance evaluation of modular functions and state databases for Hyperledger Fabric blockchain systems. *J. Supercomput.* **2023**, *79*, 2654–2690. [CrossRef]
39. Al-Sumaidaee, G.; Alkhudary, R.; Zilic, Z.; Swidan, A. Performance analysis of a private blockchain network built on Hyperledger Fabric for healthcare. *Inform. Process. Manag.* **2023**, *60*, 103160. [CrossRef]
40. Capocasale, V.; Gotta, D.; Perboli, G. Comparative analysis of permissioned blockchain frameworks for industrial applications. *Blockchain Res. Appl.* **2023**, *4*, 100113. [CrossRef]
41. Cruz, J.P.; Kaji, Y.; Yanai, N. RBAC-SC: Role-based access control using smart contract. *IEEE Access* **2018**, *6*, 12240–12251. [CrossRef]
42. Rouhani, S.; Deters, R. Blockchain based access control systems: State of the art and challenges. In Proceedings of the IEEE/WIC/ACM International Conference on Web Intelligence, Thessaloniki, Greece, 14–17 October 2019; pp. 423–428.
43. Sookhak, M.; Jabbarpour, M.R.; Safa, N.S.; Yu, F.R. Blockchain and smart contract for access control in healthcare: A survey, issues and challenges, and open issues. *J. Netw. Comput. Appl.* **2021**, *178*, 102950. [CrossRef]
44. Kamboj, P.; Khare, S.; Pal, S. User authentication using Blockchain based smart contract in role-based access control. *Peer-to-Peer Netw. Appl.* **2021**, *14*, 2961–2976. [CrossRef]
45. Zhang, L.; Li, B.; Fang, H.; Zhang, G.; Liu, C. An Internet of Things Access Control Scheme Based on Permissioned Blockchain and Edge Computing. *Appl. Sci.* **2023**, *13*, 4167. [CrossRef]

46. Yang, L.; Jiang, R.; Pu, X.; Wang, C.; Yang, Y.; Wang, M.; Zhang, L.; Tian, F. An access control model based on blockchain master-sidechain collaboration. *Cluster Comput.* **2023**, 1–21. [CrossRef]
47. Fang, W.; Chen, W.; Zhang, W.; Pei, J.; Gao, W.; Wang, G. Digital signature scheme for information non-repudiation in blockchain: A state of the art review. *EURASIP J. Wirel. Commun. Netw.* **2020**, *2020*, 1–15. [CrossRef]

electronics

MDPI

Article

Blockchain and Interplanetary File System (IPFS)-Based Data Storage System for Vehicular Networks with Keyword Search Capability

N. Sangeeta and Seung Yeob Nam *

Department of Information and Communication Engineering, Yeungnam University, Gyeongsan 38541, Republic of Korea
* Correspondence: synam@ynu.ac.kr

Abstract: Closed-circuit television (CCTV) cameras and black boxes are indispensable for road safety and accident management. Visible highway surveillance cameras can promote safe driving habits while discouraging moving violations. According to CCTV laws, footage captured by roadside cameras must be securely stored, and authorized persons can access it. Footages collected by CCTV and Blackbox are usually saved to the camera's microSD card, the cloud, or hard drives locally but there are concerns about security and data integrity. These issues may be addressed by blockchain technology. The cost of storing data on the blockchain, on the other hand, is prohibitively expensive. We can have decentralized and cost-effective storage with the interplanetary file system (IPFS) project. It is a file-sharing protocol that stores and distributes data in a distributed file system. We propose a decentralized IPFS and blockchain-based application for distributed file storage. It is possible to upload various types of files into our decentralized application (DApp), and hashes of the uploaded files are permanently saved on the Ethereum blockchain with the help of smart contracts. Because it cannot be removed, it is immutable. By clicking on the file description, we can also view the file. DApp also includes a keyword search feature to assist us in quickly locating sensitive information. We used Ethers.js' smart contract event listener and contract.queryFilter to filter and read data from the blockchain. The smart contract events are then written to a text file for our DApp's keyword search functionality. Our experiment demonstrates that our DApp is resilient to system failure while preserving the transparency and integrity of data due to the immutability of blockchain.

Keywords: blockchain; Ethereum blockchain; decentralized application (DApp); interplanetary file system (IPFS); smart contracts

Citation: N. Sangeeta; Nam, S.Y. Blockchain and Interplanetary File System (IPFS)-Based Data Storage System for Vehicular Networks with Keyword Search Capability. *Electronics* **2023**, *12*, 1545. https://doi.org/10.3390/electronics12071545

Academic Editor: Hamed Taherdoost

Received: 17 February 2023
Revised: 22 March 2023
Accepted: 22 March 2023
Published: 24 March 2023

1. Introduction and Background

CCTV camera images are a valuable source of traffic surveillance that supplements other traffic control measures. CCTV is aimed at helping in the detection and prevention of criminal activity. It can be helpful in protecting the citizens in the community. It is placed in public areas to provide evidence to appropriate law enforcement agencies. CCTV cameras can be found on busy roads, atop traffic lights, and at highway intersections. Operators detect and monitor traffic incidents using images from CCTV cameras. It may be possible to predict the duration of a traffic incident based on prior experience and traffic modeling techniques. Cameras are used to observe and monitor traffic, as well as to record traffic pattern data. Moving violation tickets are even issued using cameras.

The vehicle's event data recorder is constantly recording information in a loop while we are driving, at least until a collision occurs. Black boxes save data collected at the time of impact, as well as 5 s before and after the event. The black boxes will record all human contact with the vehicle. The data collected helps us understand the reasons for collisions and prevent them from happening again.

CCTV footage is being used in crime investigations by police officers and insurance companies [1] all over the world. Recorded footage is typically used by investigators to locate or confirm the identity of a suspect. Real-time surveillance systems allow employees or law enforcement officials to detect and monitor any threat in real time. Then, there's the archival footage record, which can be reviewed later if a crime or other issue is discovered. In these cases, the recorded footage must be securely deposited and kept for future use, making video storage a critical component of any video camera security system.

The vast majority of information collected by surveillance cameras and dashboard cameras is securely kept on hard drives as well as memory cards. The amount of storage on the MicroSD card of our security camera, on the other hand, is determined by the amount of activity recorded by our camera. This type of storage necessitates a large amount of storage space and exposes our data to risk if the device's hard drive fails or is damaged. It is critical to securely store CCTV and black box footage in order for it to be available and unaltered at all times. In many cases, the introduction and popularity of IP camera cloud storage have reduced the importance of local storage to a secondary option.

Cloud systems are an extremely good tool that offers us many advantages and functionalities. Cloud storage systems, on the other hand, have flaws such as problems with data safety [2,3], centralized data storage, and the requirement for trusted third parties. Owners are reassured of the burden of maintaining local data storage, but they end up losing direct control over storage reliability and protection. Every year, large database hacks cost millions of dollars. Furthermore, because the data is kept on an external device, the owners have no power over it; if the service provider disconnects or limits access, they will not be able to access it.

Due to the centralized nature of cloud storage data, an intruder to servers is able to view and alter it. Cloud data is untrustworthy and can be altered or removed at any time. As a result, making sure data security [4] and safeguarding users' privacy [5] are critical. Users are usually needed to cover the cost of any storage plan they select, even if they only use a portion of it.

Even the finest cloud service providers can face such a challenging problem while retaining strong maintenance standards. Centralized storage service providers occasionally fail to deliver the security service as agreed. For example, a hack on Dropbox [6] which is among the world's largest online storage companies, did result in the leak of 68 million usernames and passwords on the dark web. Well-known cloud services have started experiencing blackouts and security breaches. The mass email deletion event of Gmail [7], Amazon S3's latest shutdown [8], and the post-launch interruption of Apple MobileMe's [9] are other examples.

Blockchain technology may be able to address these issues. A blockchain is made up of a cryptographic algorithm, a timestamp, and transaction information that connects it to the preceding block. As a result, every block links to the next, forming a "chain" of blocks and producing safe and immutable records. In comparison, the blockchain is not designed for the purpose of file storage. The cost of keeping data on the blockchain is exorbitantly high. We can have decentralized as well as low-cost storage with the IPFS project [10]. Peer-to-peer networks provide greater security than centralized networks. As an outcome, they are ideal for protecting sensitive information from malicious actors.

We propose an IPFS-based distributed and decentralized storage application that offers more storage space compared to solely blockchain-based systems in this paper. Using distributed storage, information is kept on different nodes or servers on the Internet. To upload files, we use the Geth [11] software client to operate an Ethereum node and an IPFS Daemon server to operate our own IPFS node. Users will link to the DApp through the use of one's web browser as well as a blockchain wallet, Metamask [12], to connect to a blockchain in our proposed scheme. Since it is powered by Ethereum smart contracts, the decentralized application will interact with the blockchain, which will keep all the code of the application in addition to the data. The smart contracts keep track of all sources of information in IPFS files. A DApp can receive any kind of information. The hash value of the uploaded file is permanently saved on the Ethereum blockchain via smart contracts and it cannot be changed or deleted. Whenever a file is uploaded, the DApp hears the event "File Upload" and updates the DApp's user interface. We retrieve all of the smart contract events and reveal them on our DApp, which is called the "smart contract event log". The smart contract event log contains data such as the file name, file summary (including the event and location of the file), file type, size of the file, time and date of upload, Ethereum account information of the user, and the hash value of the file once it has been uploaded to IPFS. Users can also view the file by clicking on its description. The user does not need to remember and save the hash value independently, which could be dangerous if another individual has access to it. Our DApp also includes a keyword search feature to assist you in quickly locating sensitive information. Figure 1 shows an example scenario where our proposed system can be applied. When an accident occurs, our proposed system might be used to save the video taken by the dashboard camera on IPFS and the hash value of the video on the blockchain to prevent the manipulation of the video using the immutability property of blockchain.

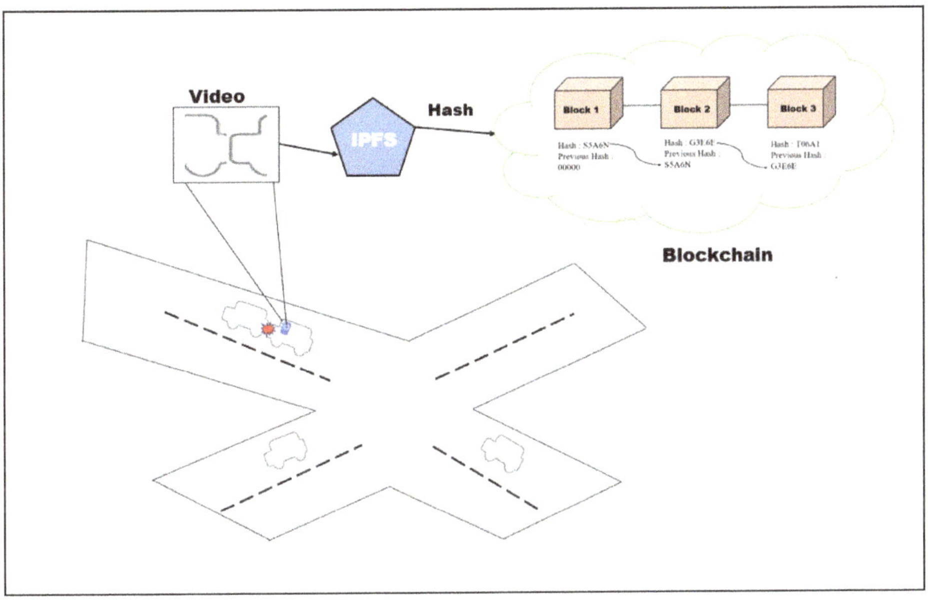

Figure 1. Example scenario to illustrate the application of the proposed system in the presence of accidents on the road.

The key contributions of our paper can be summarized as follows:

- Our proposed distributed storage application supports the storage of various file types since uploaded files are stored on IPFS and their hash values are stored in smart contracts on the Ethereum blockchain. Users need not remember the hash values since they can be retrieved from the blockchain later.
- DApp provides a keyword search feature to help users quickly find the necessary files based on Ethers.js's smart contract event listener and contract.queryFilter.
- Our experiment shows that our DApp is resilient to system failure, and our system provides better transparency than is possible with centrally managed applications.

The rest of our paper is structured as follows: Section 2 contains related work. Section 3 contains preliminary information. The proposed scheme is described in Section 4. Section 5 goes over implementation. The performance evaluation results are described in Section 6. Finally, Section 7 brings the paper to a close.

2. Related Work

Hao, J. et al. [13] studied a blockchain and IPFS-based storage scheme for agricultural product tracking. During the manufacturing, processing, and logistics processes, sensors collect real-time data on product quality as well as video and picture data, according to this study. The server parses and encapsulates the data before writing it to IPFS, and the hash address is then stored in the blockchain to complete the data storage. The collected data is not directly written to an IPFS. The authors employ a private data server, and data collected by sensors is first stored on the private data server before being directly stored on the IPFS. If the server experiences problems, such as server failure, the collected data is lost, and the server is unable to write data to IPFS. There is no keyword search function for quickly finding agricultural product information.

Rajalakshmi et al. [14] proposed a framework for access control methods in research records that manages to combine blockchain, IPFS, as well as other traditional encryption methods. The system stores the verification metadata information acquired from the IPFS on the blockchain network using Ethereum smart contracts, resulting in tamper-proof record-keeping for further auditing. There is no keyword search functionality for searching information related to research records in this proposed scheme, which only stores PDF files.

Vimal, S. et al. [15] proposed a method to improve the efficiency of the P2P file-sharing system by incorporating trustworthiness and proximity awareness during file transfer using IPFS and blockchain. Any of these hashed files can be retrieved by simply calling the hash of the file. Miners who collaborate to ensure the successful transfer of resources are compensated. This study discusses the file transfer service, as well as the security strength and some of the IPFS-based incentives.

This system is built around IPFS and Blockchain. Yongle Chen et al. [16] proposed a more efficient P2P file system scheme. The authors pointed out the high-throughput problem for individual IPFS users by incorporating the responsibility of content service providers. A novel zigzag-based storage model is utilized to improve the IPFS block storage model by taking data reliability and availability, storage overhead, and other issues for service providers into account.

Rong Wang et al. proposed a video surveillance system relying on permissioned blockchains (BCs) and edge computing in their paper [17]. Convolutional neural networks (CNN), edge computing, and permissioned BCs, as well as IPFS technology, were used in this system. Edge computing was utilized to collect and process large amounts of wireless sensor data, while the IPFS storage solution was utilized to enable huge video data storage. CNN technology was applied to real-time monitoring, and Edge computing was utilized to gather and analyze large amounts of wireless sensor data.

Sun, J. et al. [18] proposed a blockchain-based secure storage and access scheme for electronic medical records in IPFS, which ensures necessary access to electronic medical data while preserving retrieval efficiency. IPFS is a file system used in order to store encrypted electronic medical data. After receiving the hash value and encrypted hash address, the physician needs to be encrypted using the hash value and encoded hash address with a random number, hash the health information and index with the SHA256 hash function, and broadcast the hash value and encoded hash address to the blockchain. Furthermore, the system offers targeted defense against relevant keyword attacks. Medical data is not directly stored on IPFS, and electronic health data is encrypted before being stored on IPFS. It also takes time for the IPFS value to be encrypted before even being kept on the blockchain.

Most of the previous works lack a keyword search functionality for quickly locating relevant information. They do not mention how to retrieve the metadata from the blockchain. It is not possible to retrieve data from IPFS without the hash value of the file. Table 1 compares our proposed system with existing approaches.

Table 1. Comparison of existing approaches with the proposed scheme.

Constraints	Hao, J. et al. [13]	Rajalakshmi A. [14]	Sun, J. et al. [18]	Our Proposed Scheme
Delay	High delay Collected data is not directly written to an IPFS	Low delay	High delay Encryption of medical data	Low delay in uploading files to IPFS and file hash is automatically stored on BC with help of Smart contract
Tampering on the stored data	Possibilities of data tampering	No tampering	No tampering	No tampering of data as data is stored on IPFS and hash on Blockchain
Storage capacity	Less Storage capacity Stored on data server	More storage capacity	More storage capacity as the data stored on IPFS	More storage capacity as the data stored on IPFS
Heterogeneous data	Uploading only video and images on IPFS	Uploading only PDF's	Only electronic medical record	Heterogeneous data upload
Keyword Search function	No Keyword search function	No Keyword search function	No Keyword search function	Supports Keyword search function

3. Preliminaries

3.1. IPFS

The interplanetary file system is a distributed file system protocol developed by Joan Bennett in 2015 and managed by Protocol Labs. The IPFS network consists of computers running the IPFS client software. Anyone can join the IPFS network, either as an IPFS node running the IPFS client or as a network user storing and retrieving files. Any type of file can be stored, including text, music, video, and images, which is especially useful for non-fungible tokens (NFTs). In contrast to HTTP, data in IPFS is identified by content rather than location. When we upload a file to IPFS, a hash of the content is generated. This hash identifies the content uniquely and can be used to retrieve the file. If we upload a different file, the hash will be completely different, but we can always recompute the file's hash locally to ensure it matches the original IPFS hash. We selected the IPFS protocol in our proposed scheme because it is a well-known and working decentralized file storage protocol.

3.2. Ethereum

Ethereum [19] is, at its core, a decentralized global software platform that utilizes blockchain technology. It is most well-known for its native cryptocurrency, ether, abbreviated as ETH. Anyone can use Ethereum to start creating any protected digital technology. It has a token intended to be utilized by the blockchain network, but it may also be employed to pay participants for blockchain work. It is a platform for various DApps that can be deployed through smart contracts. An Ethereum Private Network is a blockchain that is completely separate from the main Ethereum network. The Ethereum Private Network is primarily used by organizations to limit blockchain read permissions.

3.3. Web3.js

Web3.js [20] is a set of libraries that allows developers to communicate with a remote or local Ethereum node via HTTP, IPC, or WebSocket. You can use this library to create websites or clients that communicate with the blockchain.

3.4. Ethers.js

Ethers.js [21] connects to Ethereum nodes using Alchemy, JSON-RPC, Etherscan, Infura, Metamask, or Cloudflare. Developers can use ethers. js to take advantage of full functionality for their various Ethereum needs.

3.5. Smart Contract

Smart contracts are programs that are implemented and stored on a blockchain when certain requirements are fulfilled. They are frequently used to automate agreement execution so that all groups have instant surety of the results even without the involvement of an additional party. They also can automate a workflow by automatically performing the next action if certain requirements are fulfilled.

3.6. Smart Contract Events

When a transaction is mined, smart contracts could also emit events and logs to the blockchain, which the front end can then process. Events are essential on any blockchain because they make connections between smart contracts, which are self-executing software programs that have the terms of the buyer's and seller's agreement straight integrated into lines of code for response with user interfaces. To use a smart contract, a user must first manually sign a transaction and interact with the blockchain. This is where automation can help users by simplifying things. Event-driven automation initiates processes without requiring human intervention. An automation tool can start a predefined process or workflow of smart contracts after detecting an event.

3.7. Decentralized Applications (DApp)

A decentralized application [22] is an application that can run autonomously, typically using smart contracts and running on a decentralized computing, blockchain, or other distributed ledger system. DApps, like traditional applications, provide some function or utility to their users.

3.8. React.js

React.js [23], also known as simply React, is a free and open-source JavaScript library. It is best to create user interfaces by combining code sections (components) into complete websites. We can use React as much or as little as we want. React enables developers to use separate software components across the client and server sides, which also speeds up development.

3.9. Dependencies

3.9.1. Node Package Manager (NPM)

The node package manager (NPM) is a command-line tool for installing, updating, and removing Node.js packages from our application. It also serves as a repository for open-source Node.js packages. A package manager is essentially a set of software tools that can be used by a developer to automate and standardize package management.

3.9.2. Node.js

Node.js is a simple programming language that can be used for prototyping and agile development, as well as to create extremely fast and scalable services.

3.9.3. MetaMask

MetaMask is a non-custodial Ethereum-based decentralized wallet that also lets users save, buy, send, transform, and swap crypto tokens, as well as sign transactions. Using Metamask in conjunction with Web3.js in a web interface simplifies communication with the Ethereum network.

3.9.4. Truffle Framework

Truffle is a set of tools that allows us to create smart contracts, write tests against them, and deploy them to blockchains. It also provides a development console and allows us to create client-side applications within our project. Truffle is the most widely used framework for creating smart contracts. It supports Solidity and Viper as smart contract languages. Truffle has three main functions: it compiles, deploys, and tests smart contracts.

4. Proposed Data Storing Scheme

Our proposed scheme divides data storage, retrieval, and searching into four steps. The system uploads a file, file hash is stored on the blockchain, monitors smart contract events, and searches for relevant information.

4.1. File Uploading

The main concept of the file uploading process is depicted in Figure 2. The file is selected from the DApp (browser) (1), and when the DApp form's submit button is clicked, the uploaded file is stored on IPFS (2). The hash of the file uploaded is returned to the DApp (3); this hash is the file's location. The file's hash is saved to a smart contract (4), which is subsequently kept on the blockchain (5), and the hash and other information of the uploaded file were also listed on the DApp (6), from which we can obtain all of the files we have uploaded to IPFS.

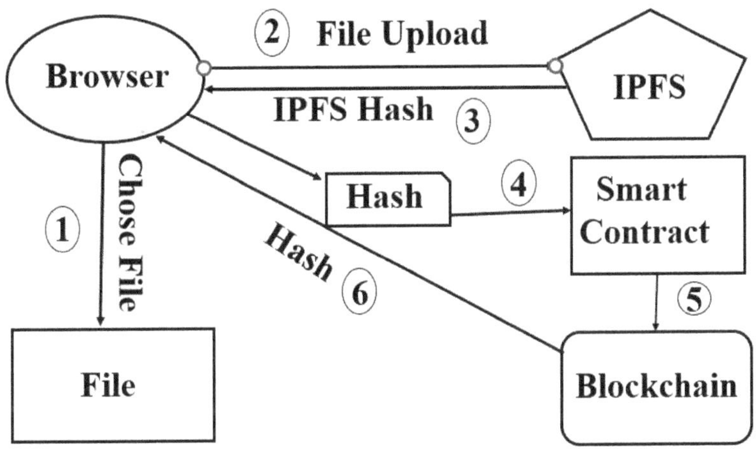

Figure 2. File Upload.

To connect to an Ethereum wallet Metamask, we used a web browser as a front end which will communicate with the blockchain and store the smart contract on it.

We will upload the file directly to an IPFS, and then IPFS will return to us a hash. We will then store this hash on the smart contract, and it will store that hash on the blockchain, allowing us to access all of the files we have created when we list them on the DApp.

A smart contract stores the hash value on the blockchain, and another smart contract lists the uploaded files on the DApp. The smart contract handles file uploading, file storage, and file listing.

Figures 3 and 4 show our smart contract. Our project's smart contract is responsible for four tasks. Define a data structure for file management, upload the files, store the file hash in the blockchain, and display the uploaded files on the DApp. We use a struct to manage the files inside Solidity. Solidity structs allow us to create more complex data types with multiple properties. By creating a struct, we can define our own type. They are useful for organizing related data. Structures can be declared outside of one contract and imported into another.

```
pragma solidity ^0.5.0;                      -----------------(1)

contract DStorage {                          -----------------(2)
   string public name = 'DStorage';          -----------------(3)
   uint public fileCount = 0;                -----------------(4)
   mapping(uint => File) public files;         ---------------(5)


struct File {                                -----------------(6)
   uint fileId;string fileHash;
   uint fileSize;string fileType;
   string fileName; string fileDescription;
   uint uploadTime; address payable uploader;}

event FileUploaded(                          -----------------(7)
  uint fileId,string fileHash,uint fileSize,
  string fileType,
  string fileName,string fileDescription, uint uploadTime,
  address payable uploader);
```

Figure 3. Solidity code for creation of a blockchain register and events to facilitate interoperability (1/2).

```
constructor() public {
}

function fileUpload(                        ------------------(8)
  string memory _fileHash,uint _fileSize,
  string memory _fileType,
  string memory _fileName,
  string memory _fileDescription)

  public {                                  -------------------(9)
   require(bytes(_fileHash).length > 0);          // Make sure the file hash exists
   require(bytes(_fileType).length > 0);          // Make sure file type exists
   require(bytes(_fileDescription).length > 0); // Make sure file description exists
   require(msg.sender!=address(0));             // Make sure uploader address exists
   require(_fileSize>0);                        // Make sure file size is greater than 0

// Increment file id
fileCount ++;                               ------------------(10)

// Add file to the contract
files[fileCount] = File(                    ------------------(11)
                   fileCount,_fileHash,_fileSize,
                  _fileType,_fileName,
                  _fileDescription,now, msg.sender);

// Trigger an event
emit FileUploaded(                         -------------------(12)
                   fileCount,_fileHash,_fileSize,
                  _fileType,_fileName,_fileDescription,
                   now,msg.sender);}}
```

Figure 4. Solidity code for creation of a blockchain register and events to facilitate interoperability (2/2).

The following steps show the tasks of a smart contract:

(i) Define data structure for the management of files:

Figure 3 shows step one in modeling the file (6). We created a file object, and inside we defined a unit id, which will be the unique identifier for the file inside our smart contract. The string will be the hash of the file, and this will be its location on IPFS, and a description of the file, which contains the location of the file and events related to the uploaded file. The address-payable uploader is the person who uploads the file, and it is the Ethereum address of that person's wallet address as they are connected to the blockchain; it is like their username on the blockchain.

(ii) Store and list the files:

Step two is to store the file on IPFS, and step three is to list the event logs on the DApp. We used mapping inside of Solidity to store the files, as shown in Figure 3. Mapping is another data structure. It can be utilized to store data as key-value pairs, with the key being any of the built-in data types but just not reference types, as well as the value being any type. We created mapping (5) as shown in Figure 3. A mapping inside of Solidity is just a key-value store. We can give it a key and a value. The data type of the key in our smart contract is an unsigned integer, and the return value is file struct (6), as shown in Figure 2. When we place a file with an id within this mapping, it will write and store it on the blockchain. Mapping is also going to give us the ability to list the files because mapping is public, and thus it gives us a function called "files" (5) that we can call, pass in the id, and fetch out each individual file. We can get back a file with all the data, such as the id, hash, file name, description, and uploader.

(iii) Upload File:

The solidity code has a function called fileUpload (8). "fileUpload" takes the following arguments: fileHash, fileSize, fileType, fileName, fileDescription. Whenever we upload a new file, we will just add a new file to the mapping. We created a new file (6) and put it inside the file's mapping (5). We are going to store the file based on the id inside the mapping, as shown in (5). We stored the file onto the blockchain as shown in (11).

Inside the smart contract, Solidity has a global variable called "msg" or "message" that has many different attributes, one of which is the person calling the function, "message sender" is the Ethereum address of the person uploading the file. We created a video struct and saved it inside the "files mapping", which we simply say "files", pass in the id, and it will be equal to a new file (11).

fileCount (4) is a variable that stores the number of files that have been created. Whenever we create the smart contract, the counter value will be zero, but we can change this value inside the function (11) as **fileCount** anytime the function is called. We could write **fileCount ++** (10) and then pass in **fileCount** in (11). **fileCount** keeps track of all the files; it is basically our ID management system, and we save it inside the file mapping, which acts like our database.

(iv) Creating an Event:

The event allows us to know when the file was uploaded. We can create events from the Solidity code. We define an event called "fileUpload" and we pass in the same arguments as the struct (7); this is going to allow us to subscribe to the event whenever it is triggered from our application. We can trigger the upload event (12). We use the emit keyword, then **FileUploaded** which has the same name as the event (7) and we pass in the arguments file count, fileHash, fileSize, fileType, filename, file description, and now, msg.sender.

Next, we added some requirements to the function to make it robust. We can use Solidity's **require** function (9). The **require** function checks that a set of parameters is true before the rest of the function executes. Table 2 shows the list of variables used in our smart contract.

Table 2. Smart Contract variables.

Variables	Why It Is Used
fileCount	Keeps track of how many files have been added to the current smart contract.
mapping File	key value store and lists the files
struct	Manage the files
event FileUploaded	Allows us to know when the file was uploaded
function fileUpload	Uploads new file
emit FileUploaded	Trigger an event

Recently, diverse types of formal methods are investigated to enhance the security of smart contracts, since the compromise of smart contracts can lead to a catastrophic monetary loss [24]. However, our smart contract codes have not been analyzed using those formal methods yet, and we will verify our codes in our future work.

Our first project element is a private Ethereum blockchain that will act as the back end for our DApp. Ethereum nodes maintain an archive of the blockchain's code. The information is dispersed throughout the network. The Geth is utilized to run an Ethereum node.

By running a node on the Ethereum network, we could also perform transactions as well as communicate with smart contracts. The uploaded file's hash is saved in a smart contract, and then immutably stored also on the Ethereum blockchain.

The next component is IPFS, which enables us to keep files in a distributed fashion. Because files are large, storing megabytes and gigabytes of files on the blockchain may not be feasible. This is where IPFS comes into play. It has nodes, just like Ethereum, and we distribute files that cannot be tampered with across the network. IPFS uses hashes. When you upload a file to IPFS, it will be stored somewhere and identified by its hash. We run our own IPFS node, which supports an IPFS gateway for file retrieval and storage and runs the IPFS Daemon server. We cannot store or retrieve data unless the Daemon server is up and running, or unless we link to public gateways such as Infura [25].

When a user uploads CCTV footage to our DApp, they can specify the location as well as event details such as whether it was an accident or a traffic violation. This information is fed into the DApp as a file description. This information is critical when uploading a video to the DApp because users can quickly search for location and event information using the DApp's keyword search function.

We first must import and link our Ethereum blockchain account to Metamask before we can use the DApp. Our web browser now supports blockchain networks, and we can upload files to IPFS using our custom-designed DApp user interface (UI). First, we must select the file, enter its description (such as file event and location), and then click the submit button. When we click the submit button, the file is sent to IPFS and we receive the IPFS result, which contains the hash value and path of the file. Metamask directs us to accept the transaction, save the hash in a smart contract, and store the smart contract on the blockchain via a confirmation pop-up. To store the hash on the blockchain, we should pay some gas in the manner of ethers. When we confirm the Metamask transaction, the hash of the uploaded file is preserved on the Ethereum blockchain.

The DApp monitors the "file upload" event and updates the DApp's User interface automatically. The event log of the smart contract is generated by retrieving and displaying all events from the smart contract within our DApp. The smart contract event log includes the file no, file description, type of file, file size, timestamp, Ethereum information of the uploaded person, and the hash value of the file after it has been stored in IPFS. By clicking on the file description, individuals may view the uploaded files in their web browser. The hash value does not need to be remembered or stored separately by the user.

4.2. Keyword Searching

Users of the blockchain network can view transaction details but cannot identify the individuals who made the transactions. On our DApp, we can see the transactions and use the data for keyword searching.

(i) Read information from the blockchain:

When events occur in the smart contract, the smart contract emits events in order to communicate with DApps and other smart contracts. When we invoke a smart contract function, it has the ability to generate an event. It is critical for us to be able to listen to these events in real time when developing DApps.

To listen for smart contract events, we used Ethers.js smart contract event listener. To communicate with a smart contract using Ethers.js, we must first create a new contract object with Ethers as shown in step (1) of Figure 5.

```
Creating Instances
 new ethers.Contract( address , abi , signerOrProvider ) -------------- (1)

const abi = [] -----------(2)
const address = ""  ------------(3)
const provider = new ethers.providers.JsonRpcProvider('HTTP://localhost:8545'); ------(4)
const contract = new ethers.Contract(address, abi, provider); ----------(5)

contract.queryFilter( 'event') ------- (6)
contract.queryFilter( event [ , fromBlockOrBlockHash [ , toBlock ] ) ⇒ Promise< Array< Event > > ---------- (7)
   |Return Events that match the event.
```

Figure 5. Ethers.js filter to read events from blockchain.

As shown in steps (2), (3), and (4) of Figure 5, we need the blockchain address for the smart contract, the ABI of the smart contract, and the signer or provider (4). The ABI is a JSON object that describes how the smart contract works; it describes the interface, which essentially means what functions the smart contract has, what function arguments it accepts, and what it responds to when we try to read data from it. Ether.js allows us to store ABIs as an array and only pull in the parts we want when we are setting up a smart contract object. We require file upload information for our project, so we included ABI, which is related to the file upload event. Then, we need a provider or a signer; in our project, we have a provider. A provider is an abstraction of an Ethereum network connection that provides a concise, consistent interface to standard Ethereum node functionality. We take our smart contract ABI and create a new contract address ABI, and then we provide all of the required information as shown in step (5) of Figure 5.

We used contract.queryFilter to filter the information, as shown in step (6) of Figure 5. Using this command, we will examine every single FileUploaded event that has ever occurred on our blockchain. We include this filter to reduce the search space inside the Ethereum blockchain. Ethers.js allows us to examine the FileUploaded events and specify which blocks we want to examine as shown in step (7) of Figure 5.

(ii) Keyword search text file creation:

We can create a text file for keyword searches once the events are retrieved from the blockchain. The smart contract events are written into a text file. We store only necessary information in the text file, e.g., information such as file name, event type, location, Ethereum account number, and smart contract.

Figure 6 shows how to retrieve data from the blockchain and conduct keyword searches. To listen to smart contract events, we used a command prompt to send requests to the blockchain (1). Blockchain responded with a filtered smart contract event log containing all of the information about the uploaded file, including the smart contract address, file name, file hash and description, Ethereum address of the uploaded file, and so on (2). When we received a smart contract event log, we saved some of the event logs in a text file (3). We wrote code in react.js to filter the results and search for keywords on the DApp. When a user searches for a keyword on the DApp, the request is sent to a text file containing smart contract events, which is then filtered, and the result is returned to the DApp (4). Users can look up a word or an alphabet.

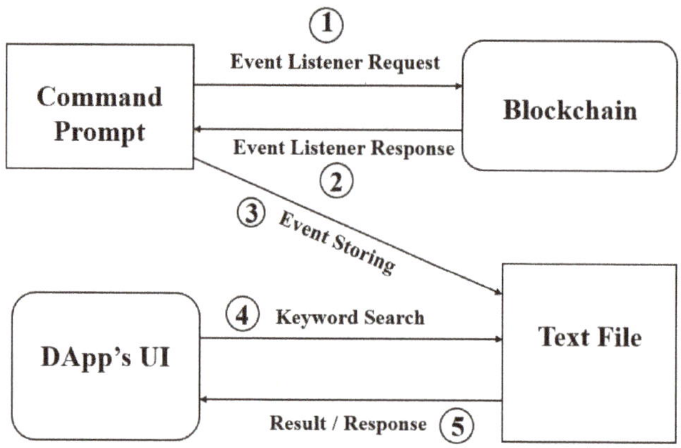

Figure 6. Keyword Search Function of the proposed DApp.

5. Implementation

On the Windows 10 operating system, we used a private Ethereum blockchain to implement a proposed scheme. The Ethereum core network is not connected to a private Ethereum network. Organizations primarily use it to limit blockchain read permissions. Installing geth/parity allows the current node to join the Ethereum network and download the blockchain to local storage. We used Go Ethereum to create our Ethereum blockchain (Geth).

5.1. Steps to Create Private Ethereum Network

The following steps show how we built our private Ethereum network:

5.1.1. Download "Geth"

Go Ethereum (Geth) can be directly downloaded and installed from **geth.ethereum.org**, accessed on 16 February 2023. Because Geth is a command line interface, we execute all commands from the command line. After installing Geth on our system, we typed geth and pressed enter in a command prompt and obtained the output as shown in Figure 7.

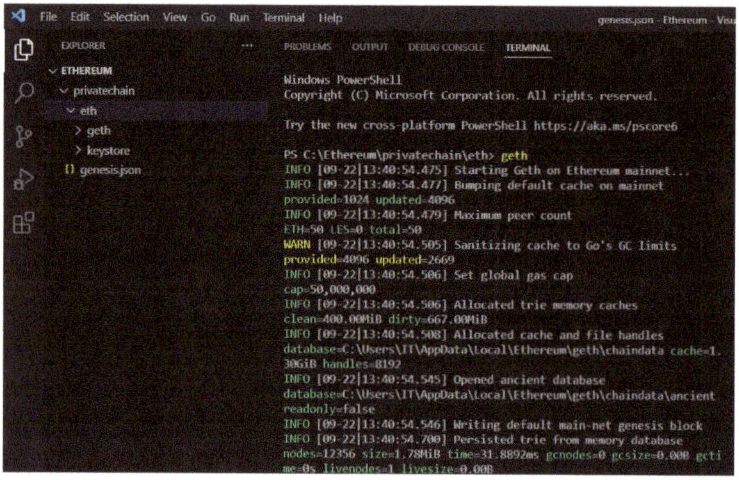

Figure 7. Geth command.

We used the geth command to connect to a blockchain, and the geth command will run in fast sync mode. Fast sync is Geth's current default sync mode. Fast Sync nodes download the headers of each block and retrieve all the nodes beneath them until they reach the leaves. Instead of reprocessing all transactions that have ever taken place, fast sync downloads the blocks but only validates the affiliated proof-of-works (which could take weeks). When we stop and restart the geth, it will operate in full sync mode. Full sync needs to download all blocks and incrementally generate the blockchain state by running each block since genesis. The data size of the Ethereum blockchain is currently around 800–1000 gigabytes, and we do not need to download the entire Ethereum blockchain on our system.

5.1.2. Make a Folder for Our Private Ethereum Network

For the private Ethereum network, we created a separate folder called "Private Ethereum". This folder separates the Ethereum private network files from the public files.

5.1.3. Construct a Genesis Block

In blockchain, all transactions are recorded in the form of blocks in sequential order. There are an infinite number of blocks, but there is always one distinct block that gives rise to the entire chain, known as the genesis.

The genesis block, also known as Block 0 or Block 1, is the first block ever recorded on its respective blockchain network. There are no transactions. The genesis block is used to initialize the blockchain, as shown in Figure 8. A genesis block is required to create a private blockchain. The genesis block can be created with any text editor and saved with the JSON extension in the **Private Ethereum** folder. Figure 9 shows the genesis block file.

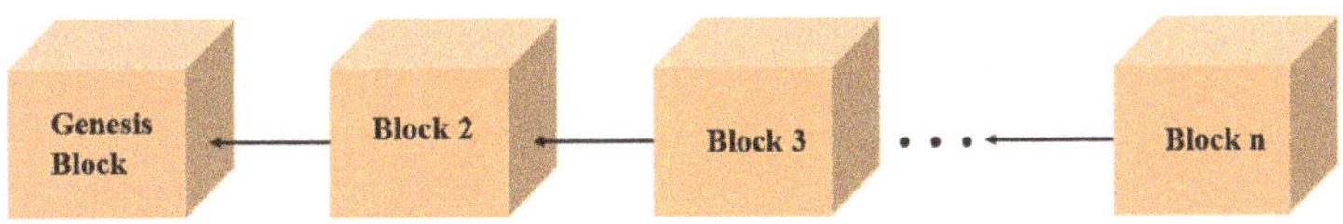

Figure 8. Genesis block in a blockchain.

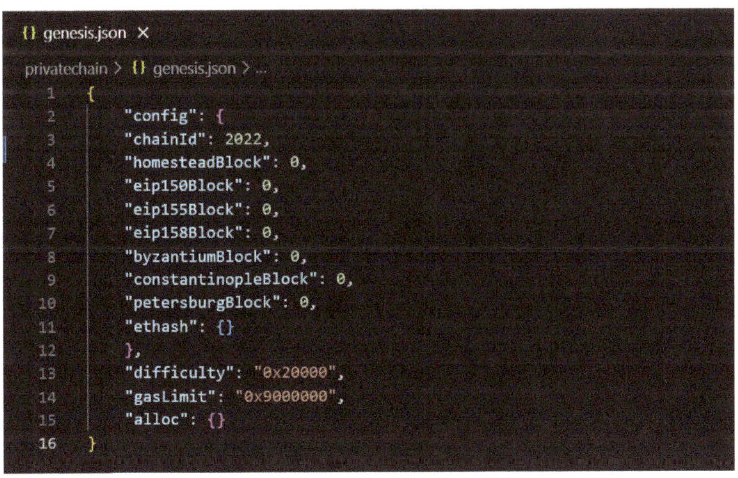


```
{
    "config": {
    "chainId": 2022,
    "homesteadBlock": 0,
    "eip150Block": 0,
    "eip155Block": 0,
    "eip158Block": 0,
    "byzantiumBlock": 0,
    "constantinopleBlock": 0,
    "petersburgBlock": 0,
    "ethash": {}
    },
    "difficulty": "0x20000",
    "gasLimit": "0x9000000",
    "alloc": {}
}
```

Figure 9. Genesis block file.

5.1.4. Run the Genesis File

To extract the genesis file, we open the **Private Ethereum** folder in Visual Studio Code and run the command **geth init ./genesis.json -datadir eth.** Eth is the name of a folder. Geth is connected to the genesis file after running the above command.

5.1.5. Set Up the Private Network

We created a private network in which multiple nodes can add new blocks. We must use the command **geth –datadir ./eth/ –nodiscover** to accomplish this. When–**nodiscover** is used to start a geth node, it prevents the node from being discovered by the network's bootnode. Every time the private network chain is needed, commands in the console must be executed to connect to the genesis file and the private network. A private Ethereum network and a personal blockchain are now available. Figure 10 shows the running status of a private Ethereum network.

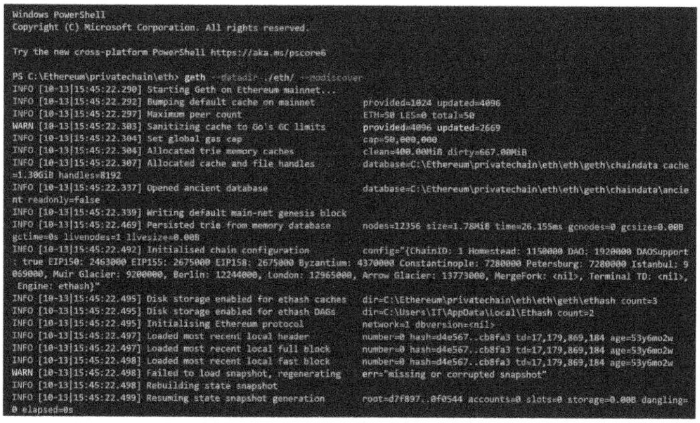

Figure 10. Private Ethereum network.

5.1.6. Make Externally Owned Account (EOA)

EOAs are controlled by users who have access to the account's private keys. These accounts, which can both send transactions and trigger contract accounts, are typically used in conjunction with a wallet. To manage the blockchain network, EOA is required. To make it, we launched Geth in two windows. One terminal to run Geth as shown in Figure 10 and another terminal to create EOA. We entered the command **geth attach \\.\pipe\geth.ipc** in the second terminal (console window). This will connect the second terminal to the private Ethereum network established in Figure 10. We used the command **personal.newAccount()** to create a new account. After executing this command, we entered our password to obtain our account number and saved it for future use as shown in Figure 11.

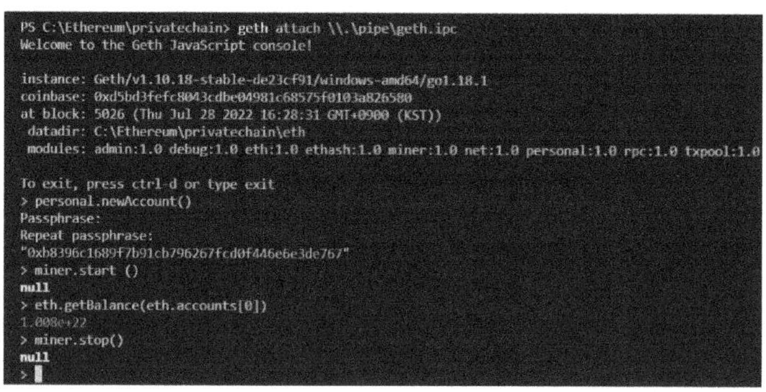

Figure 11. Externally owned account, Mining Start and Stop.

5.1.7. Ethereum Mining on Our Private Chain

If we mine on the Ethereum main chain, we will need expensive equipment with powerful graphics processors. ASICs are typically used for this but high performance is not required in our private network, and we can begin mining with the command **miner.start ()** as shown in Figure 11.

After a few seconds, some ether was found in the default account if the balance status is checked as shown in Figure 11. To check the balance, we used the command **eth.getBalance(eth.accounts[0]).** Figure 12 shows the mining process. We used the command **miner.stop()** to stop mining as shown in Figure 11.

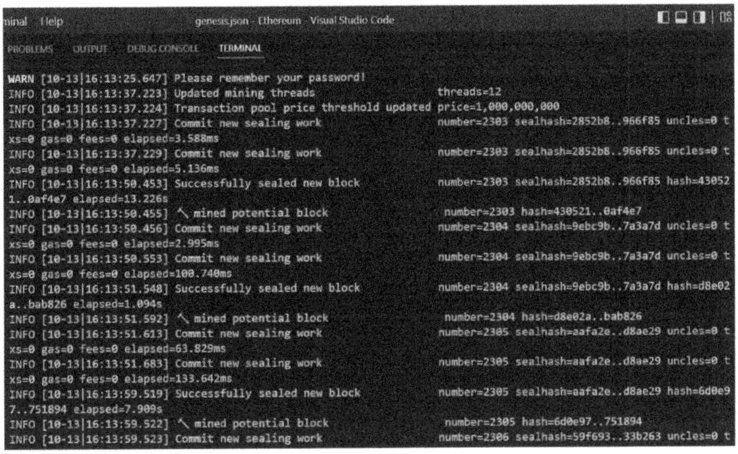

Figure 12. Mining Process.

5.1.8. Connecting the Private Ethereum Network to Metamask

We closed the terminal in which our private network was running and opened a new terminal and typed the command **geth –datadir ./eth/ –nodiscover –http –http.addr "localhost" –http.port "8545" –http.corsdomain="*" –http.api web3,eth,debug,personal,net –ws.api web3,eth,debug,personal,net –networkid 7777 –allow-insecure-unlock,** as shown in the Figure 13 and now our private Ethereum is connected to Metamask.

Explanation of the used commands as follows:

- http.addr value:
  Listening interface for HTTP-RPC servers (default: "localhost").
- http.port value:
  Listening port for HTTP-RPC server (default: 8545).
- http.corsdomain value:
  A list of domains separated by commas that will accept cross-origin queries (browser enforced). Because the HTTP server can be accessed from any local application, the server includes additional safeguards to prevent API abuse from web pages. The server must be configured to accept Cross-Origin requests in order to allow API access from a web page. The —http.corsdomain flag is used to accomplish this. The —http.corsdomain command accepts wildcards, allowing access to the RPC from any location: —corsdomain '*'.
- http.api value:
  APIs accessible via the HTTP-RPC protocol.
- ws.api value: APIs accessible via the WS-RPC interface.
- nodiscover:
  The peer discovery mechanism is disabled.
- networkid value:
  Sets network id explicitly.
- allow-insecure-unlock:
  When account-related RPCs are exposed via http, this allows for insecure account unlocking.

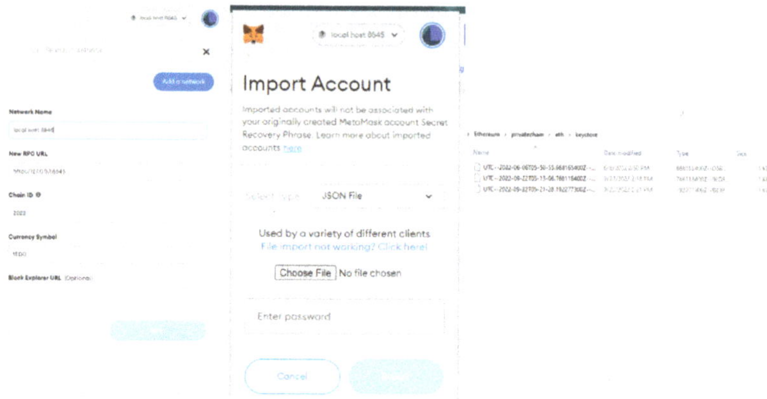

Figure 13. Importing Ethereum account in Metamask.

We launched Metamask and added the Network "Local Host 8545" with the Chain ID "2022". It is the chain ID we specified in our private Ethereum network's genesis block. By importing a JSON file from our private Ethereum folder, we imported a private Ethereum account. The JSON file can be found in the keystore's Private Network folder. Figure 13 depicts how to add a Private Ethereum account to Metamask.

5.2. Running Our Own IPFS Node

To store information on IPFS, we must run an IPFS Daemon server on our own IPFS node. To use IPFS, we must first download and install the Go language from the golang website, then go to the IPFS command line install page and download "install go-ipfs". We navigate to the download path, extract the files to C drive, and then run ipfs.exe to start the Daemon server, as shown in Figure 14.

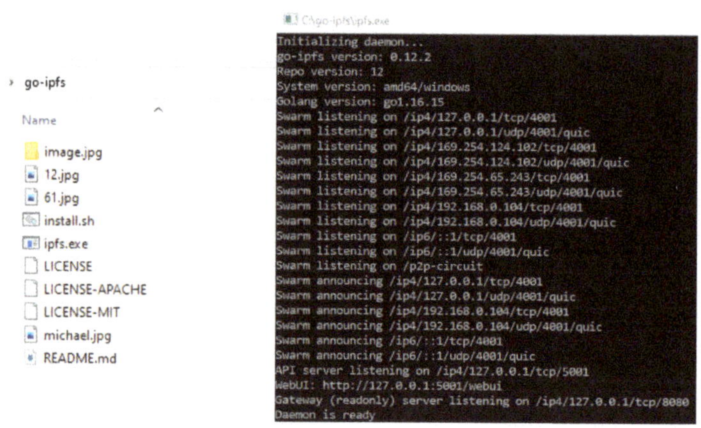

Figure 14. IPFS Execution and Daemon server.

5.3. Deploying Smart Contract

A smart contract stores the hash of the uploaded file. To make smart contracts in the Solidity programming language, the Truffle framework is used. The Truffle Suite is a collection of tools specifically designed for Ethereum blockchain development. The suite includes three pieces of software. Truffle is capable of helping compile and deploy smart contracts in addition to injecting them into web apps and building DApp front ends. Truffle is now a popular Ethereum Blockchain IDE.

5.4. File Uploading and Retrieving

After writing the smart contract, deploying, and publishing it to our Ethereum blockchain, we then utilize Metamask to connect our DApp to the Ethereum blockchain. A Metamask is required to communicate with the blockchain. The client-side application, which is also going to communicate with IPFS, was built with React.

Figures 15 and 16 show how we initially deployed the smart contract to Ethereum, then launched the DApp with the command **npm run start**, imported an Ethereum account into Metamask, and linked Metamask to our DApp. Figures 17 and 18 show how to submit a file to IPFS, deposit the file's hash in a smart contract, record the smart contract on the Ethereum blockchain, and successfully retrieve the file using our DApp.

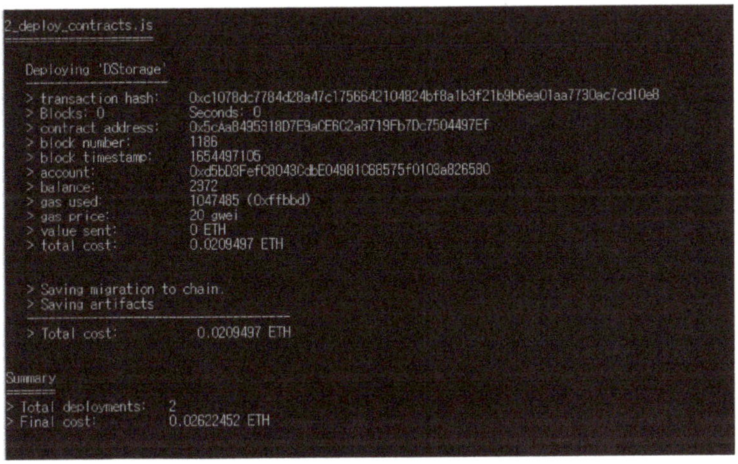

Figure 15. Smart contract deploy.

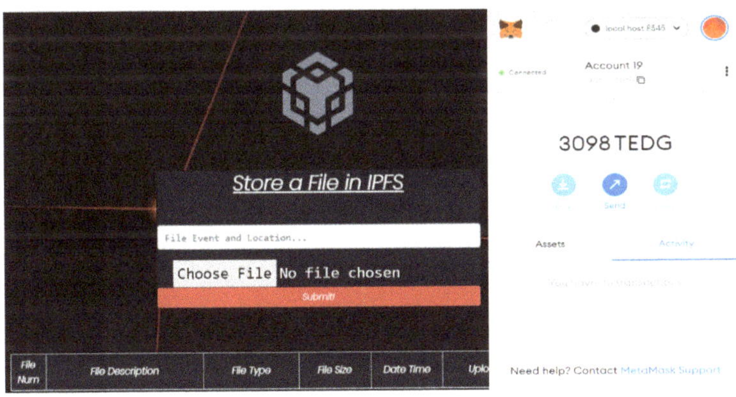

Figure 16. Connecting Metamask to the DApp.

We chose the file and entered the location as well as the location of the file in the user interface of DApp after logging into Metamask, then clicked the submit button also confirmed the transaction of Metamask, as shown in Figure 17. To deploy the smart contract, upload files, and store hash values on the blockchain, we start and maintain mining.

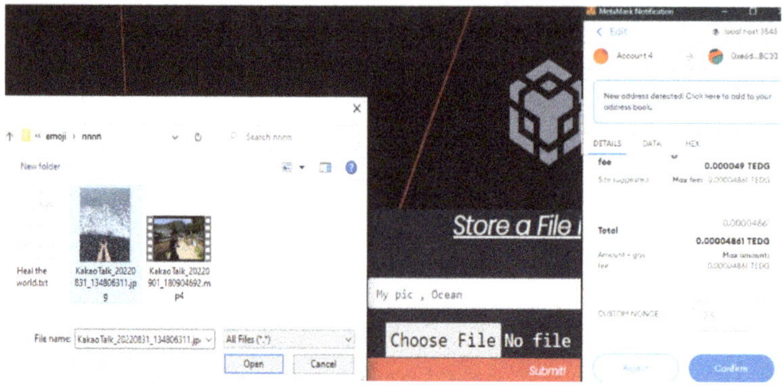

Figure 17. Choosing a file and confirming Metamask transaction.

As the transaction is confirmed, the DApp listens for the event "File Upload" and updates the DApp's user interface automatically. Whenever a transaction has been mined, smart contracts generate events and logs to the blockchain, which can then be processed by the front end. Our DApp retrieves and displays all smart contract events. It is referred to as a "smart contract event log". The event log of the smart contract contains the file number, file description (which includes an event and location of the file), type of the file, file size, date and time, the uploader's Ethereum account details, and the hash of the file. By having to click on the file's file details, users are able to view the uploaded files through their web browser. Figure 18 depicts a smart contract event log and various file types retrieved.

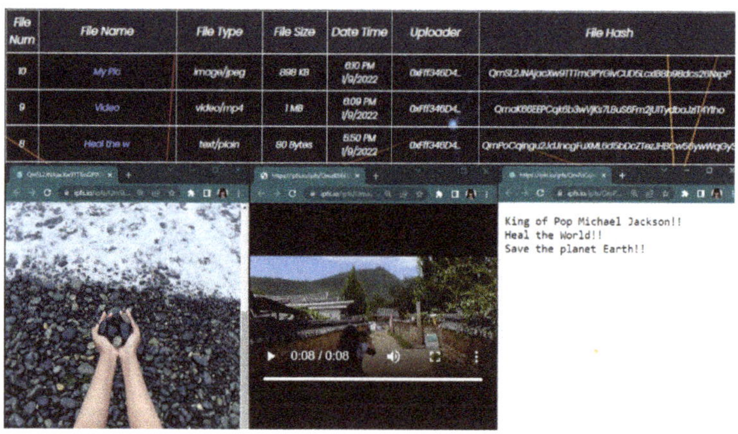

Figure 18. Event log and file retrieve.

5.5. *Keyword Searching*

Our DApp supports the keyword search method. In order to conduct keyword searches, we obtain event information from Blockchain. Smart contracts could even emit logs as well as events to the blockchain whenever an Ethereum transaction is mined, which the front end can then process. An event broadcasts information about a file upload, and we could have access to all of the events so that we could listen to them in real time, or we could just use them to obtain all of the most recent file uploads on the blockchain.

We can read smart contract events outside of the DApp's user interface by using Web3.js or Ethers.js. In our implementation, Ethers.js is used to read smart contract events. We only have one event in our smart contract, so we use a filter to retrieve information from that event, which is File Upload. A smart contract event log is shown in Figure 19.

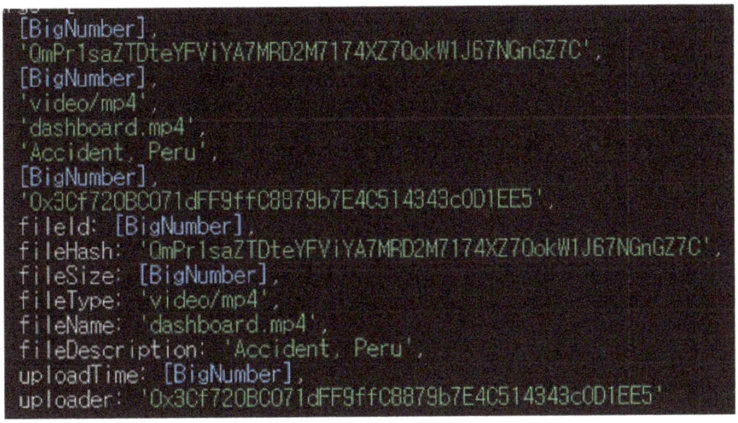

Figure 19. Smart contract events.

The smart contract events are then written to a text file, allowing our DApp to conduct keyword searches. We store only necessary information in the text file, e.g., information such as file name, event type, location, Ethereum account number, and smart contract. When looking for sensitive information on the DApp, keyword searching is essential. Entering an alphabet or a keyword into keyword searching will filter the results to show only the keyword we entered. In the case of an alphabet search, the DApp will display all events that include the letter we typed into the search box. This method makes navigating an event easier and more efficient. Keyword searching is shown in Figure 20.

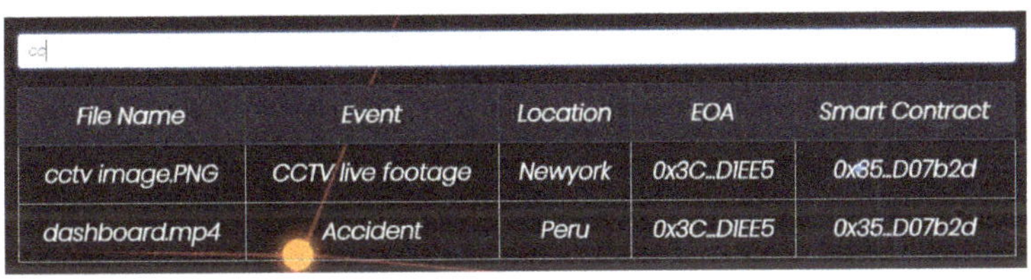

Figure 20. Keyword Searching.

6. Performance Evaluation

The majority of applications we use today are centralized, which means they are managed by a single authority. Google [26] and Facebook [27], for example, retain complete ownership of their respective products, running their apps and storing user data on private servers and databases. While this gives Google and Facebook control over their applications and user experiences, it can also be discouraging to users. Users of centralized apps have little control over their data or experience within the app. They must have faith in the app's developer to listen to their feedback, provide product services, and treat them and their data with dignity. However, with other centralized applications facing backlash over privacy and the monetization of user data, many users are wary of relying on them.

Centralized applications run programs and store critical user information on centralized servers. The entire application may fail if a single, central server is compromised. DApps enable users to complete transactions, verify claims, and collaborate in real time without relying on a centralized intermediary.

Our DApp operates on a peer-to-peer network, similar to a distributed ledger, with each network member contributing to the program. Each of the roles that a central server would normally provide, from computing power to storage, is distributed across the network. We do not need to keep and secure a central server, and users can directly participate in the app's operation. Our system is robust to system failure. There is no single point of failure in our DApp and is distributed across a network of public nodes, with copies of critical information distributed among them. The application is unaffected if one or more IPFS nodes are compromised. Even if there is a virus attack, a hardware failure, or the system is turned off, the user can still retrieve the uploaded files and perform keyword searches.

When a user uploads data to IPFS, it is chopped into smaller chunks, hashed, and assigned a unique content identifier (CID), which serves as a fingerprint. This makes it faster and easier to store small amounts of data on the network. A cryptographic hash (CID) is generated for each piece of data, making each upload to the network unique and resistant to security breaches or tampering.

The experiment we conducted demonstrates that our DApp is resistant to system failure, robust, and transparent.

The experiments we carried out are listed below.

Scenario 1:

In Scenario 1, the system unexpectedly shuts down, and when it is restarted, the DApp's event log vanishes, as illustrated in Figure 21. We can retrieve the event log outside of the DApp using smart contract event listeners. In Figure 19, we used Ethers.js to retrieve the event log. The data associated with the uploaded file is included in the event log. As a result, system failure has no effect on the uploaded data.

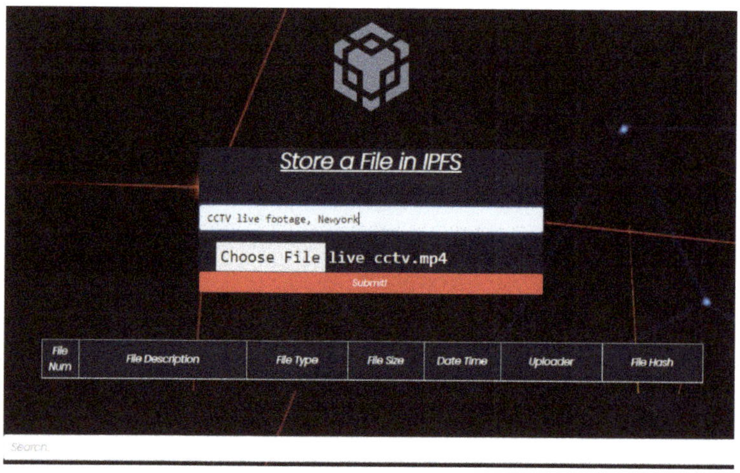

Figure 21. No event log listed on the DApp.

Scenario 2:

The information in the keyword search text file was accidentally deleted in Scenario 2 as shown in Figure 22, and we were unable to perform the keyword search on the DApp. As demonstrated in Scenario 1, we recreated the keyword search text file using information retrieved from the smart contract event log and performed a keyword search as illustrated in Figure 23. Table 3 summarizes the scenarios of performance evaluation.

Table 3. Performance Evaluation Scenario Summarization.

Scenario #	Description
1	The system unexpectedly shuts down. When the system restarted, the DApp's event log vanished. We used ether.js to retrieve the event log.
2	The information in the keyword search text file was accidentally deleted. We recreated the keyword search text file by using information retrieved from the smart contract event log and performed a keyword search.

Figure 22. Text file with no data.

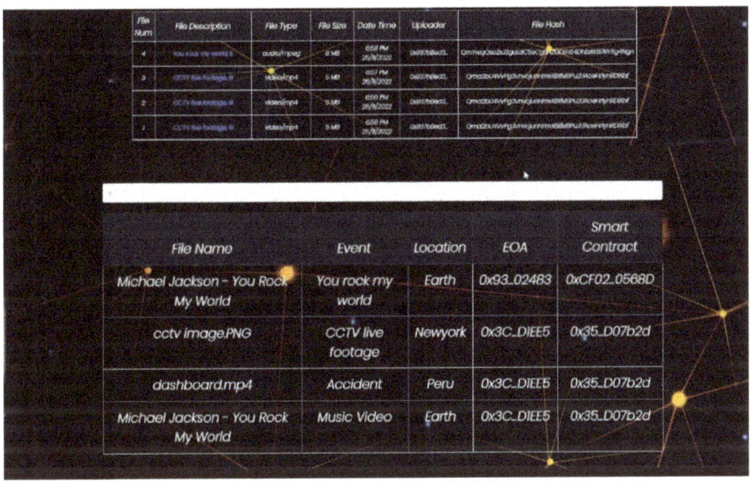

Figure 23. Keyword Search.

If a malicious actor manages to compromise the blockchain network, any changes are visible on a public network, allowing both users and developers to respond quickly. Our DApp operates on a public ledger, which means that anyone with internet access can participate in the application and network. As a result, anyone can view the transaction record and any changes made to those records. Therefore, this system can provide better transparency than centralized applications can provide. On a publicly distributed ledger, no central entity can revoke transparency, limit viewership, or censor participation.

7. Conclusions

In this paper, we present the design and implementation of a decentralized application that uses Ethereum blockchain and IPFS to store CCTV and black box footage securely and efficiently. The DApp allows users to easily manage their storage. For scalability, only hashes of the files are stored on the blockchain via smart contracts. Our proposed scheme works in a decentralized manner. When a file is uploaded, the DApp listens for the event File Upload and automatically updates the DApp's user interface. All smart contract events are fetched and displayed on our DApp. The extracted information is called a smart contract event log, and it includes information about the file, timestamp, the uploader's account information, and the hash of the IPFS file returned. By clicking on the file's description, users can gain access to it. The selected file is then displayed in the web browser. DApp also includes a keyword search feature to help us find any information quickly. To filter and read data from the blockchain, we used ether.js' smart contract event listener and contract.queryFilter. We used the smart contract address as well as the smart contract's ABI. The smart contract events are then written into a text file. The text file only contained necessary information, such as the file name, event type, location, Ethereum account number, and smart contract. Our experiment shows that our DApp is not affected by system failure. We can secure an application by managing the data in a decentralized manner. Because our DApp runs on a public ledger, anyone with internet access can participate in the application and network. As a result, anyone can view and modify the transaction record. As a result, unlike centrally managed applications, this system provides greater transparency. We anticipate that our DApp can be used in a variety of fields, such as for keeping records of student research securely at universities, the medical information of patients at hospitals, and customer information at banks due to its ability to store various file types.

In our current system, the access control function is not included in the smart contract yet, and thus, the hash values of one's files can be exposed to anyone who knows his or her smart contract address. We will investigate the access control scheme for the smart contract to resolve this issue in our future work. In addition, we will also verify the source code of our smart contract using well-known formal methods.

Recently, Ethereum has been upgraded by changing its consensus mechanism from proof-of-work (PoW) to proof of stake (PoW), and this new version is also known as Ethereum 2.0. However, this new consensus mechanism has not been verified intensively compared to the PoW mechanism, and thus, we used an old version of Ethereum and its corresponding Ethereum Virtual Machine (EVM) environment in this paper. We will implement and investigate our proposed system on the new version of Ethereum in our future work.

Author Contributions: Conceptualization, N.S. and S.Y.N.; data curation, N.S.; formal analysis, N.S.; methodology, N.S.; project administration, N.S. and S.Y.N.; resources, N.S.; software, N.S.; supervision, S.Y.N.; validation, N.S.; visualization, N.S.; writing—original draft, N.S.; writing—editing and review, N.S. and S.Y.N. All authors have read and agreed to the published version of the manuscript.

Funding: This research was supported in part by the National Research Foundation of Korea (NRF), with a grant funded by the Korean government (MSIT) (2020R1A2C1010366). This research was supported in part by Basic Science Research Program through the National Research Foundation of Korea (NRF) funded by the Ministry of Education (2021R1A6A1A03039493).

Data Availability Statement: No new data were created or analyzed in this study. Data sharing is not applicable to this article.

Conflicts of Interest: The authors declare no conflict of interest.

References

1. Mateen, A.; Khalid, A.; Nam, S.Y. Management of Traffic Accident Videos using IPFS and Blockchain Technology. *KICS Summer Conf.* **2022**, *1*, 1366–1368.
2. Singh, A.; Chatterjee, K. Cloud security issues and challenges: A survey. *J. Netw. Comput. Appl.* **2017**, *79*, 88–115. [CrossRef]
3. Shin, Y.; Koo, D.; Hur, J. A survey of secure data deduplication schemes for cloud storage systems. *ACM Comput. Surv.* **2017**, *49*, 1–38. [CrossRef]
4. Yinghui, Z.; Dong, Z.; Deng, R.H. Security and privacy in smart health: Efficient policy-hiding attribute-based access control. *IEEE Internet Things J.* **2018**, *5*, 2130–2145.
5. Zhang, Y.; Chen, X.; Li, J.; Wong, D.S.; Li, H.; You, I. Ensuring attribute privacy protection and fast decryption for outsourced data security in mobile cloud computing. *Inf. Sci.* **2017**, 379, 42–61. [CrossRef]
6. Dropbox. Available online: https://www.theguardian.com/technology/2016/aug/31/dropbox-hack-passwords-68m-data-breach (accessed on 17 February 2023).
7. Arrington, M. Gmail Disaster: Reports of Mass Email Deletions. December 2006. Available online: https://techcrunch.com/2006/12/28/gmail-disaster-reportsof-mass-email-deletions/ (accessed on 17 February 2023).
8. Amazon. Amazon s3 Availability Event: 20 July 2008. Available online: https://simonwillison.net/2008/Jul/27/aws/ (accessed on 17 February 2023).
9. Krigsman, M. Apple's MobileMe Experiences Post-Launch Pain. July 2008. Available online: https://www.zdnet.com/article/apples-mobileme-experiences-post-launch-pain/ (accessed on 17 February 2023).

10. Benet, J. Ipfs-Content Addressed, Versioned, p2p File System. 2014. Available online: https://arxiv.org/abs/1407.3561 (accessed on 17 February 2023).
11. Geth. Available online: https://geth.ethereum.org/ (accessed on 17 February 2023).
12. Metamask. Available online: https://metamask.io/ (accessed on 17 February 2023).
13. Hao, J.; Sun, Y.; Luo, H. A Safe and Efficient Storage Scheme Based on BlockChain and IPFS for Agricultural Products Tracking. *J. Comput.* **2018**, *29*, 158–167.
14. Rajalakshmi, A.; Lakshmy, K.V.; Sindhu, M.; Amritha, P. A blockchain and IPFS based framework for secure Research record keeping. *Int. J. Pure Appl. Math.* **2018**, *119*, 1437–1442.
15. Vimal, S.; Srivatsa, S.K. A new cluster P2P file sharing system based on IPFS and blockchain technology. *J. Ambient. Intell Hum. Comput.* **2019**, 1–8. [CrossRef]
16. Chen, Y.; Li, H.; Li, K.; Zhang, J. An improved P2P file system scheme based on IPFS and Blockchain. In Proceedings of the 2017 IEEE International Conference on Big Data (Big Data), Boston, MA, USA, 11–14 December 2017; pp. 2652–2657. [CrossRef]
17. Wang, R.; Tsai, W.-T.; He, J.; Liu, C.; Li, Q.; Deng, E. A Video Surveillance System Based on Permissioned Blockchains and Edge Computing. In Proceedings of the 2019 IEEE International Conference on Big Data and Smart Computing (BigComp), Kyoto, Japan, 27 February–2 March 2019; pp. 1–6. [CrossRef]
18. Sun, J.; Yao, X.; Wang, S.; Wu, Y. Blockchain-Based Secure Storage and Access Scheme For Electronic Medical Records in IPFS. *IEEE Access* **2020**, *8*, 59389–59401. [CrossRef]
19. Ethereum. Available online: https://ethereum.org/ (accessed on 17 February 2023).
20. Web3. Available online: https://web3js.readthedocs.io/en/v1.8.0/ (accessed on 17 February 2023).
21. Ethers. Available online: https://docs.ethers.io/v5/ (accessed on 17 February 2023).
22. Cai, W.; Wang, Z.; Ernst, J.B.; Hong, Z.; Feng, C.; Leung, V.C.M. Decentralized Applications: The Blockchain-Empowered Software System. *IEEE Access* **2018**, *6*, 53019–53033. [CrossRef]
23. React. Available online: https://reactjs.org/ (accessed on 17 February 2023).
24. Krichen, M.; Lahami, M.; Al-Haija, Q.A. Formal Methods for the Verification of Smart Contracts: A Review. In Proceedings of the 15th International Conference on Security of Information and Networks (SIN), Sousse, Tunisia, 11–13 November 2022; pp. 1–8.
25. Infura. Available online: https://infura.io/ (accessed on 17 February 2023).
26. Google. Available online: https://www.google.com/ (accessed on 17 February 2023).
27. Facebook. Available online: https://www.facebook.com/ (accessed on 17 February 2023).

electronics

MDPI

Article

Implementing Blockchain-Based Traceable Certificates as Sustainable Technology in Democratic Elections

Robin Singh Bhadoria [1,*], Arka Prabha Das [2], Abul Bashar [3] and Mohammed Zikria [3]

[1] Department of Computer Engineering & Applications, GLA University, Mathura 281406, Uttar Pradesh, India
[2] Indian Institute of Information Technology (IIIT) Bhopal, Bhopal 462003, Madhya Pradesh, India
[3] College of Computer Engineering and Science, Prince Mohammad Bin Fahd University, Al-Khobar 31952, Saudi Arabia
* Correspondence: robin19@ieee.org

Abstract: A democratic election is a crucial event in any country. Therefore, the government of the country is concerned with creating more competitive and fairer elections. This paper discusses the survey and scope of Blockchain technology adoptions in conducting elections. A distributed digital ledger is used in the Blockchain technology that is utilized for recording transactions happening between two parties. Ledger conducts this processing in an efficient and effective manner with latest secure mechanism of encryption algorithms. Therefore, the data stored in several blocks in each transaction is secure, transparent, and tamper-proof, which ultimately improves the transparency and voter confidentiality. This paper demonstrates how the benefits of the Blockchain technology such as immutability, transparency and end-to-end verifiability can be utilized by the national governments around the world to ensure fair democratic elections. In short, we aim to present a rigorous mechanism of a Blockchain based e-voting system, its efficiency based on different consensus algorithms and the overall progress and analysis based on some critical parameters to anticipate the feasibility of the successful implementation of the proposed e-voting system.

Keywords: Blockchain technology; e-voting system; smart contract; distributed ledger; transparency and confidentiality

Citation: Bhadoria, R.S.; Das, A.P.; Bashar, A.; Zikria, M. Implementing Blockchain-Based Traceable Certificates as Sustainable Technology in Democratic Elections. *Electronics* **2022**, *11*, 3359. https://doi.org/10.3390/electronics11203359

Academic Editor: Hamed Taherdoost

Received: 6 September 2022
Accepted: 29 September 2022
Published: 18 October 2022

1. Introduction

The most common means of vote casting is through ballot papers. This method has been widely criticized because of fraudulent voting and booth capturing witnessed across various countries worldwide. Thus, manually casting the vote has been replaced with electronic machines to record the vote for individual citizens of the country. The machines saved paper costs and reduced time and replaced the manual exercise involved in conventional counting and resulted in dumping of fake votes. Such voting machines introduced more transparency and verifiability to its voters [1].

Even after all these replacements, several concerns still remain for voters. The Distributed Ledger Technology (DLT) can be combined with such voting machines to make the electoral process more robust and error-free. DLT is secured and immutable through the use of complex encryption algorithms. In simple terms, Blockchain is defined as a distributed database whose copy is issued to everyone involved in the transaction process [2]. One can add records in the database but cannot alter them. Therefore, data stored inside the Blockchain is secure, transparent, and tamper-proof.

A distributed digital ledger is used in Blockchain technology that is utilised for recording transactions happening between two parties. This task is achieved by the ledger in a very efficient manner. In creating chain of blocks, each block comprises of data and its associated hash value of previously created block in such a chain [3]. The data stored inside such blocks may depend on the type of Blockchain, especially its version. "Hash" is the second element that is always unique, very similar to a human fingerprint that can be identified amongst trillions of hands. Hash is calculated just after the creation of a block and the Hash identifies the block along with its contents. Any manipulations in the block will automatically cause changes in its associated hash value [4]. Thus, the role of hash is very significant in identification of any block, if it is modified. This gives unique characteristic for Blockchain, and each block is linked or chained in sequence to one another. The first block is an initial block, and thus it does not have any hash value and is also known as the *genesis block*. When anyone tries to modify or alter the data in any block, the hash value associated with the blocks also gets modified which helps in identification of such block and make it as "*invalid*". This scenario makes a chain of blocks as more secure and immutable.

The conceptualization of Blockchain Technology is creating a number of records, namely blocks, that hold data and its associated generated hash value (always *unique*). It creates a distributed ledger that keeps a record of all data for every transaction [5]. Three major pillars of the Blockchain are immutability, decentralized and transparency. Such a technique is also known as Distributed Ledger Technology (DLT) as demonstrated in Figure 1.

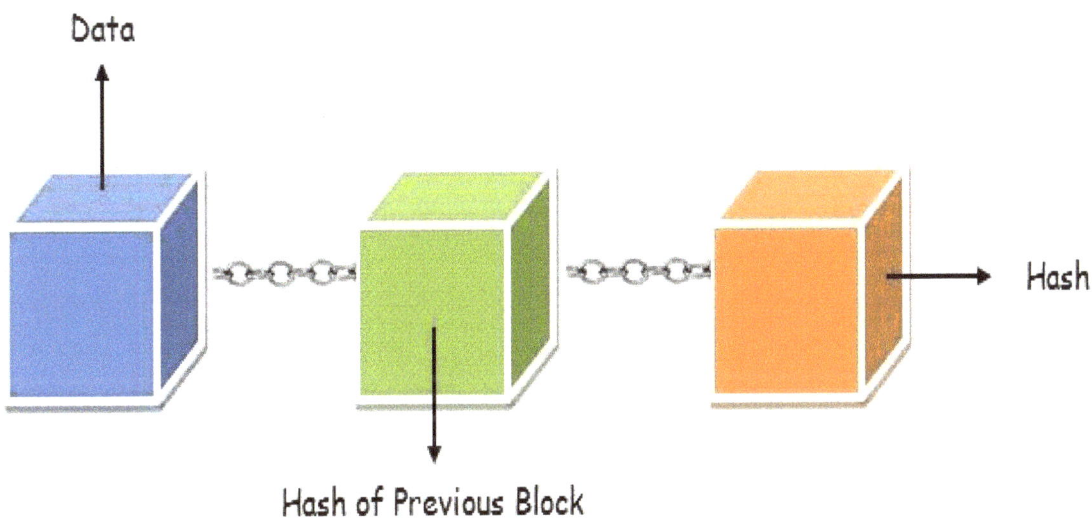

Figure 1. Blockchain based Traceable Certificates in a Distributed Ledger.

1.1. Three Pillars of Blockchain Technology

1.1.1. Decentralization

The need of the decentralized system can only be understood when we are aware about the vulnerabilities of a centralized system that is used in a traditional fund transfer system. Banks and client-server model are examples of the centralized system in which bank as a central authority controls the entire transaction process [6].

To address such limitations, the idea of a decentralized system is introduced in which data has been utilized to store, record and synchronize transactions at different nodes. In decentralization, every node can make transaction associated with the data. Blockchain Technology has been established with the aid of distributed networks, digital signature and encryption/decryption techniques from the security domain. A decentralized system uses peer-to-peer (P2P) networks in which every node can own the copy of the complete data in the chain of blocks [7].

1.1.2. Transparency

Generally, Blockchain transactions are not encrypted. Current block stores the hash of the previous block. The encryption technique is used in Blockchain which ultimately secures the data. Thus, this characteristic enables Blockchain technology to maintain transparency and privacy in the entire network nodes of peer connection. The identity of an individual node is kept hidden through the use of complex cryptographic unique alphanumeric characters and usually symbolized only by its public identifier/address [8].

1.1.3. Immutability

This term is used to depict something that has entered into the chain of blocks and can never be modified or altered in anyway. Even though the data can be added to the chain, but already existing blocks of data cannot be altered. Due to cryptographic hash function, such property is exhibited by Blockchain Technology. Moreover, hashing is a methodology or technique in which the input data length is a variable quantity whereas the output length is fixed [9].

2. Role of Blockchain in Overall Governance

Blockchain can not only be utilized in elections but can also be used to improve overall governance by incorporating it in property registry systems [10], public sector banking [11], healthcare [12] and building smart cities [13]. All the sectors which are prone to cyber-crimes can be made secure with the help of Blockchain. Thus, the novelty of the proposed research lies in developing theoretical approach for sustainable development of the society on the basis of the Blockchain-based Traceable Certificates. The major benefits of Blockchain adoption specifically into an e-Election can be listed below [14]:

- The first benefit that Blockchain can bring about is transparency. Decentralized ledger of Blockchain records result—in accuracy and safety thus ensuring trust at every stage of the voting process;
- Immutable public ledger enables the tracking and counting of votes while being visible to everyone. This feature of Blockchain provides legitimacy of the voting;
- Blockchain and its distributed ledger provides an unhackable system as there is no involvement of fallible or corruptible central body;
- Blockchain allows for anonymity during voting by providing private keys to the voters. These applications of such private keys keep the votes polled by the voters anonymous;
- Processing time is reduced in Blockchain because results can be gathered and processed quickly soon after the completion of the voting phase.
- Blockchain as an Ultimate Solution for Securing Elections

This idea of Blockchain in conducting secure election has previously been implemented by companies such as Agora and Polys but the former was not able to justify its presence and had a controversy with the Sierra Leone government while the latter never achieved scalability in any state election. There have been several challenges which need to be kept in mind while designing system for e-Election conduction [14,15]:

- Difficulty in integration with legacy systems;
- Complexity and lack of Blockchain talented personnel;
- Lack of scalability;
- Lack of interoperability;
- Lack of good governance;
- Lack of user experience and education.

Several noteworthy attempts have been made recently, but none of them have achieved the scalability which is required for Blockchain based voting to be successful as depicted in Table 1 [16,17].

Table 1. Adoption of Blockchain Technology in e-Election and the reported vulnerabilities.

Year	Country	Consequences
August, 2018	Tsukuba, Japan	Tested only for social purposes but not for elections. (State Sponsored)
November, 2018	West Virginia, USA	All vulnerabilities are not covered. (Boston based Voatz named app)
March, 2018	Sierra Leone	Officially not accepted. (Switzerland based company name Agora)
June, 2019	Russia	Moscow City election conducted
June 2020	African Nations	Flexibility and adequate security to the election procedure.

Due to the presence of the above-mentioned properties and after looking at the frauds that occurred in digital electoral systems as discussed in Table 2, it is recommended that the Blockchain technology is used in Electronic Voting Machines to make them more intelligent and secure [18,19].

Table 2. Consequences of electoral frauds in various countries.

Country	Issue
India	Booth capturing and rigging
United States of America	Rigging via hacking
Russia	Ballot stuffing
United Kingdom	Proxy voting
Nigeria, South Africa	Voter impersonation and booth capturing
Germany	EVMs have been prone to hacking
Netherlands	EVMs lack of transparency
Ireland	EVMs lack of transparency and trust

The perception associated with the casting of a vote by authorized voters can be visualized in Figure 2, that justifies the usage of Blockchain technology in such a system [20,21]. When a fraud voter penetrates the system through fake credentials it can immediately be reported at the zonal office by authorities. Such malicious activities can easily be determined through Blockchain technology [22]. The first block called *genesis block* is created with legitimate data associated with transaction identity, source/destination address, voter/candidate details, etc.

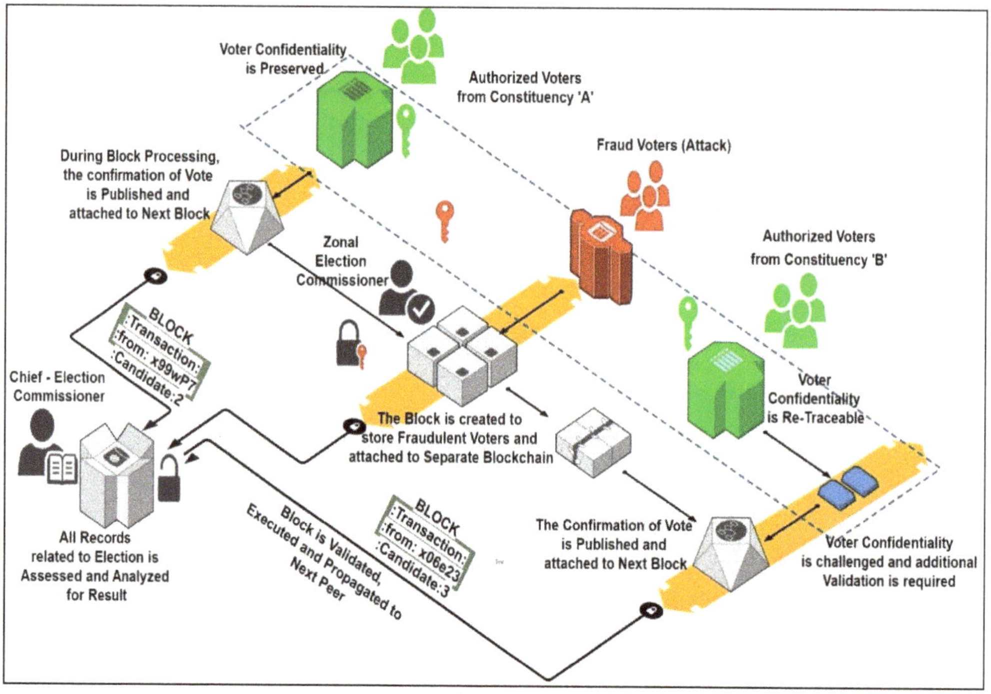

Figure 2. Casting of Electronic Vote through Blockchain based Traceable Certificates.

3. Material & Methods

Before studying the working of Blockchain technology in electronic voting, it is important to know the vulnerabilities in today's election in detail. Several parameters may influence any e-voting processes that are as follows [23,24]:

Hacked voter registration databases: Cyber-attack on voter's registration database can also threaten people's ability to vote. A registration database consists of information such as voter's name, phone number, address, etc. Such information is known as Personally Identifiable Information (PII). Hackers can exploit the stored information by selling it on the dark web and use it to target potential voters with disinformation and to gain benefits.

Hacked voting hardware: Any type of electronic device or software used in the machine is subject to cyber-attacks. Results stored in these devices can be vulnerable to hacking. Hackers need only one single point to breach an entire model of the voting machine. Attackers may also inject malware into machines developed by reputed companies to cause a dangerous effect on the votes of millions of voters [25,26].

Compromised election reporting systems: Reporting systems could be manipulated to announce false results. If automated data streams are used to inform the results to news organizations, then attackers may manipulate data streams and trick news organizations to announce the wrong winner. In this context, highly realistic fake videos can be created announcing bogus winners using Generative Adversarial Networks (GANs). It is a type of Neural Network used to carry out unsupervised learning. GANs can be utilized by attackers to fabricate audio, video, and image content, which seem realistic and plausible.

Post-election audits: The procedure used to count the votes and the equipment are checked for their correctness. If any bug or error is found during the audit, election officials are informed, and they can act as a deterrent against fraud. However, experts believe that voting machines that only record votes electronically are not suitable for ensuring election integrity.

A glimpse of Election Security: There are three stages in an election process: pre-election, election, and post-election. There are several steps that are followed in an election process, as shown in Figure 3. The process also contains several vulnerabilities which need to be identified to prevent future attacks.

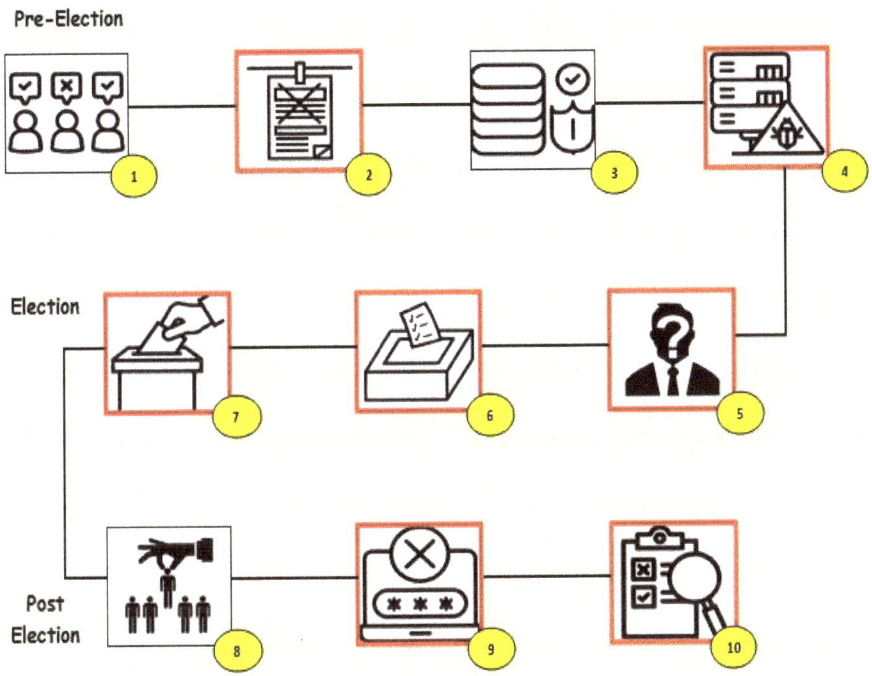

Figure 3. Election process and its vulnerabilities.

Sept.1 Voter forms a political opinion;
Sept.2 Disinformation campaign against the voter;
Sept.3 Voter enters their name in a voter registration database;
Sept.4 Hackers attack the voter registration database and alter the records;
Sept.5 Voter is unable to find their record because of altered voter record;
Sept.6 If a voter casts a vote, their vote could be changed by a hacked voting machine;
Sept.7 Voter's vote could be miscounted due to tampering caused in the machine;
Sept.8 A winner is declared;
Sept.9 Reporting systems are compromised to spread alternative results;
Sept.10 Mismatch in the results causes dispute over election's integrity which prompts a post-election audit that can be vulnerable to inaccuracies.

Steps shown in Figure 3 as highlighted into red color rectangular depicts the vulnerabilities and possible breach into security [27].

3.1. Consensus Protocol for a Common Understanding in Generating Certificates

There is a need for a common point of understanding in a decentralized consensus mechanism. This can be termed as Proof of Work (PoW) in which a certain procedure is used to validate the transaction in a given peer-network and creates a new block for consortium Blockchain [28,29]. Consensus is a kind of agreement that must be taken up by each participating node in consortium. There can be major algorithms for consensus protocol for different features as depicted in Table 3.

Table 3. Various Consensus Algorithms with Major Features.

Feature	Proof of Work (PoW)	Byzantine Fault Tolerance (BFT)	Proof of Capacity (PoC)	Proof of Burn (PoB)	Proof of Stake (PoS)
Consistency	Y	Y	Y	Y	N
Scalability	Y	Y	Y	Y	Y
Partition Tolerance	N	Y	N	N	N
Efficient	N	Y	N	Y	Y

3.2. Algorithms for Voting & Publishing Schemes

During the processing of each block after the casting of the vote by the authorized voter using Blockchain based election, the data associated with the elected candidate and voter itself is stored within the block. Such block is published and attached to the next block that creates a chain in series [30,31]. The smart contract is created by the chief election commissioner (administrator) in respective blocks.

If a voter wishes to REGISTER for casting their vote, then the voter must ensure to SETUP for predefined system software possessed by Chief Election Commissioner ed authority) [32]. The voter should use CREDENTIALS to cast their vote through e-ballot. This can be recorded with a digital signature with mentioned VALIDITY. The job of a legitimate voter is specified in Algorithm 1.

Algorithm 1: Voting Scheme for individual voters

Voting Scheme for individual voters
Initially, SETUP the device as per the requirement of system software
If the Voter is not REGISTERED then
Use CREDENTIALS to REGISTER with verification
Cast a Vote with DIGITAL SIGNATURE
VALIDITY of the e-ballot for particular session END

After casting a vote by the authorized voter, it is the duty of the election commissioner (administrator) to PUBLISH the vote [33]. This should be verified first by VALIDATE and then APPEND it to the next block in the series. This process is depicted in Algorithm 2.

Algorithm 2: Publish Scheme for Vote by Election Commissioner

Publish Scheme for Vote by Election Commissioner
Firstly, CHECK the VALIDITY for the e-ballot
If VALIDITY is FALSE then
e-ballot can be CANCELLED
Otherwise,
e-ballot can be PUBLISHED and APPENDED to the next block

3.3. Blockchain as the Solution to Vulnerabilities

Let us understand how Blockchain affects the voting process as shown in Figure 4 [34,35].

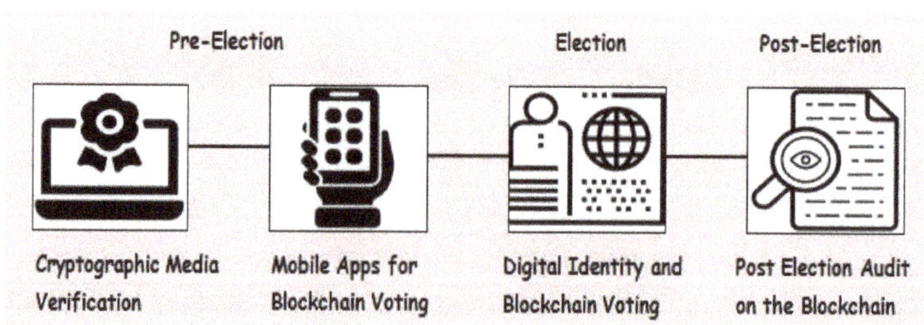

Figure 4. Securing Election Process through Blockchain based Traceable Certificate.

- Cryptographic Media Verification: Cryptographic techniques would help to determine the trusted and accountable sources of information. Voters would only consume the information that is stamped with a unique cryptographic identifier. In this work, the practice of "Cryptographic media verification" is based on previous existing unique cryptographic identifier created by authorized persons (in the government).
- Mobile Apps for Blockchain Voting: Voting through mobile apps would increase voter's participation in an election process and adding Blockchain to the application would help in securing mobile internet voting.
- Digital Identity and Blockchain Voting: Biometric identity such as iris and face data has been used to match a voter's identity with his/her identity stored in the voter's registration database at the time of his or her registration. This technique has been adopted to verify the identity of the person.
- Post-Election Audit on the Blockchain: Each voter would be allowed to examine each ballot to confirm the accuracy of the counted votes without revealing his or her identity.

4. Use of Blockchain in Electronic Voting for Certificates

Indian electoral arrangements currently utilize the EVM (Electronic Voting Machine), wherein the person casting his vote presses a button corresponding to the candidate they wish to vote. However, there have been recent modifications after the emergence of VVPAT (Voter Verified Paper Audit Trail) through which the voter can also verify whether his vote has been received by the candidate to whom he has casted his vote to [36]. The addition of VVPAT to EVM has simplified the process but added some serious issues regarding security. To remove these bottlenecks, Blockchain would prove to be an effective solution. Once Blockchain is induced in this electoral process, the threats of booth capturing would no longer exist, and the results would be full of trust [37].

To reform the electoral process in the biggest democracy is not easy, but in the long run it would be beneficial. To begin with, the Chief Election Commission of India should devise a Blockchain-based electronic voting system. All eligible voters must be allowed to vote only after their biometric verification is successful. Once verified, voters must select the candidate to whom they want to vote for, and this vote would be converted to a block [38]. This block will then be verified and will contain all the information necessary such as the candidate who received the vote, identity of the voter (into hidden format such as ****), timestamp, etc. This would then become an indispensable part of the Blockchain. Similarly, all the voters would then follow the same process and create such blocks. The duty of the Chief Election Commission would then be to verify the identity and display it for everyone to see. Since blocks are connected via the hash of the previous block, changing any one block would lead to tampering of the complete information which is not possible.

The process of casting vote and counting the votes of a particular candidate into peer of network, there must be a set of specific functions (RANGE) as mentioned in Algorithm 3. These functions can rely on a particular smart contract generated between e-voter and the corresponding candidate as discusses in Algorithm 3.

Algorithm 3: Smart Contract for e-Voters and Candidate Function

```
Add Candidate into the Peer Network
ADD Candidate as per the requirement of system software
IF CONFIRM e-voter COUNT does not exceed the RANGE
CREATE or APPEND the COUNT
END IF
CHECK the VALIDITY for the e-Voter
If VALIDITY is FALSE then
e-voter can be CANCELLED
Otherwise,
CAST the vote and APPENDED it to the next block
Increment the vote COUNT
```

Case Study: Indian Electoral System

India has a vibrant electoral democracy governed by the Constitution of India through which fundamental rights and the country's citizen duties can be configured. Such elections are conducted by distinct roles from the election commission of India [38]. As such conducting elections in India is a tedious and cumbersome process because the country holds the position of the world's most populated democracy. Indian states have been subjected to allegations from various political and non-political organizations regarding malfunctioning of the currently used system for elections i.e., VVPAT (Voter Verified Paper Audit Trail) and EVM (Electronic Voting Machine) [39,40].

These systems have been upgraded and made better than the ballot paper system to reduce paper wastage and time; however, it has also brought some severe issues with it such as being prone to electronic faults, hacking, etc [41,42]. Moreover, transportation of these machines from a central control unit to polling stations has led to wear and tear. Thus, to avoid all these added issues security personnel trained Election Commission officials, etc. are appointed to take care of the machine. However, with the emergence of Blockchain technology, expenditure on such avoidable factors would decline. This will improve the overall governance and the electoral process of the country. The Indian government spent about 3426 crore INR for conducting elections in 2014 [43] which witnessed a 131% rise in the costs as compared to the 2009 elections.

For any voting system, there can be a number of parameters that need to be considered while designing an automated, secure and trusted e-voting system [44]. Firstly, it should majorly focus on events such as register (create), poll, validity, verify and publish. This can be well-conceptualized from three-point of views, i.e., voter's view, candidate's view and chief election commissioner's view as depicted in Figure 5 [45].

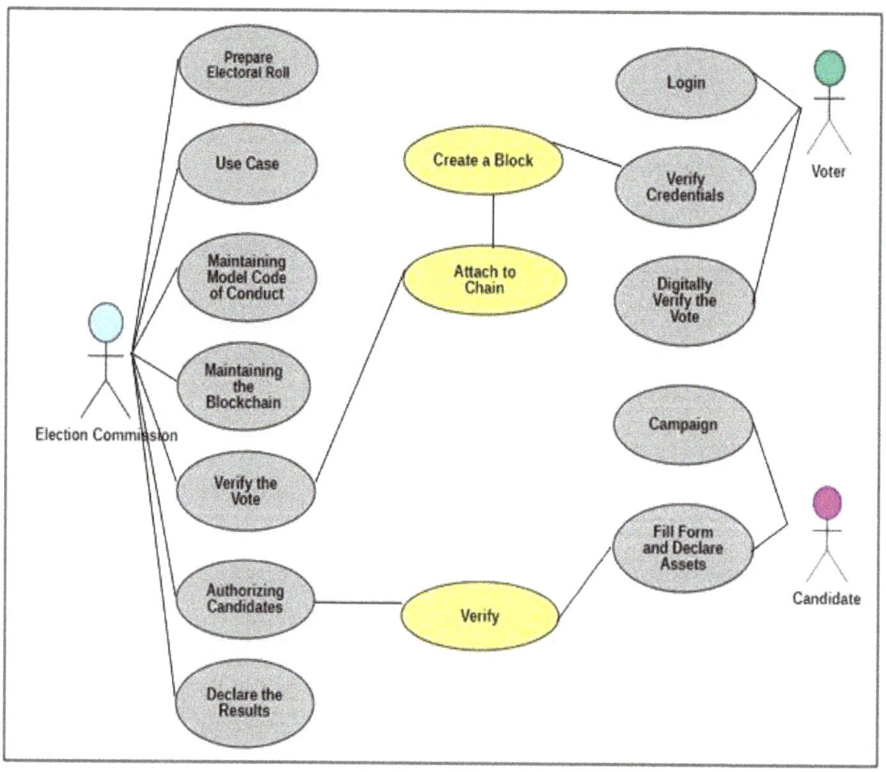

Figure 5. Use case for e-Electoral Voting System.

5. Results & Discussions

Decentralization with security and privacy-preserving features can be of primordial importance for its application in activities of mass participation as a general election [46,47]. The statistical analysis of different parameters of any public Blockchain should be considered in order to facilitate the process more efficiently. One of the key challenges which is ubiquitous to such public Blockchain is the cost of deployment. However, in this case the principal aim is to achieve optimized security and reliability. In Ethereum Blockchain, all the programmable transactions require some charges for ensuring safety in the networks and to overcome computational challenges. All operations such as computations, smart contract deployment and storage on the EVM require fees to complete the tasks. In our case with some initial fluctuations, we have observed throughout consistency in the chain length and the transaction energy dissipation. However, the charges are expected to increase with the deployment of more complex smart contracts, which in turn would result in making the entire process comparatively expensive, as shown in Figure 6.

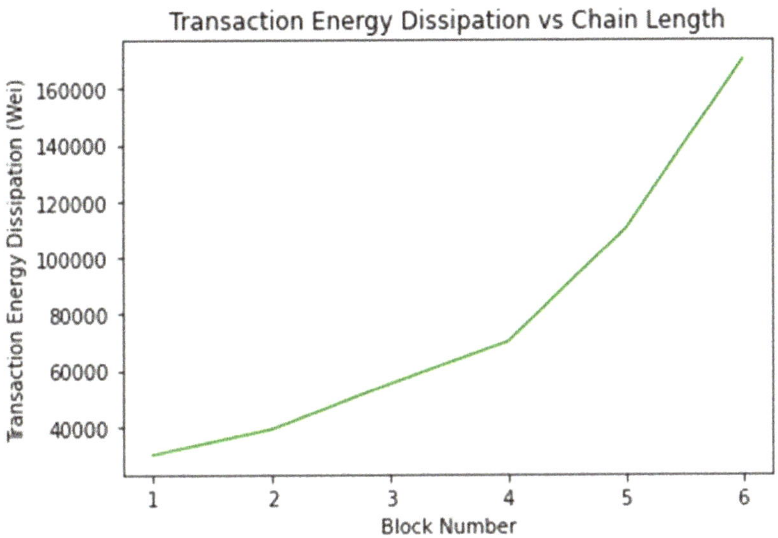

Figure 6. Comparison of transaction energy dissipation and block number.

Block time is the length of time it takes to create a new block in a Blockchain. In an election process block time could be one of the decisive factors for the successful implementation and adoption of such a system. We have observed that the block time is expected to increase exponentially with the chain length by measuring it at an interval of one block, as shown in Figure 7. One of the main factors which influences the block time is the difficulty level of the network. In Ethereum homestead released Blockchain the level of difficulty is calculated using the following procedure: where//denotes integer division and 2** denotes the two to the power operation. The int function returns the largest integer less than or equal to a given number:

$$block\_time = current\_block\_timestamp - parent\_block\_timestamp \tag{1}$$

$$\begin{aligned} current\_block\_difficulty = {} & parent\_block\_difficulty + (parent\_block\_difficulty/2048) \\ & *\max(1-(block\_time/10), -99) \\ & +\mathrm{int}(2**((current\_block\_number/100000)-2)) \end{aligned} \tag{2}$$

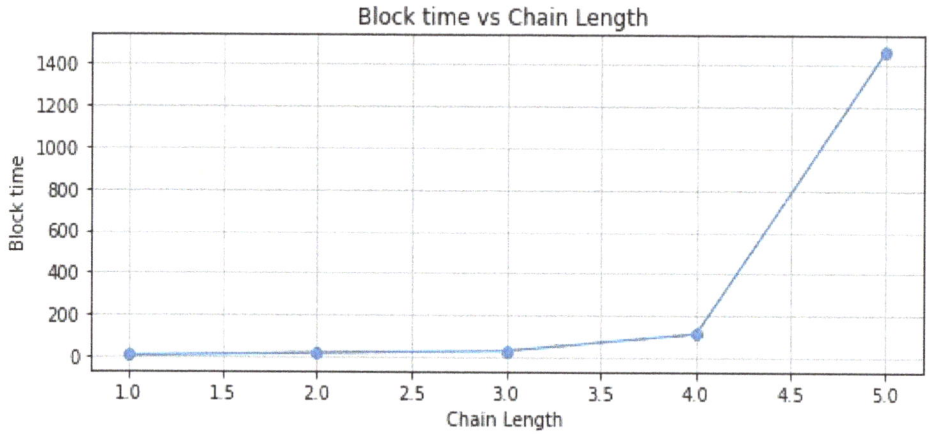

Figure 7. Comparison of block time and chain length.

The proposed mechanism can be deployed to any other Blockchain with lower gas fees to make the process more cost effective, provided it is open-source, reliable and meets the protocol and security standards for the execution of a general election. We have also made an attempt to estimate the variation of throughput (transactions per second, tps) and average latency of the Blockchain with the send rate (tps). Transaction throughput may be defined as the measure of how fast a Blockchain can process a particular transaction. Blockchain network latency is the time between submitting a transaction to a network and the first confirmation of acceptance by the network. An analysis of such parameters can be a decisive factor, particularly when the chain length is very large. In our case we have found a strong correlation in the variation of throughput and average latency with the send rate (tps) of the chain. We evaluated the performance of the system over different transaction sending rates (10–130 tps). Although the average latency showed a steady increase with the increase in the transaction send rate, the throughput increased till the transaction send rate increased to 100 tps and then the growth rate slowed down, as shown in Figure 8.

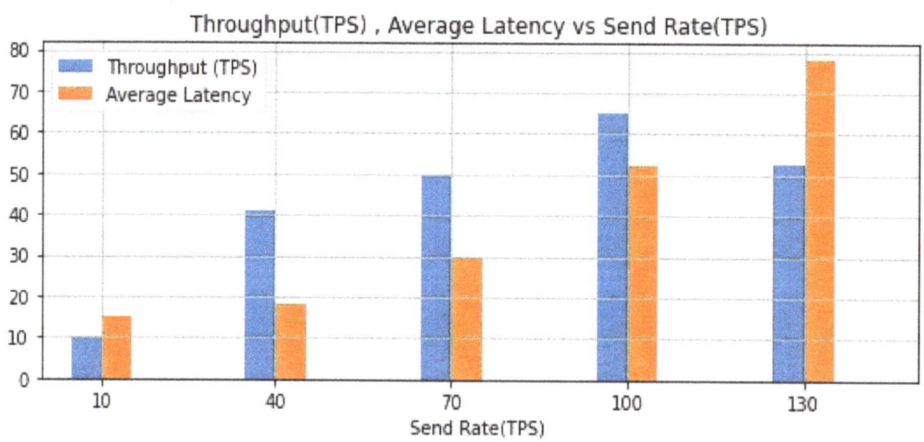

Figure 8. Variation of throughput and average latency with the send rate.

So, the analysis of the system over the aforementioned parameters reveals that adoption of decentralization for an event of mass participation such as an election process is a viable option. As from the point of feasibility of cost, it is clear that although the process has dependency on gas fees, the cost is economical and governments and organizations would find it affordable.

Overall, the system performed as per our expectations. The accuracy over varying transaction send rates (tps) over the network has been found to be considerably better in comparison to the existing centralized voting system, as shown in Figure 9. The analysis shows strong correlation with the desired outcomes in terms of cost and security, and we conclude that the adoption of Blockchain based traceable certificates in democratic elections would ensure transparency, confidentiality and security of the process.

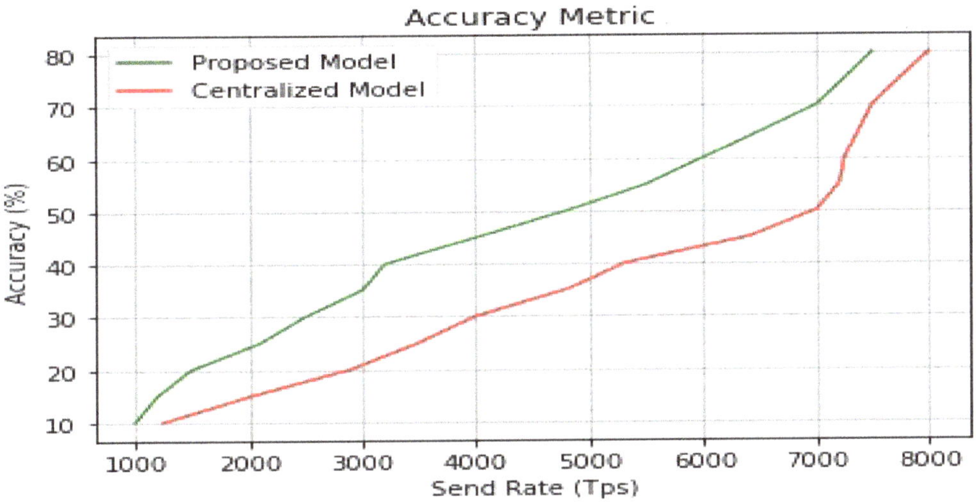

Figure 9. Measure of accuracy with the send rate.

Analyzing the Feasibility of Proposed Mechanism for Achieving ESG-Goals in the Context of a Democratic Society

The proposed mechanism is dependent on factors such as the block processing time (for syncing with other nodes) and transactions processing time of the network used. The following chart represents the gas used for the network transactions in the proposed system. A and V represents the operations dependent on the administrator and the voters, respectively.

Table 4 discusses the parameters that affects the public blockchain networks over cost of development of the system. The total cost of implementing the system would be attained by adding the costs for deploying, maintaining and monitoring the system across public or enterprise blockchains.

Table 4. Cost comparison for different blockchain networks over consensus algorithms.

Networks	Affecting Parameters for Cost Issues	Consensus Algorithms
Ethereum	high gas fee	PoW or PoS
Hyperledger Fabric	data storage in private database, reliance on authorized organizations	CFT or BFT
Stellar	small circulation, risks of volatility	BFT
Quorum	low scalability	Practical BFT
Hedera Hashgraph	not open-source, partially decentralized	Asynchronous BFT

However, the system can be made more efficient through the use of more low-cost networks. The possibility of the development of more energy efficient and scalable private networks and protocols in future would further enhance the feasibility of the usage of the system. Further, a second layer can be used on top of a main network with high gas cost networks, for faster response and low gas cost. Transaction verification mechanisms such as the Proof-of-Work consensus protocol require high processing power and hence are energy consuming which might negatively impact the climate change mitigation efforts since a considerable proportion of electricity is obtained from combustible fossil fuels worldwide [48–50].

6. Conclusions

This article presents the need for a secured voting system based on the Blockchain technology. Such a technology has a bright future and would capture the market in the coming years through its security features such as immutability, transparency, distributed nature and end-to-end connection through smart contracts. For events such as elections, voter's confidentiality and transparency are the major characteristics in a democratic country. To conduct such an election through online or digital means, Blockchain technology plays a vital and prominent role in securing this event. Our observations reveal that the implementation of the Blockchain technology in elections would not only be feasible but also will be very effective in terms of both cost and security. The government should ensure the choice of consensus algorithms, parameters such as block size, difficulty of the chain etc. based on the number of voters and available time. The smart contract is a legal event or action which would get automatically executed whenever it is intended to be included by its developers. Such contract binds the integrity of the voter with the created block and would then append it to the next processing block to form the sequence or chain which would be immutable. This guarantees confidentiality and transparency for voter's rights. Strong network connectivity and reliable hardware infrastructure and software services for mining, security, processing power and memory would be required to maintain the constant throughput in the Blockchain during the entire electoral process. There is a strong possibility of an exponential rise in block time with the increase in the difficulty of the chain if power consuming consensus algorithms such as Proof-of-work are used. In this regard, it should be noted that the consensus algorithm goes towards achieving enhanced security, transparency and scalability. In 'Proof-of-work', the primary intention is to mine the coin whereas in 'Proof-of-stake', the intention is to validate the transaction. For more energy efficient mechanisms, such as the 'Proof-of-stake' can be also used, however, in case of 'Proof-of-stake' mechanism, inconsistencies in the governance issues remain such as excessive influence of validators with maximum holdings on transaction verification, possibilities of induced centralization in the process through double spending etc. Additionally, certain security challenges such as advanced spear phishing attacks on the voters by cybercriminals, threat of natural disasters which might bring in severe interruptions in the process, hardware vulnerabilities etc., continue to exist. We conclude that the proposed Blockchain based e-voting system would, however, be effective in achieving integrity and security in any democratic election around the world.

Our future work would comprise of a more comparative evaluation of the system over various private networks and thorough an analysis of performance through its implementation using different consensus algorithms. Based on our findings from this paper, we also aim to investigate the possibility of a dedicated Blockchain which would meet the criteria for the low consumption of energy and security standards to become more relevant for its implementation in a real national election.

Author Contributions: Conceptualization, R.S.B., A.P.D. and A.B.; methodology, R.S.B., A.B. and M.Z.; software, R.S.B. and A.P.D.; validation, R.S.B. and A.B.; formal analysis, R.S.B., A.B. and M.Z.; investigation, R.S.B., A.P.D. and A.B.; resources, R.S.B., A.B. and M.Z.; data curation, R.S.B. and A.P.D.; writing—original draft, R.S.B., A.P.D. and A.B.; preparation, R.S.B., A.B. and M.Z.; writing—review and editing, R.S.B., A.B. and M.Z.; visualization, A.B. and M.Z.; supervision, R.S.B., A.B. and M.Z.; project administration, R.S.B., A.B. and M.Z.; funding acquisition, A.B. and M.Z. All authors have read and agreed to the published version of the manuscript.

Funding: This research was funded by Cybersecurity Center, Prince Mohammad Bin Fahd University, Khobar, Saudi Arabia with grant number PCC-Grant-202121.

Data Availability Statement: The data used to support the findings of this study have been retrieved from: US Election 2020Race to Presidential Election 2020 by County (URL: https://www.kaggle.com/datasets/unanimad/us-election-2020) (accessed on 27 August 2022) and Indian Election Dataset: State and National Level Election Data from 1977–2015. (https://www.kaggle.com/datasets/awadhil23/indian-election-dataset) (accessed on 27 August 2022).

Acknowledgments: The work conducted in this research has received the grant from Cybersecurity Center, Prince Mohammad Bin Fahd University, Khobar, Saudi Arabia (https://pmu.edu.sa/cybersecurity/) (accessed on 27 August 2022) under Project Number: PCC-Grant-202121.

Conflicts of Interest: The authors declare no conflict of interest.

References

1. Zaghloul, E.; Li, T.; Ren, J. *d*-BAME: Distributed Blockchain-Based Anonymous Mobile Electronic Voting. *IEEE Internet Things J.* **2021**, *8*, 16585–16597. [CrossRef]
2. Nakamoto, S. Bitcoin: A Peer-To-Peer Electronic Cash System. 2008. Available online: https://bitcoin.org/bitcoin.pdf (accessed on 5 June 2022).
3. Park, S.; Specter, M.; Narula, N.; Rivest, R.L. Going from bad to worse: From Internet voting to blockchain voting. *J. Cybersecur.* **2021**, *7*, tyaa025. [CrossRef]
4. Hu, W.; Li, H. A blockchain-based secure transaction model for distributed energy in Industrial Internet of Things. *Alex. Eng. J.* **2020**, *60*, 491–500. [CrossRef]
5. Dinh, T.N.; Thai, M.T. AI and Blockchain: A Disruptive Integration. *IEEE Comput.* **2018**, *51*, 48–53. [CrossRef]
6. Eyal, I. Blockchain Technology: Transforming Libertarian Cryptocurrency Dreams to Finance and Banking Realities. *IEEE Comput.* **2017**, *50*, 38–49. [CrossRef]
7. Febriyanto, E.; Triyono; Rahayu, N.; Pangaribuan, K.; Sunarya, P.A. Using Blockchain Data Security Management for E-Voting Systems. In Proceedings of the 2020 8th International Conference on Cyber and IT Service Management (CITSM), Pangkal, Indonesia, 23–24 October 2020; pp. 1–4.
8. Shabnam, S.; Sayyad, F. Voting Using Blockchain Technology. In *Intelligent Computing and Networking*; Springer: Singapore, 2021; pp. 285–291.
9. Mustafa, M.K.; Waheed, S. An E-Voting Framework with Enterprise Blockchain. In *Advances in Distributed Computing and Machine Learning*; Springer: Singapore, 2021; pp. 135–145.
10. Xie, J.; Yu, F.R.; Huang, T.; Xie, R.; Liu, J.; Liu, Y. A Survey on the Scalability of Blockchain Systems. *IEEE Netw.* **2019**, *33*, 166–173. [CrossRef]
11. Khalfan, M.; Azizi, N.; Haass, O.; Maqsood, T.; Ahmed, I. Blockchain Technology: Potential Applications for Public Sector E-Procurement and Project Management. *Sustainability* **2022**, *14*, 5791. [CrossRef]
12. Singh, L.; Kumar, A.; Singh, Y. Internet of Healthcare Things (IoHT) and Blockchain: An Efficient Integration for Smart Health Care Systems. In *Healthcare and Knowledge Management for Society 5.0*; CRC Press: Boca Raton, FL, USA, 2021; pp. 135–149. [CrossRef]
13. Rejeb, A.; Rejeb, K.; Simske, S.J.; Keogh, J.G. Blockchain technology in the smart city: A bibliometric review. *Qual. Quant.* **2022**, *56*, 2875–2906. [CrossRef]
14. Maesa, D.D.F.; Mori, P. Blockchain 3.0 applications survey. *J. Parallel Distrib. Comput.* **2020**, *138*, 99–114. [CrossRef]
15. Umeh, J. Beyond Bitcoin and the Blockchain. *ITNOW* **2018**, *60*, 48–49. [CrossRef]
16. Abayomi-Zannu, T.P.; Odun-Ayo, I.; Tatama, B.F.; Misra, S. Implementing a Mobile Voting System Utilizing Blockchain Technology and Two-Factor Authentication in Nigeria. In *Proceedings of the First International Conference on Computing, Communications, and Cyber-Security (IC4S 2019)*; Springer: Singapore, 2020; pp. 857–872. [CrossRef]
17. Alvi, S.T.; Uddin, M.N.; Islam, L.; Ahamed, S. From Conventional Voting to Blockchain Voting: Categorization of Different Voting Mechanisms. In Proceedings of the 2020 2nd International Conference on Sustainable Technologies for Industry 4.0 (STI), Dhaka, Bangladesh, 19–20 December 2020; pp. 1–6. [CrossRef]
18. Rubtcova, M.; Pavenkov, O. Implementation of Blockchain technology in electronic election in Sierra Leone. In Proceedings of the 2018 Conference: "Re-Thinking Regions in Global International Relations", Davao City, Philippines, 23–24 March 2018; Volume 23.
19. Pawlak, M.; Poniszewska-Marańda, A. Trends in blockchain-based electronic voting systems. *Inf. Process. Manag.* **2021**, *58*, 102595. [CrossRef]
20. Khan, K.M.; Arshad, J.; Khan, M.M. Empirical analysis of transaction malleability within blockchain-based e-Voting. *Comput. Secur.* **2020**, *100*, 102081. [CrossRef]
21. Berdik, D.; Otoum, S.; Schmidt, N.; Porter, D.; Jararweh, Y. A Survey on Blockchain for Information Systems Management and Security. *Inf. Process. Manag.* **2020**, *58*, 102397. [CrossRef]

22. Hassan, N.U.; Yuen, C.; Niyato, D. Blockchain Technologies for Smart Energy Systems: Fundamentals, Challenges, and Solutions. *IEEE Ind. Electron. Mag.* **2019**, *13*, 106–118. [CrossRef]
23. Taş, R.; Tanrıöver, Ö.Ö. A Systematic Review of Challenges and Opportunities of Blockchain for E-Voting. *Symmetry* **2020**, *12*, 1328. [CrossRef]
24. Li, C.; Xiao, J.; Dai, X.; Jin, H. AMVchain: Authority management mechanism on blockchain-based voting systems. *Peer—Peer Netw. Appl.* **2021**, *14*, 2801–2812. [CrossRef]
25. Baudier, P.; Kondrateva, G.; Ammi, C.; Seulliet, E. Peace engineering: The contribution of blockchain systems to the e-voting process. *Technol. Forecast. Soc. Chang.* **2020**, *162*, 120397. [CrossRef]
26. Yang, X.; Yi, X.; Nepal, S.; Kelarev, A.; Han, F. Blockchain voting: Publicly verifiable online voting protocol without trusted tallying authorities. *Futur. Gener. Comput. Syst.* **2020**, *112*, 859–874. [CrossRef]
27. Sadia, K.; Masuduzzaman, M.; Paul, R.K.; Islam, A. Blockchain-Based Secure E-Voting with the Assistance of Smart Contract. In *IC-BCT. 2019*; Springer: Singapore, 2020; pp. 161–176. [CrossRef]
28. Alam, M.; Yusuf, M.O.; Sani, N.A. Blockchain technology for electoral process in Africa: A short review. *Int. J. Inf. Technol.* **2020**, *12*, 861–867. [CrossRef]
29. Li, K.; Li, H.; Wang, H.; An, H.; Lu, P.; Yi, P.; Zhu, F. PoV: An Efficient Voting-Based Consensus Algorithm for Consortium Blockchains. *Front. Blockchain* **2020**, *3*, 11. [CrossRef]
30. Kshetri, N.; Voas, J. Blockchain-Enabled E-Voting. *IEEE Softw.* **2018**, *35*, 95–99. [CrossRef]
31. Benabdallah, A.; Audras, A.; Coudert, L.; El Madhoun, N.; Badra, M. Analysis of Blockchain Solutions for E-Voting: A Systematic Literature Review. *IEEE Access.* **2022**, *10*, 70746–70759. [CrossRef]
32. Specter, M.A.; Koppel, J.; Weitzner, D. The ballot is busted before the Blockchain: A security analysis of voatz, the first internet voting application used in us federal elections. In Proceedings of the 29th {USENIX} Security Symposium ({USENIX} Security 20), Boston, MA, USA, 12–14 August 2020; pp. 1535–1553.
33. Dimitriou, T. Efficient, Coercion-free and Universally Verifiable Blockchain-based Voting. *Comput. Netw.* **2020**, *174*, 107234. [CrossRef]
34. Pawade, D.; Sakhapara, A.; Badgujar, A.; Adepu, D.; Andrade, M. Secure Online Voting System Using Biometric and Blockchain. In *Data Management, Analytics and Innovation*; Springer: Berlin/Heidelberg, Germany, 2020; pp. 93–110. [CrossRef]
35. Roh, C.H.; Lee, I.Y. A study on electronic voting system using private Blockchain. *J. Inf. Process. Syst.* **2020**, *16*, 421–434.
36. Wolchok, S.; Wustrow, E.; Halderman, J.A.; Prasad, H.K.; Kankipati, A.; Sakhamuri, S.K.; Yagati, V.; Gonggrijp, R. Security analysis of India's electronic voting machines. In Proceedings of the 17th ACM conference on Computer and communications security, Chicago, IL, USA, 4–8 October 2010; pp. 1–14.
37. Krishnamurthy, R.; Rathee, G.; Jaglan, N. An enhanced security mechanism through blockchain for E-polling/counting process using IoT devices. *Wirel. Netw.* **2019**, *26*, 2391–2402. [CrossRef]
38. Soni, Y.; Maglaras, L.; Ferrag, M.A. Blockchain Based Voting Systems. In *European Conference on Cyber Warfare and Security*; Academic Conferences International Limited: Athens, Greece, 2020; pp. 241–248.
39. Goyal, M.; Kumar, A. Sustainable E-Infrastructure for Blockchain-Based Voting System. In *Digital Cities Roadmap: IoT-Based Architecture and Sustainable Buildings*; Scrivener Publishing: Beverly, MA, USA, 2021; p. 221. [CrossRef]
40. Roopak, T.M.; Sumathi, R. Electronic Voting based on Virtual ID of Aadhar using Blockchain Technology. In Proceedings of the 2020 2nd International Conference on Innovative Mechanisms for Industry Applications (ICIMIA), Bangalore, India, 5–7 March 2020; pp. 71–75.
41. Schaub, M.; Phares, H.B. Cryptocurrency value changes in response to national elections: Do they behave like money or commodities. *Appl. Econ. Lett.* **2019**, *27*, 1135–1140. [CrossRef]
42. Rathor, S.; Agrawal, A.; Yadav, D.P. The Efficient use of Blockchain for Reducing Frauds in Parental Property Distribution. In Proceedings of the 2020 International Conference on Power Electronics & IoT Applications in Renewable Energy and its Control (PARC), Mathura, India, 28–29 February 2020; pp. 46–49.
43. Peck, M.E. Blockchain world—Do you need a blockchain? This chart will tell you if the technology can solve your problem. *IEEE Spectr.* **2017**, *54*, 38–60. [CrossRef]
44. Ma, X.; Zhou, J.; Yang, X.; Liu, G. A Blockchain Voting System Based on the Feedback Mechanism and Wilson Score. *Information* **2020**, *11*, 552. [CrossRef]
45. Rathor, S.; Agrawal, A. A robust verification system for recruitment process by using blockchain technology. *Int. J. Blockchains Cryptocurrencies* **2020**, *1*, 389–399. [CrossRef]
46. Huang, J.; He, D.; Obaidat, M.S.; Vijayakumar, P.; Luo, M.; Choo, K.K.R. The Application of the Blockchain Technology in Voting Systems: A Review. *ACM Comput. Surv. CSUR* **2021**, *54*, 1–28. [CrossRef].
47. Khan, K.M.; Arshad, J.; Khan, M.M. Investigating performance constraints for blockchain based secure e-voting system. *Futur. Gener. Comput. Syst.* **2019**, *105*, 13–26. [CrossRef]
48. Hussain, A.A.; Emon, M.; Tanna, T.A.; Emon, R.I.; Onik, M.; Hassan, M. A Systematic Literature Review of Blockchain Technology Adoption in Bangladesh. *Ann. Emerg. Technol. Comput. AETiC* **2022**, *6*, 1–30. [CrossRef]

49. Seifelnasr, M.; Galal, H.S.; Youssef, A.M. Scalable open-vote network on ethereum. In *International Conference on Financial Cryptography and Data Security*; Springer: Berlin/Heidelberg, Germany, 2020; pp. 436–450.
50. Jain, P.K.; Pamula, R.; Yekun, E.A. A multi-label ensemble predicting model to service recommendation from social media contents. *J. Supercomput.* **2021**, *78*, 5203–5220. [CrossRef]

electronics

Review

Blockchain-Based E-Voting Systems: A Technology Review

Mohammad Hajian Berenjestanaki [1,*], Hamid R. Barzegar [1], Nabil El Ioini [2] and Claus Pahl [1,*]

[1] Faculty of Engineering, Free University of Bozen-Bolzano, 39100 Bolzano, Italy; hamidreza.barzegar@unibz.it
[2] School of Computer Science, University of Nottingham Malaysia, Semenyih 43500, Selangor, Malaysia; elioini.nabil@nottingham.edu.my
* Correspondence: mhajian@unibz.it (M.H.B.); claus.pahl@unibz.it (C.P.); Tel.: +39-0471-016000 (M.H.B.); +39-0471-016177 (C.P.)

Abstract: The employment of blockchain technology in electronic voting (e-voting) systems is attracting significant attention due to its ability to enhance transparency, security, and integrity in digital voting. This study presents an extensive review of the existing research on e-voting systems that rely on blockchain technology. The study investigates a range of key research concerns, including the benefits, challenges, and impacts of such systems, together with technologies and implementations, and an identification of future directions of research in this domain. We use a hybrid review approach, applying systematic literature review principles to select and categorize scientific papers and reviewing the technology used in these in terms of the above key concerns. In the 252 selected papers, aspects such as security, transparency, and decentralization are frequently emphasized as the main benefits. In contrast, although aspects like privacy, verifiability, efficiency, trustworthiness, and auditability receive significant attention, they are not the primary focus. We observed a relative lack of emphasis on aspects such as accessibility, compatibility, availability, and usability in the reviewed literature. These aspects, although acknowledged, are not as thoroughly discussed as the aforementioned key benefits in the proposed solutions for blockchain-based e-voting systems, whereas the considered studies have proposed well-structured solutions for blockchain-based e-voting systems focusing on how blockchain can strengthen security, transparency, and privacy, in particular, the crucial aspect of scalability needs attention.

Keywords: blockchain; digital transformation; e-voting systems; security; scalability; systematic review

Citation: Hajian Berenjestanaki, M.; Barzegar, H.R.; El Ioini, N.; Pahl, C. Blockchain-Based E-Voting Systems: A Technology Review. *Electronics* **2024**, *13*, 17. https://doi.org/10.3390/electronics13010017

Academic Editor: Hamed Taherdoost

Received: 10 November 2023
Revised: 7 December 2023
Accepted: 13 December 2023
Published: 19 December 2023

1. Introduction

Blockchain technology has been recognized as a potential solution for secure and transparent e-voting systems. By leveraging the decentralization, immutability, and transparency of blockchain technology, e-voting systems can prevent fraud and manipulation, improve voter anonymity, and increase trust in the electoral process. Moreover, blockchain-based e-voting systems can reduce the cost and time associated with traditional voting systems.

Traditional voting mechanisms commonly rely on centralized entities, which can give the opportunity for vulnerabilities such as the tampering of results or electoral fraud. The decentralized and immutable features inherent in blockchain technology offer a promising solution to the vulnerabilities related to traditional and other e-voting approaches. Blockchain technology has the ability to create a tamper-proof and transparent platform for conducting e-voting. Blockchain-based e-voting systems provide secure, verifiable, and auditable voting procedures through the integration of cryptographic techniques and consensus protocols.

The growing interest in blockchain-based e-voting systems indicates the importance of a comprehensive and systematic evaluation of the current knowledge in this domain. One of the aims of this review is to identify the main benefits of e-voting systems based on blockchain technology through an in-depth review of the previous research. These benefits include heightened security, transparency, decentralization, and privacy. Additionally, we intend to identify the challenges and limitations that come with these systems, which include privacy and security concerns, scalability issues, and technical limitations.

Moreover, a comprehensive understanding of the technologies and implementations involved in blockchain-based e-voting platforms is imperative in order to evaluate their feasibility and functionality. Furthermore, this systematic review provides technical insight into common blockchain frameworks, consensus algorithms, and security and privacy-enhancing techniques used in these systems. In addition, we aim to conduct an examination of the impacts of proposed blockchain-based e-voting systems in the literature on various aspects of the voting process, including security, privacy, efficiency, and scalability.

Overall, the purpose of this review is to conduct an extensive review of the current state of the literature related to blockchain-based e-voting systems. We look into the benefits, challenges, technological aspects, impacts, and potential research and development areas in the context of e-voting systems using blockchain technology. We conduct a combined review method, employing the principles of systematic literature review to choose and classify scientific papers. Additionally, we examine the technology implemented in these with respect to the already mentioned key concerns. The evaluation follows the PRISMA guidelines [1], which guarantee a rigorous and transparent methodology for the synthesis of available research data. The PRISMA protocol (Preferred Reporting Items for Systematic Review and Meta-Analysis Protocols) is a reporting guideline designed to aid researchers in the preparation and documentation of systematic review and meta-analysis protocols.

2. Voting System Types and Requirements

We first categorize the types of voting systems before defining relevant requirements for them.

2.1. Voting Systems

Voting systems have been combined with advancements in information technology, making them increasingly efficient and accessible. There are a number of voting system types that can be differentiated from a technical standpoint.

1. Traditional voting: the conventional method where voters either mark paper ballots manually or use mechanical lever machines. The ballots, whether marked remotely or at a polling station, are collected and counted by election officials. Within traditional voting, there are two main categories:
    - Paper-based voting: In this method, voters typically mark their choices on the ballot paper by hand next to the candidate or option they wish to vote for, and then the ballots are counted manually [2]. It can be further categorized into remote and on-site voting. Remote paper-based voting refers to the process of casting a vote by mail or other means of delivery, whereas on-site paper-based voting refers to the process of casting a vote in person at a polling station [3].
    - Mechanical lever machines: They were first used in the 1890s and are operated by the voter indicating their choice by pressing a lever next to the preferred candidate. Once the voter is finished, the voter pulls the large lever again, which causes the counters associated with their choice to be incremented by one and the machine prepared for the next voter [4].
2. E-Voting: A voting method that uses electronic devices to record, cast, or count votes. In general, e-voting systems can be divided into four subcategories, as follows:
    - Punch-card: Developed in the 1960s, utilized modified Hollerith cards where voters used a stylus to punch out chads corresponding to their candidate choices. After voting, the punched card was deposited in a ballot box. These cards were later counted using a card reader [2].
    - Direct Recording Electronic (DRE): An electronic system that presents ballots and records voter selections directly into computer memory. Voters interact

with DREs using push-buttons, touchscreens, or dials. Some DREs feature Voter Verified Paper Audit Trail (VVPAT) printers, allowing voters to confirm their choices on a paper record, which can be used for post-election audits or recounts [5].
- Optical scanning systems: Specialized computer hardware and software are used to read and interpret votes. Voters mark their choices on machine-readable ballots by filling in symbols next to their preferred candidates. Once marked, these ballots can either be scanned directly at the polling place or collected and scanned at a central location [6].
- Ballot-Marking Devices (BMDs): Presents ballots electronically, lets voters make selections, and then produces a human-readable paper ballot without storing the vote electronically. Introduced after the Help America Vote Act of 2002 to aid voters with disabilities, BMDs can either mark pre-existing ballots or print summaries, sometimes with barcodes or QR codes. From 2016 onwards, some areas expanded BMD usage to all voters, becoming more common in 2020 [7].
- I-voting: Internet voting denotes a subset of e-voting methodologies wherein ballots are transmitted and registered via the Internet [8,9]. Terms such as "remote e-voting", "mobile voting", and "online voting" are often used in the literature to describe these systems. All of the terms outlined above are, however, grouped under the broader conceptual framework of i-voting systems, which is itself an instance of an e-voting paradigm. Furthermore, Blockchain-based e-voting systems are a type of i-voting that relies on the internet by using a peer-to-peer computer network that employs blockchain technology to cast and count votes in an election [10–12].

2.2. Voting Systems Requirements

A requirement is a need or constraint on the software or system to be developed. We can distinguish the properties of these systems into functional requirements (FR) and non-functional requirements (NFR). According to [13–20], an e-voting system is required to comply with a number of requirements if considered as an alternative to traditional voting systems.

Based on the above references, we propose here a division of requirements into different categories, namely functional and non-functional non-security requirements on the one hand and security as a functional and non-functional requirement type on the other hand. Our categorization forms a structured base set of properties that we will refer to in the discussion of benefits, challenges, impacts, and future research directions later.

2.2.1. Non-Security Requirements

- Functional Requirements
  - User-Centric Voting Design: The concept that a voting system should be easy for all people to use. This means that it should have a user-friendly interface and show choices without giving any candidate an advantage.
  - Flexibility: It refers to the ability of the system to adapt to a variety of formats, languages, and voting ballots, making it compatible with different platforms and technologies. To provide a flexible and adaptable electronic voting experience, this phrase emphasizes adapting to changes, complying with deadlines, and permitting numerous ballot question types, including open-ended questions.
- Non-Functional Requirements
  - Equality: It assigns priority to equitable and consistent voter access, ensuring that regardless of the process of voting, all voters have equal voting rights and opportunities and receive the same information and opportunities.
  - Accessibility: This term highlights the importance of providing individuals with functional limitations or disabilities with the necessary access to vote, ensuring

voters have undiscriminating access to the voting infrastructure, and enabling entities to have logical and/or physical access to the voting system.
- Openness: For an e-voting system, the functioning of the system (hardware and software) should be transparent to citizens, and the people should be able to understand and verify how the voting system works.
- Auditability: It refers to the necessity of being able to verify that all votes in the final election tally are precisely accounted for, along with having reliable and authentic election records with a (possibly) physical but always permanent audit trail that ensures voter secrecy.
- Cost-effectiveness: It addresses the need for essentially affordable and reusable systems with implementation and maintenance costs that are acceptable and competitive with traditional voting methods.
- Interoperability: In order to ensure smooth integration and compatibility with different components and technologies, it makes sure that voting system data are imported, exported, or reported in an interoperable format using widely accepted, openly available interfaces and communications protocols.

2.2.2. Security Requirements

- Functional Requirements
  - Authentication and eligibility:
    * Voter authenticity: requires voter identification based on the voter registration database and ensures that only eligible voters cast their votes.
    * Uniqueness: the voter can only submit a vote once, and the final result of that vote can never be altered.
    * Eligibility: guarantees that only legitimate voters are able to vote and that their identities are confirmed precisely.
  - Anonymity and secrecy:
    * Anonymity: the voter's identity remains unlinked to their vote, and personal information or identity should remain concealed.
    * Secrecy: ensuring that no one involved in the voting process can link a specific ballot to a particular voter, preserving voter anonymity; in addition, the content of their vote remains confidential.
  - Uncoercible ballot assurance:
    * Uncoercibility: the fundamental principle of an e-voting system is to prevent any external influence, coercion, or vote-selling, ensuring that voters cannot prove or reveal their voting decisions, thereby safeguarding the integrity of the voting process and obstructing attempts at manipulating or pressuring voters for electoral gain.
    * Non-valid voting capability: voters should be able to cast ballots that they know are invalid if they so desire without compromising the integrity of the election in any way.
- Non-functional Requirements
  - Integrity and reliability:
    * Data protection: guarantee that each vote is reliably recorded and remains tamper-proof, while also applying rigorous data protection measures to prevent unauthorized access to or manipulation of voting data.
    * System integrity: ensure resistance against security failures or vulnerabilities, the voting system needs to maintain its functionality by preventing reconfiguration during operation and using multiple levels of controls.
    * Reliability: ensure the system functions robustly without losing any votes, even in the presence of multiple failures, including those related to voting machines and network communication, and prevent malicious code or bugs,

thus providing voters with the utmost confidence in its secure and efficient operation under anticipated physical conditions.

- Detection and monitoring:
  * Testing: The principle that electoral authorities, political parties, and social organizations should have the ability to put the voting systems to the test to ensure they meet the established criteria. This testing process should be thorough and conducted by experts to evaluate and verify that the systems meet the required security standards.
  * Monitoring: record important activities through event logging mechanisms in a format suitable for automated processing while also generating, storing, and reporting error messages in real time as they occur during the voting process.
- Fairness: the importance of maintaining a fair voting environment by avoiding biased or misleading information, ensuring that the voting system does not provide evidence about any voter's intention before the end of the voting phase, and remaining neutral so that the system does not influence the eligible voter's intention during the voting process.
- Verifiability and accuracy: allowing voters and election officials, parties, and independent observers to verify that the votes are accurately recorded and counted, ensuring the system can securely record votes, enabling them to use control mechanisms accurately with direct control of ballot changes and selections, providing voters the ability to verify their intentions in the vote without alterations, and offering sound and independently verifiable evidence that each authentic vote is accurately reflected in the election results.
- Availability: the system's ability to remain consistently available to all eligible voters, protect against denial of service attacks, establish redundant communication paths, ensure continuous availability during the election, have alternative support and election sites ready in case of failures, maintain a minimum Mean Time Between Failures (MTBF), have updated backups readily available for disaster recovery, and protect sensitive information.

Integrating blockchain technology into e-voting can satisfy some of these requirements. However, we will see that multiple challenges remain to be addressed to establish a reliable and trustworthy voting system.

3. Background, Related Work, and Objectives

We introduce blockchain basics before summarizing related work on blockchain for e-voting. From this, we will identify gaps and define objectives for this review.

3.1. Blockchain Technology

A blockchain is a decentralized and distributed ledger made of a sequence of blocks linked to each other. Each block contains a list of transactions, and each transaction is a record of an event or action. The block header, which includes the previous block hash, timestamp, nonce, and Merkle root, identifies each block. The previous block hash links the current block to the previous one. The timestamp verifies the data in the block and assigns a time or date of creation for digital documents. The nonce, a number used only once, is a central part of the proof of work in the block. The Merkle root, a type of data structure frame for different blocks of data, stores all the transactions in a block by producing a digital fingerprint of the entire set of transactions. This structure provides assurance that once data are recorded in a block, they cannot be altered in the future without modifying all subsequently recorded blocks, making blockchain transactions immutable and secure. Figure 1 represents an overview of the blockchain structure with the chain of blocks that encapsulate the transactions and secure them with hashes and other data. These blocks are broadcasted and replicated across a network of peers. This method is characterized by its robust security measures through cryptographic principles, which effectively mitigate the risks of manipulation and fraudulent activities. The decentralized nature of blockchain enables universal accessibility of the distributed database to all participants in the network, which is governed by a consensus algorithm. Therefore, blockchain data are immutable; it additionally traces and validates transactions based on their origins. This technique makes digital transactions transparent, secure, and tamper-proof. Considering these unique characteristics, blockchain is an appropriate mechanism for integration with e-voting systems.

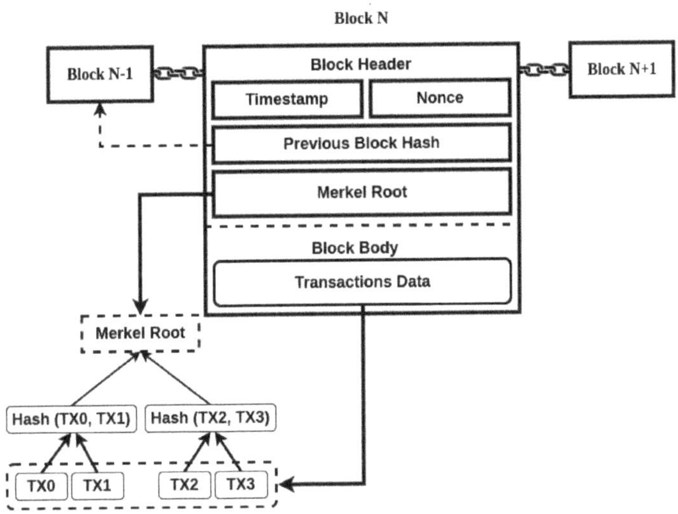

Figure 1. The blockchain structure.

3.2. *Blockchain Applications Across Domains*

Blockchain technology has emerged as a revolutionary trend across various domains, and whereas blockchain technology application in e-voting systems attracts interest in enhancing electoral integrity and transparency, it is equally valuable in other domains, each with distinct requirements and objectives. This section aims to provide a comparison and analysis of blockchain applications in different domains such as healthcare, financial services, supply chain management, cloud computing, education, and IoT (Internet of Things) [21], highlighting their parallels and contrasts with their use in e-voting systems.

- Blockchain in healthcare: In healthcare, blockchain is employed for secure data sharing, patient privacy, and interoperability among different healthcare systems [22]. Its application in healthcare shares some aspects of e-voting, such as the emphasis on data security and privacy. However, whereas blockchain in healthcare deals with continuous data flow and personal health records, in e-voting, it addresses the singular event of casting and recording votes.
- Blockchain in financial services: In financial services, blockchain technology revolutionizes transactions and trust mechanisms. Similar to e-voting, where blockchain brings transparency and verifiability to the voting process, in financial services, it introduces a new concept of trust and efficiency in transactions [23]. The key difference lies in blockchain's role in handling continuous financial transactions as opposed to the discrete event of voting.
- Blockchain in supply chain management: blockchain technology in supply chain management focuses on improving transparency, reducing fraud, and enhancing efficiency [24], whereas both supply chain management and e-voting systems benefit from blockchain's immutability and transparency, supply chain management uniquely utilizes blockchain for continuous tracking of goods and transactions, in contrast to the periodic nature of elections.

- Blockchain in cloud computing: In cloud computing, blockchain enhances security, data provenance, and creates new service models like Blockchain-as-a-Service (BaaS). The integration of blockchain in cloud computing shares similarities with e-voting in terms of improving security and reliability. However, the use cases in cloud computing are more varied and continuous, focusing on service enhancement and data integrity across diverse cloud-based applications [25].
- Blockchain in education: Blockchain technology in education mainly focuses on enhancing data security, credential verification, traceability, and record management. Through its immutable feature, blockchain technology not only ensures the integrity of educational records and certificates, consequently creating trust in academic credentials, additionally, it effectively secures and tracks the progress of academic patents, copyrights, and research innovations, significantly enhancing the management and protection of property within the educational domain [26–28]. Compared to its application in e-voting, where blockchain ensures vote integrity and transparency, in education, it serves to preserve academic achievements and automate administrative processes.
- Blockchain in IoT: Blockchain technology in IoT includes enhancing security, scalability, and trustworthiness in diverse applications like smart cities. The decentralized nature of blockchain in IoT addresses issues similar to those in e-voting, like ensuring security and scalability [29]. However, IoT applications deal with a broader range of data types and greater scalability challenges than electronic voting systems.

3.3. Related Work

Studies exploring potential applications of blockchain technology in the domain of e-voting aim to evaluate its feasibility, security, and efficiency in enhancing the transparency and integrity of the election process.

Taş and Tanrıöver [30] reviewed in 2020 the state of blockchain-based voting research, identifying potential challenges and forecasting future directions. They presented a conceptual description of the desired blockchain-based e-voting application and conducted a review of 63 research papers. The articles that were examined were categorized into five main categories: general, integrity, coin-based, privacy, and consensus. They concluded that, whereas blockchain-based voting systems can prevent data manipulation and integrity issues, the most frequently highlighted issues are scalability, cost-effectiveness, authentication, privacy, and security in blockchain-based e-voting systems.

Jafar et al. [31] presented a conceptual description of a blockchain-based e-voting application in addition to an introduction to the blockchain's fundamental structure and characteristics in relation to e-voting. They mentioned that whereas blockchain systems could help solve some of the issues that currently affect election systems, the authors conclude that the most frequently mentioned issues in blockchain applications are scalability, user identity, transactional privacy, energy efficiency, immatureness, acceptableness, and political leaders' resistance.

In [32], Pawlak et al. indicated the remaining problems like security attacks, coercion, cost efficiency, and privacy that still need to be solved. The paper serves as a valuable resource for understanding the current trends and challenges in blockchain-based electronic voting systems.

Huang et al. [33] in 2021 provided a comprehensive review of blockchain-based voting systems, discussing their advantages, challenges, and technical innovations. They also provide a taxonomy of blockchain and identify key challenges in blockchain-based voting systems such as authentication, anonymity, coercion-freeness, and auditability.

Jafar and Ab Aziz in [34] emphasized the benefits and challenges of blockchain-based e-voting systems, providing useful details on probable future applications of this technology with regard to democratic processes. They demonstrated how blockchain technology offers security, transparency, and a reduced risk of fraud. However, they brought up issues with scalability, transactional privacy, and immaturity for these systems.

Devi and Bansal [35] provided a comprehensive review of the security requirements and potential threats in e-voting systems. They discuss various cryptographic techniques that can be used to secure these systems.

Benabdallah et al. [36] presented a comprehensive analysis of blockchain solutions for e-voting. They discussed the challenges faced by e-voting systems and how blockchain technology can address these issues. They also provide a comparison of several blockchain-based e-voting solutions, identifying their strengths and weaknesses. The paper also addressed the limitations and issues raised by this technology, such as scalability, unpredictable attacks, weakness of the identification system, new issues raised using blockchain technology, efficiency and decentralization, the digital divide, and vulnerabilities in smart contracts.

Jafar et al. in their systematic literature review [37] discussed the challenges and solutions for scalable blockchain-based electronic voting systems, in addition to anticipating future developments. To evaluate cost and time, they identified well-known proposals, their implementations, verification methods, and various cryptographic solutions in previous research. They analyzed performance parameters, the primary benefits and limitations of different systems, and the most common approaches to blockchain scalability.

In [38], Vladucu et al. provided a thorough overview of blockchain-based e-voting systems currently in use by various countries and companies, as well as those proposed for academic research. The authors discussed the challenges that blockchain e-voting systems face and identified areas for future research to improve their trustworthiness. Furthermore, they included a detailed explanation of the terminology used in blockchain-based e-voting systems, such as consensus algorithms, cryptography, and system characteristics.

Despite this number of reviews, a comprehensive and comparative analysis is still required, as we will justify below.

3.4. Implementations of Blockchain-Based E-Voting Systems

In the following, we present several projects that are currently being developed or have already implemented e-voting on blockchain.

- Luxoft: Luxoft Holding Inc., a global IT service provider of technology solutions, is developing an e-voting infrastructure that will enable the world's first consultative vote on blockchain in Zug, Switzerland. Hyperledger Fabric was used to create an authorized blockchain that included a network, applications, and algorithms. In order to allow voters to cast their ballots, Zug's digital ID registration app based on Ethereum was authorized through uPort. Luxoft announces its intention to open source this technology and creates a Government Alliance Blockchain to encourage blockchain use in public institutions [39].
- Votem: A company specializing in election management, its main product is the CastIron platform. This platform is built on blockchain technology and offers several distinctive features, including a distributed database, immutability, permission-based access, and an audit trail. Votem has successfully handled over 13 million voters, serving both government elections and various associations in the United States and around the world. Notably, their track record boasts zero instances of fraud, compromise, attacks, or hacking, highlighting the security and reliability of their system [40]
- Voatz: A blockchain-based mobile voting tool that was launched in 2018 in West Virginia for overseas military voters participating in the 2018 midterm elections in the United States. Voatz includes biometric validation, such as fingerprints or retinal scans, so that voters validate their applicants and themselves on the application. A recent study found Voatz has major security flaws that allow attackers to monitor votes and edit or block ballots in large amounts [41].
- POLYAS: In the summer of 1996, Finland held the first POLYAS online election, with 30,000 voters participating in three languages. The company uses blockchain technology to offer an electronic voting system to the public and private sectors. Germany's

Federal Office for Information Security granted the first online election certification in 2016. The online voting system satisfies anonymity, accuracy, singularity, verifiability, and auditability. In Europe and the USA, several important companies employ POLYAS to manage their electronic voting systems [42].

- Polys: An online voting system that increases confidence in the voting process and results. Because it is based on blockchain technology, it is secure and transparent. Both the voting procedure and the results are immutable. Transparent cryptographic techniques are employed on the top of the blockchain to protect voter anonymity. Voters can check at any moment to ensure that their vote is valid and unmodified [43].
- DecentraVote: A blockchain-based solution for virtual meetings was originally developed by a team at the iteratec location in Vienna. DecentraVote uses a public Ethereum network based on Proof of Authority consensus with permissioned validator nodes. The smart contract constructed a Merkle tree of all voting rights on-chain, and the Zero-Knowledge Succinct Non-Interactive Argument of Knowledge (zk-SNARK) generated a proof for every voting right off-chain. DecentraVote does not address national political elections [44].

3.5. Research Gap and Objectives

Our systematic analysis of blockchain-based e-voting systems is guided by identified gaps in the current literature and specific objectives we aim to achieve. Despite ongoing research in this domain, existing studies often focus on the limitations of blockchain-based e-voting, lacking a comprehensive comparison with traditional and electronic voting systems in terms of benefits and challenges. The primary objectives of this systematic analysis are therefore:

1. To conduct a comprehensive comparison of blockchain-based e-voting systems against traditional and e-voting systems, focusing on understanding their relative benefits and challenges.
2. To review and analyze the concrete implementation techniques of blockchain in e-voting systems, identifying how they address existing challenges.
3. To provide the potential implications of blockchain-based e-voting systems for addressing existing challenges in the blockchain-based e-voting systems.
4. To establish an up-to-date roadmap for future research, emphasizing areas that require further investigation in the rapidly evolving landscape of blockchain-based e-voting.

This study aims to fill these gaps by offering a comprehensive and holistic review of blockchain-based e-voting systems. This involves an in-depth exploration of current challenges and potential areas for future research, thereby contributing to a more thorough understanding of blockchain technology's role in enhancing the integrity and efficiency of voting processes.

3.6. Contribution of the Review

To address the research gaps, this review conducts a comprehensive analysis of the existing literature on blockchain-based e-voting systems by, firstly, selecting papers using systematic literature review principles and, secondly, analyzing their technology aspects systematically. Specifically, the research aims to achieve the following objectives:

- Identify and analyze the benefits and challenges of blockchain-based e-voting systems in comparison to traditional voting and other e-voting systems, identifying the impact of blockchain-based e-voting systems on various aspects of the voting process.
- Explore the implementation technologies utilized in blockchain-based e-voting systems.
- Provide summarizing observations and recommendations for future research and development in this field.

In order to address the aforementioned objectives, the following research questions guide this systematic review:

- Benefits: What are the benefits of using blockchain technology in e-voting systems over other implementation approaches? The benefits are expressed in terms of requirements met by blockchain-based e-voting systems but not by other voting and e-voting types.
- Challenges: What are the challenges faced in implementing blockchain-based e-voting systems? These are expressed in terms of requirements that are already satisfied by other types of voting and e-voting systems but generally not yet met by blockchain-based e-voting systems.
- Impact: What are the impacts of proposed blockchain-based e-voting systems on different qualities? Impacts are expressed in terms of requirements that have been shown as satisfied (becoming a benefit of these) or not satisfied (becoming a challenge for blockchain-based e-voting systems).
- Technologies: what are the common technologies and implementations used in blockchain-based e-voting systems, including popular blockchain frameworks, consensus algorithms, security and privacy enhancing techniques?
- Future Research: based on the challenges identified and technologies reviewed, what future research and development directions should be explored in blockchain-based e-voting systems to enhance their functionality and quality?

Our results and observations aim to provide insights to legislators, researchers, and practitioners regarding the essential technical challenges that need to be tackled to establish widespread and secure blockchain-based e-voting systems. In addition, this study aims to provide guidance for future research by recognizing areas where research is lacking and indicating potential possibilities for future studies. Finally, this review shall provide insights into the potential solutions for implementing secure and ubiquitous blockchain-based e-voting systems, which can contribute to the practical implementation of such systems.

4. Methodology

This review follows the PRISMA protocol to ensure a transparent and rigorous review process and applies systematic literature review principles to selected papers. This systematic approach includes a structured review of the current literature on blockchain-based e-voting systems. The objective of this review is to provide a fair analysis of the available information using a systematic approach designed to minimize bias by following common selection, analysis, and validation procedures.

The hypothesis of this study is that by applying the distinct features of blockchain technology, such as decentralization, immutability, and transparency, it is possible to address the weaknesses and constraints related to traditional voting systems. This idea suggests integrating blockchain technology, and this hypothesis implies that this leads toward enhanced democratic procedures.

A search technique is used to discover relevant research, which includes utilizing precise keywords and concepts that relate to electronic voting, such as e-voting, i-voting, evoting, ivoting, electronic voting, internet voting, and election. Furthermore, the search approach encompasses blockchain-related terms such as blockchain, distributed ledger, and DLT. Boolean operators, in particular ("OR", "AND") are used to combine keywords and filter search results, ensuring that only papers that address both subjects are retrieved.

- Search query: *(evoting OR ivoting OR e-voting OR i-voting OR ((electronic OR internet) AND (voting OR vote OR election))) AND (blockchain OR "distributed ledger" OR DLT)*

The literature search was conducted using reputable databases (ACM, IEEE, Elsevier, Springer, and Scopus). The process of searching for relevant studies involves initially screening titles to identify potentially relevant ones. This is followed by a thorough review of the full text of the articles to determine whether they answer any of the research questions.

A number of exclusion and inclusion criteria can be established. Inclusion criteria are:

- Papers that are directly related to or contribute to the comprehension of blockchain-based e-voting systems are relevant to the title.
- Papers should be available in English to ensure accessibility and comprehension.

- Papers with an available full-text version, which allows for a comprehensive analysis and extraction of data.

Exclusion criteria are:

- To avoid repetition and ensure a unique set of papers, it is necessary to remove any duplicate titles.
- Exclude papers that are not written in English, as they can hamper comprehension and analysis.
- Exclude book chapters and focus on research articles and conference papers.
- To ensure the inclusion of valid and reliable research, papers that are officially retracted are excluded.
- Exclude papers if their topic does not align with the blockchain-based e-voting systems.

Figure 2 indicates the approach employed to conduct database analysis and, afterward, the inclusion and exclusion of publications for the purpose of our study.

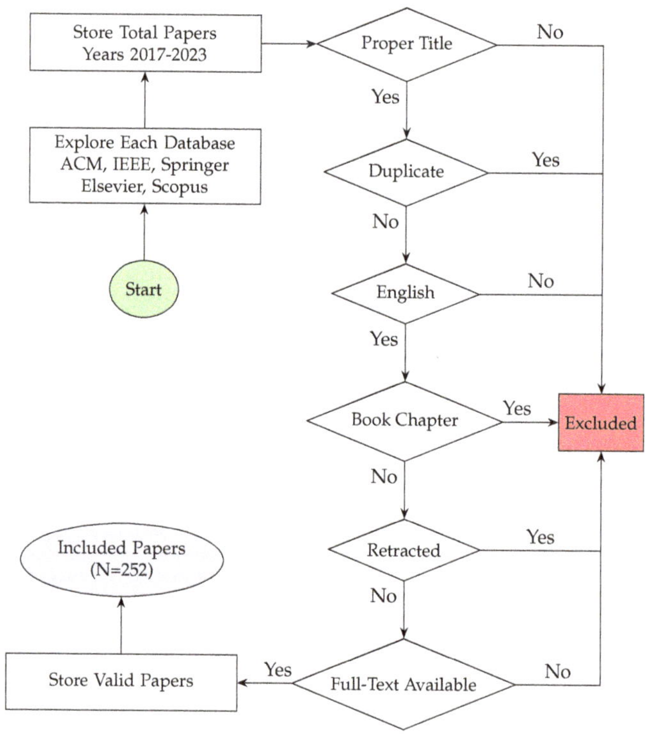

Figure 2. Procedure for database examination and paper inclusion.

The process of certainty assessment includes the evaluation of the level of certainty in the research outcomes. That confidence depends on the quality of the included studies and the cohesiveness of their results. High certainty indicates strong and reliable evidence, whereas low certainty indicates the need for further investigation or the existence of significant limitations in the currently available set of data. In order to ensure an efficient and rigorous assessment, separate reviewers are responsible for conducting an accurate assessment for each study that was randomly chosen. In cases where disagreements occur between the reviewers, these disagreements are resolved through broad consideration or, if determined essential, by requesting the perspective of an additional reviewer in order to attain a consensus.

5. Results—Benefits, Challenges, and Impacts

In this section, we present results derived from the selection process indicated earlier. Through the analysis of the data collected, our objective is to explore the research questions and construct findings from the outcomes of the systematic review. We identified the final number of publications from each database that should be included in the systematic review by applying these criteria to the corresponding databases. The results of this procedure are presented in Table 1.

Table 1. Overview of paper categories across databases.

Category	ACM	IEEE	Elsevier	Springer	Scopus	Total
Total	34	187	20	142	250	633
Inappropriate Title	18	80	0	30	2	130
Duplicate	0	1	9	42	176	228
Not English	0	2	0	0	2	4
Book Chapter	0	0	2	0	0	2
Retracted	0	0	0	0	1	1
Not Available	0	1	0	1	14	16
Included Papers	16	103	9	69	55	252

Figure 3 illustrates the publication trend of academic research literature that passed the inclusion and exclusion criteria, showing an increasing academic interest within this domain over time.

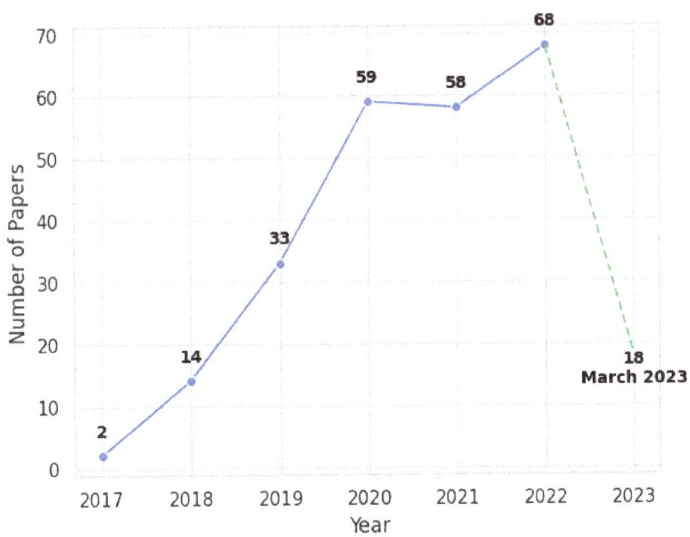

Figure 3. Publication trend in blockchain-based e-voting research.

We present the results for each of the research questions as follows:

- We address benefits, challenges, and impacts before looking at implementation technologies and summarizing future research in the following sections.
- For each, we comment on all properties mentioned in relation to the specific blockchain perspective.

- We also list the properties in the order of their frequency for the specific concern across the selected study papers, summarizing total occurrences and normalized numbers for better comparison.

5.1. Results—Benefits of Blockchain-Based E-Voting Systems

Various studies recommend blockchain-based e-voting systems due to their benefits. We compare here the benefits associated with blockchain-based e-voting systems with those of traditional (e-)voting systems, in terms of the requirements listed above for e-voting.

We categorize these benefits into major requirement categories, each further decomposed into several more detailed specific properties, if needed. In order to extract these benefit properties, we employed a hybrid strategy that includes both syntactic and semantic selection methods. We extracted the properties from relevant sections (Abstract, Introduction, and Related Work), thereby ensuring a targeted assessment of the content. These properties were identified as general benefits of blockchain technology and advantages offered by proposed blockchain-based e-voting systems, as discussed in the related work sections of the respective literature in comparison to conventional election systems.

We now list properties identified as benefits in the literature over traditional voting system types. We provide further comments on sources and explanations on each indicating how blockchains can achieve the benefits. Note that we order the benefits based on their frequency of occurrence across the selected study papers.

1. Security: a major benefit of blockchain-based e-voting systems, where subcategories highlight a unique perspective:
    - Integrity: holistic assurance of security aligned with the design [45].
    - Immutability: once a vote is recorded, it cannot be altered, ensuring the voting process's finality [46].
    - Durability: robust against data loss and ensures the permanency of stored data.
    - Stability: Resistance to disruptions or manipulations like hacking. Stability is enhanced by strong encryption systems, often inherent in blockchain technology [47].
    - Non-repudiation: a voter cannot dispute the validity of their cast vote [48].
2. Transparency: The blockchain-based e-voting system's inherent design encourages open voting, recording, management, and counting procedures. It facilitates independent audits [49] and ensures that all transactions (votes) on the blockchain are visible to all participants and can be independently verified.
3. Privacy: the ability of blockchain-based e-voting systems to protect voters' personal information and the confidentiality of their voting choices.
    - Anonymity: protecting a voter's identity [50].
    - Confidentiality (secrecy): the voters' choices are private, and outcomes are not presented ahead of time [51].
    - Untraceability: prevent the tracing of a vote back to its individual voter [50].
    - Pseudoanonymity: voters' actual identities are masked, but their voting activities are linked to unique identifiers similar to pseudonyms or addresses [52,53].
4. Verifiability: the ability to confirm that votes have been cast as intended, stored, and counted.
    - Public verifiability: the ability of all to verify the entire election process [54].
    - Individual verifiability: the ability for every voter to verify that their vote was precisely recorded and counted [54].
5. Auditability: ensure the voting process accuracy and truthfulness [55].
6. Accessibility: provide every eligible voter with an equal opportunity to participate in the voting process.
    - Availability: blockchains generally ensure that voters are able to cast their votes anytime within the stipulated period without facing any issue.
    - Broad turnout: technology allows substantial participation of eligible voters.

- Universal access: the ability of the system to be used effectively by all eligible voters.

7. Decentralization: Refers to the distribution of voting system authority, responsibility, and operations across a network compared to a central entity. This property is fundamental to blockchain technology and is essential for enhancing confidence among citizens by minimizing control of a potentially corrupt third party [36].
8. Usability: facilitate an extensive number of voters casting votes in accordance with their choices in an effective way while being satisfied with the process [56].
    - Simplicity: how simple and straightforward the system is to operate.
    - Understandability: clarity in system operation ensures that voters cast their votes as intended.
9. Efficiency: ability of an e-voting system to allow voters to cast votes in a swift and inexpensive manner.
    - Cost efficiency: The system's capacity to carry out voting operations at a cost that is affordable. This can involve a lower-cost setup and maintenance, material distribution, and human expenses.
    - Time efficiency: the system's ability to speed up voting and vote tallying.
    - Performance efficiency: the ability to handle massive amounts of data (votes), process, and count votes accurately, securely, and swiftly.
10. Trustworthiness: Secure, transparent, and fair system that ensures the accurate tracking and integrity of each vote. It is a balance of rigorous security measures, prompt results, and scalability, all of which are critical to preserving trust in the voting process [57].
    - Eligibility: only eligible voters can participate [58].
    - Fairness: election results are not exposed before the voting process finalizes [58].
    - Accountability: ability to determine whether or not the official vote record is inaccurate is facilitated by the blockchain [59].
    - Uniqueness: each eligible voter merits one and only one vote.
    - Accuracy: each vote is precisely accounted for, ensuring there is no modification, omission, or unauthorized inclusion [14].
    - Credibility: how much voters, politicians, and the general public trust and believe in the e-voting system.
    - Reliability: the system's consistency in performance through time ensures accurate, error-free function and availability [60].
11. Compatibility: ability of the e-voting system to operate in conjunction with various types of hardware, software, protocols, and legislation.
    - Adaptability: ability of an e-voting system to alter or adjust in order to accommodate various circumstances or necessities that may emerge [61,62].
    - Flexibility: ability to adapt to different frameworks, election types, voting methods, and voter interfaces.
12. Resistance to coercion: capacity of an e-voting system to shield voters from potential manipulations or coercions [36,63].

We enumerate in Table 2 the papers that mention the above properties as benefits of blockchain-based systems, ordered by the number of occurrences within the 252 selected papers. These properties are referred to as benefits either in the abstract, the introduction, or the related works sections of these papers.

Table 2. Distribution of papers mentioning the benefits of blockchain-based e-voting systems.

Benefit Category	No. of Papers	Normalized (%)
Security	224	88.89
Transparency	180	71.43
Decentralization	139	55.16
Privacy	96	38.10
Verifiability	85	33.73
Efficiency	67	26.19
Trustworthiness	63	25.00
Auditability	58	23.02
Accessibility	44	17.46
Usability	7	2.78
Compatibility	5	1.98
Resistance to Coercion	3	1.19

Normalized Percentage $= \frac{\text{Number of Papers in a Category}}{\text{Total Number of Papers}} \times 100.$

Blockchain-based e-voting systems offer first and foremost security, transparency, and decentralization, as mentioned in 224, 180, and 139 papers, respectively. Moreover, 96, 85, and 67 papers mention privacy, verifiability, and efficiency as significant benefits. Although less frequently discussed, trustworthiness, auditability, and accessibility also have significant advantages. The least frequently discussed factors are usability, compatibility, and resistance to coercion.

5.2. Results—Challenges in Blockchain-Based E-Voting Systems

Despite the properties of blockchain technology and the benefits it offers, these systems are not inherently applicable across all voting contexts due to some barriers. Our objective is an understanding of the obstacles and challenges associated with using blockchain technology for e-voting systems, specifically identifying properties that traditional e-voting systems have but blockchain-based ones do not.

As before, we arranged them into groups, ordered according to their frequency.

1. Privacy: It encompasses efforts to protect the secrecy of everyone who casts a vote, keep sensitive voter information from leaking out, and minimize the risk of tracking individual voters. However, ensuring privacy in e-voting causes challenges due to the conflicting objectives of auditability and transparency with privacy [64,65].
2. Security: It is a crucial aspect of blockchain-based e-voting systems, as it encompasses various measures to maintain the voting process's integrity, and availability. Defensive measures against cyber-attacks, Zero-Day exploits, and smart contract vulnerabilities are challenges for the blockchain security fundamental qualities. In [66], several types of attacks on blockchain such as hash-based attack, centralization attack, traffic attack, network level attack, injection attack, integrity attack, and private key leakage attack are discussed. It is necessary to mitigate such threats and prevent fraudulent use or disclosure of sensitive voter data without authorization [67,68].
3. Scalability: As the number of participants and transactions increases, it becomes crucial to maintain high performance and throughput. The inherent characteristics of blockchain, such as the need for consensus among distributed nodes and the necessity of storing every transaction on the blockchain, present scalability challenges. The decentralized nature of blockchain can lead to slow transaction processing times and increased resource requirements. In order to reach scalability in blockchain-based e-voting systems, it is necessary to address transaction throughput, network bandwidth, and data storage capacity. To ensure that blockchain-based e-voting

systems can accommodate an increasing number of participants and transactions while maintaining the security and decentralization nature of blockchain, scalability concerns need to be dealt with [36,69].

4. Technical aspects: various implementation challenges for blockchain-based e-voting systems arise, encompassing algorithm restrictions, technical complexity of consensus algorithms, hardware platform compatibility, integration with existing systems, complexity of technology, interoperability (including protocol interoperability), technical limitations, transparency in certain implementations, implementation challenges, complexity of implementation, complex design requirements, automating configuration, and limitations of authentication schemes [70–73].
5. Efficiency and feasibility: This encompasses various factors, including computation resource efficiency, energy consumption, performance efficiency, cost efficiency, and feasibility. Computation resource efficiency includes minimizing computational overhead associated with the consensus protocol and effectively allocating resources to handle the increasing workload. For minimizing the operational costs of blockchain-based e-voting systems, energy efficiency is crucial. The development of energy-efficient protocols, algorithms, and hardware can help reduce energy consumption [31,74–76].
6. Acceptability and immaturity: It refers to the level of trust and confidence stakeholders have in blockchain-based e-voting systems. To address this, it is necessary to achieve security, privacy, transparency, and reliability, thus building an environment that encourages the acceptance of blockchain-based e-voting systems. The immaturity of blockchain technology in e-voting leads to a lack of real-world experiments, extensive testing, stakeholder engagement, and comprehensive evaluation [11,34,38,77,78].
7. Usability: it is necessary to achieve a balance between a user-friendly interface and the security and integrity of the voting process [38,79].
8. Coercion freeness: it refers to challenges to protect voters from external pressures or coercive influences that could compromise their right to vote freely [33,64,80].
9. Accuracy and reliability: Ensuring accuracy is paramount to guaranteeing that each vote is recorded and counted correctly, without any errors or omissions. Blockchain technology has the potential to enhance accuracy by creating a transparent and tamper-proof record of all voting transactions. However, to achieve a reliable and credible e-voting system, it is crucial to design a protocol that is fair, prevents double-voting, and avoids reliance on a central authority [81,82]. By developing and implementing robust cryptographic techniques, secure consensus algorithms, and comprehensive auditing mechanisms, blockchain-based e-voting systems can enhance accuracy, reliability, and credibility, ensuring the integrity and fairness of the electoral process [83,84].
10. Accessibility: Access to voting opportunities is a fundamental principle. Limited internet access in certain locations presents a significant challenge to accessibility in blockchain-based e-voting systems. Providing a method such as offline voting that is consistent with the overall system is complex [85–87].
11. Regulatory and governance: Implementing blockchain-based e-voting systems requires adherence to legislation as well as adjusting to a constantly evolving legal landscape. Addressing regulatory and legal difficulties entails managing jurisdictional requirements, data privacy legislation, and electoral laws, and ensuring legal standards are challenging.

    Furthermore, ensuring interoperability and compatibility across different e-voting systems and platforms needs to establish common standards and protocols for blockchain-based e-voting, as it can provide seamless integration and collaboration among various stakeholders. Addressing regulatory and governance challenges, including the establishment of standards, is a significant challenge for blockchain-based e-voting systems [88–90].
12. Decentralization and consensus mechanisms: The distribution of authority, control, and decision-making power throughout the e-voting process, from registration to

result calculation, is referred to as decentralization at all stages. Achieving the appropriate level of decentralization is a challenge for ensuring transparency, avoiding central points of failure, and increasing system trustworthiness. Furthermore, for reaching a proper level of decentralization, selecting a suitable consensus mechanism to securely and quickly validate and confirm transactions is a related issue [91]. Consensus techniques are crucial for assuring network participant agreement and defending against fraudulent operations. Choosing the best consensus mechanism necessitates careful consideration of variables such as scalability, security, energy efficiency, and the specific needs of the e-voting system [92,93].

In Table 3, we provide a summary of papers that identify the above features as challenges of blockchain-based e-voting systems. These items are selected from various sections, primarily the Abstract, Introduction, and Related Works, applying a hybrid technique combining syntactic and semantic selection techniques. This approach signifies that these features are acknowledged either as inherent challenges to blockchain technology or as specific issues introduced by proposed blockchain-based e-voting systems.

Table 3. Distribution of papers mentioning the challenges of blockchain-based e-voting systems.

Challenge Category	No. of Papers	Normalized (%)
Privacy	108	42.86
Security	104	41.27
Scalability	87	34.52
Technical Aspects	40	15.87
Efficiency and Feasibility	36	14.29
Acceptableness and Immaturity	32	12.70
Coercion Freeness	21	8.33
Usability	18	7.14
Accuracy and Reliability	16	6.35
Accessibility	8	3.17
Regulatory and Governance	8	3.17
Decentralization and Consensus Mechanisms	3	1.19

Normalized Percentage = $\frac{\text{Number of Papers in a Category}}{\text{Total Number of Papers}} \times 100$.

Some advancements addressing the challenges in blockchain-based e-voting systems can be observed.

1. Enhanced privacy: Recent advances in cryptographic techniques, such as zero-knowledge proofs and homomorphic encryption, blind signatures, ring signatures, and mix networks, have significantly enhanced the privacy aspect of blockchain-based e-voting systems. These methods enable the verification of votes without revealing the voter's private information, simultaneously balancing privacy with the necessary transparency and auditability.
2. Enhanced security: In response to security challenges, there have been significant developments in both blockchain architecture and cryptographic defenses. In addition, enhanced consensus algorithms, like Proof of Stake (PoS) and Practical Byzantine Fault Tolerance (PBFT), have been implemented to counteract various blockchain-specific attacks. Additionally, the integration of advanced security protocols and mechanisms could become standard methods, improving these systems against cyber threats
3. Scalability improvement: To address scalability issues, innovative solutions such as off-chain transactions, sharding, optimized consensus protocols, and layer-2 scaling solutions like Lightning Networks have been introduced. These technologies have

proven effective in increasing transaction throughput, allowing for more scalable e-voting systems.

4. Technical improvement: to address the technical complexities, approaches for optimizing the chosen consensus algorithm for efficiency, simplifying technical complexities, ensuring hardware platform compatibility, ensuring interoperability with existing systems and protocols, implementing automation for configuration, and constantly seeking feedback for refinement are some of the steps taken or that need further research to evolve the system.
5. Energy and cost efficiency: The shift towards more energy-efficient consensus mechanisms, like Delegated Proof of Stake (DPoS), has notably reduced the operational costs and energy consumption of blockchain networks. Further, ongoing research into optimizing blockchain infrastructure and in other layers (on-chain and non-chain) can lead to the economic feasibility of blockchain-based e-voting systems.
6. Increasing acceptability: Experimental projects and real-world evaluations can play an important role in building trust and demonstrating the viability of blockchain-based e-voting systems. By developing educational resources and engaging stakeholders, this technology can be accepted and understood by a broader audience.
7. User-friendly interfaces: Significant efforts can be made to develop interfaces that are both simple for voters and secure. These interfaces often include guiding instructions and reliable verification mechanisms to ensure a seamless and secure voting experience.
8. Provide coercion-resistant: To achieve this aim in a blockchain-based e-voting system, there are several methods in the literature: implementing strong end-to-end encryption, utilizing zero-knowledge proofs, enforcing receipt-freeness, using blind signatures, employing multi-step authentication, securing physical components, maintaining a transparent blockchain, implementing auditing and monitoring, and ensuring user-friendly interfaces. Together, these strategies ensure the integrity of the voting process, prevent coercion, and enable voters to participate freely and without fear of repercussions.
9. Accuracy and reliability enhancements: By adopting robust cryptographic techniques and providing a decentralized ledger with transparent, auditable transactions, accuracy and reliability can be enhanced. By using identity verification mechanisms and smart contracts to ensure fairness, double voting can be prevented, whereas decentralized oracles and on-chain storage of critical data can reduce reliance on centralized sources. Consensus mechanisms and regular security testing are key to overall reliability. In all these cases, blockchain-based e-voting systems become more accurate and reliable.
10. Improved accessibility: Efforts to expand accessibility include developing offline voting mechanisms and protocols in mobile voting apps and establishing remote voting centers in areas with limited internet access. These centers can be equipped with the necessary technology to ensure that mobile voting applications are accessible to voters. Provide features for people with disabilities, such as screen readers, voice-guided interfaces, etc. Consider having backup plans in place in case of technical failures or disruptions in areas with limited internet access.
11. Regulatory compliance and governance: establishing legal frameworks and standards is a key focus, ensuring that these systems comply with the regulatory challenges associated with blockchain-based e-voting.
12. Decentralization and consensus mechanism optimization: customized consensus mechanisms that adjust to the unique requirements of e-voting systems can enable achieving a balance between speed, security, and decentralization.

5.3. Results—Impacts of Blockchain-Based E-Voting Systems

In this section, we discuss the identified impacts of different proposed systems. This extraction process involves retrieving the data from various sections of the studies, including evaluation and results, discussion, and conclusion. The impact categories follow those for benefits and challenges stated in the preceding sections.

Table 4 presents a quantitative description of the impacts of proposed systems across various categories.

The attributes that have the most notable relative impacts are security (41.67%), efficiency (34.52), and privacy (18.65%). These three attributes play a key role in maintaining the integrity, performance, and secrecy of the e-voting procedure.

Table 4. Impacts of proposed systems in various categories.

Impact Category	No. of Papers	Normalized (%)
Security	105	41.67
Efficiency	87	34.52
Privacy	47	18.65
Reliability	35	13.89
Scalability	27	10.71
Verifiability	22	8.73
Usability	16	6.35
Transparency	14	5.56
Accessibility	13	5.16
Resistance to Coercion	10	3.97
Auditability	8	3.17
Acceptableness	3	1.19

Normalized Percentage = $\frac{\text{Number of Papers in a Category}}{\text{Total Number of Papers}} \times 100$.

5.4. In-Depth Analysis of Results

The analysis, particularly focused on the data presented in Sections 5.1 and 5.2 and their respective tables, revealed insights. Section 5.1, as indicated by its table, shows broad agreement on blockchain's role in enhancing security and integrity, with a majority of the papers emphasizing these advantages. This trend emphasizes blockchain's potential to increase trust and participation in electoral processes. Furthermore, Section 5.2 indicates scalability and voter privacy as leading concerns, with a significant percentage of studies highlighting these issues. This suggests an urgent need for developing scalable blockchain architectures and integrating advanced privacy-preserving techniques in e-voting systems.

Section 5.3, supported by its respective table, further enriches our understanding. A notable percentage of studies in the impacts section report significant improvements in the efficiency and speed of voting processes facilitated by blockchain technology. This highlights blockchain's role not just in security, but also in optimizing and automating electoral procedures.

6. Results—Technologies and Implementation of Blockchain-Based E-Voting Systems

E-voting systems based on blockchains use a variety of concepts and technologies to enable secure and trustworthy elections. Blockchain frameworks like Ethereum and Hyperledger Fabric, consensus algorithms like Proof of Work, Proof of Stake, and Practical Byzantine Fault Tolerance, and privacy-enhancing techniques like homomorphic encryption and zero-knowledge proofs are among these technologies. Furthermore, authentication mechanisms such as biometric verification and identity management systems are critical in confirming voter legitimacy and maintaining the voting system's integrity.

In this section, we present a technology summary in five broader categories:

- Blockchain platforms;
- Consensus algorithms;

- Security and privacy techniques;
- Authentication and identity verification techniques;
- Other techniques (cryptography, development, testing).

6.1. Blockchain Platforms

The blockchain frameworks and technologies domain includes a variety of platforms and tools used in the design and implementation of blockchain-based systems. Blockchain frameworks such as Ethereum, Hyperledger Fabric, Bitcoin, and Multichain provide the foundation required for developers to create decentralized apps.

Figure 4 includes a range of widely used blockchain frameworks, including the proposed blockchain e-voting systems context. In all of the frameworks mentioned, Ethereum is the most popular choice, as evidenced by the 34.91% portion of utilized frameworks. Although particular papers mentioned specific frameworks, there are further studies, and no specific blockchain framework is explicitly stated. Instead, they proposed customized systems that are based on the general concept of blockchain technology.

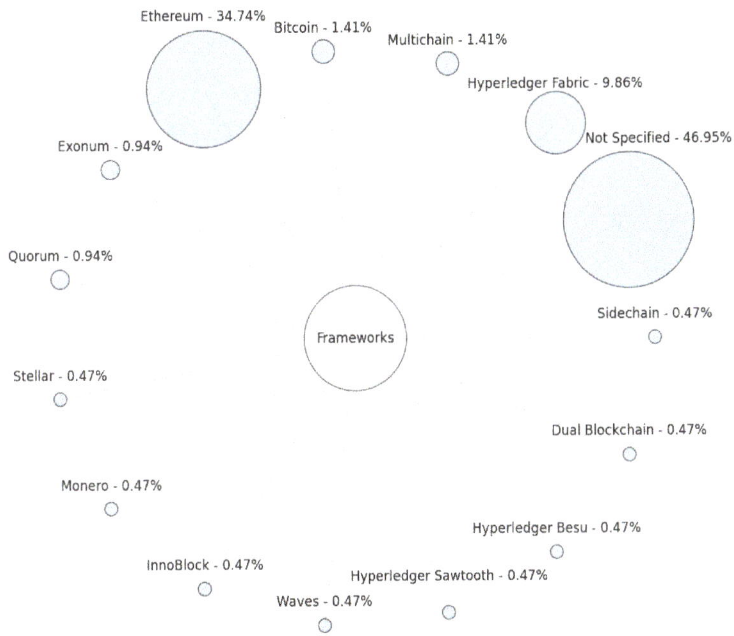

Figure 4. Blockchain frameworks distribution of proposed blockchain-based e-voting systems.

6.2. Consensus Algorithms

The consensus algorithms that were mentioned are illustrated in Table 5. Although a substantial number of papers do not explicitly mention the consensus algorithm used, it is reasonable to assume that for most proposed systems that use Ethereum as their framework, the consensus algorithm can be considered as Proof of Work (PoW). The following and most substantial protocol is referred to as "Proof of Work (PoW)", resulting in approximately 5.2% portion of used consensus algorithms. In the following, we provide a brief definition for each of these consensus algorithms:

Table 5. Adoption of consensus algorithms in blockchain-based e-voting systems (if mentioned).

Consensus Algorithm	No. of Papers	Normalized (%)
Proof of Work (PoW)	11	100
Proof of Stake (PoS)	6	54.55
Proof of Authority (PoA)	6	54.55
Byzantine Fault Tolerance (BFT)	6	54.55
Practical Byzantine Fault Tolerance (PBFT)	4	36.36
Raft consensus algorithm	3	27.27
Delegated Proof of Stake (DPoS)	2	18.18
Crash Fault Tolerant (CFT)	1	9.09
Stellar consensus protocol (SCP)	1	9.09
Hybrid (PoC combined with PoS)	1	9.09

Normalized Percentage $= \frac{\text{Number of Papers in a Category}}{\text{Max Number of Papers in any Category}} \times 100.$

1. Proof of Work (PoW): Commonly used consensus algorithm, including Bitcoin. It is a technique that requires members, known as miners, to solve computationally demanding puzzles in order to secure the network and validate transactions [94].
2. Proof of Stake (PoS): a consensus process in which block creators (validators) are selected depending on their wealth or stake in the network, and their possessions act as a guarantee, inciting honesty and network security [95].
3. Proof of Authority (PoA): A consensus approach used with authorized entities or individuals as block validators. Unlike other consensus methods, PoA is based on a predetermined set of reliable validators who proved their credibility in the network [96].
4. Byzantine Fault Tolerance (BFT): A technique that obtains agreement among participants even in the presence of malfunctioning or malicious nodes. BFT consensus algorithms are designed for dealing with Byzantine failures, in which nodes behave unexpectedly and inconsistently [97].
5. Practical Byzantine Fault Tolerance (PBFT): A specific algorithm that provides BFT in distributed systems. A leader node is selected to propose a block of transactions, which the other nodes, called replicas, validate and agree on [98].
6. Raft consensus algorithm: Developed for fault-tolerant log management to handle replicated logs. The Raft algorithm elects a leader to replicate logs across all nodes. The leader logs client requests and replicates them to cluster nodes. After a majority of nodes acknowledge log entries, the leader commits them and informs the followers [99,100].
7. Delegated Proof of Stake (DPoS): A PoS consensus algorithm variant. DPoS relies on the PoS concept by delegating block creation and validation commitments to a selected number of trusted delegates elected through vote [101].
8. Crash Fault Tolerant (CFT): A type of consensus method established for distributed systems that can endure crash failures, in which nodes in the system stop responding or crash. In it, a simple majority voting method is frequently used, in which nodes vote on the proposed state or decision. The system considers a value or decision to be acceptable if a majority of nodes agree on it [102].
9. Stellar consensus protocol (SCP): It combines the principles of federated agreement and Byzantine agreement to offer the Stellar network with a decentralized and fault-tolerant consensus mechanism. It enables nodes to agree on the state of the blockchain and keep the security and integrity of system transactions [103].
10. Hybrid (Proof of Credibility (PoC) combined with Proof of Stake (PoS): The weight of each vote in the consensus process is determined by the value of the tokens staked by

validators through the Proof of Stake (PoS) mechanism. The method brings Proof of Credibility (PoC) to address the issue of coin collapse in the PoS consensus mechanism. This combination of PoS and PoC is a safe hybrid structure that ensures full security when deployed in e-voting systems [104].

6.3. Security and Privacy Techniques

The use of blockchain-based e-voting systems needs to take security and privacy into consideration. Since it is decentralized and transparent, blockchain offers the possibility to boost the trustworthiness and credibility of e-voting systems. The use of security and privacy techniques in blockchain-based e-voting systems could assist in alleviating concerns about vote tampering, manipulation, and privacy violations.

Table 6 shows the number of studies that deploy security and privacy techniques. Data collection covers a broad spectrum of concepts and techniques. We list the number of publications and a normalized value in order to indicate the magnitude relative to other techniques.

The acronyms for each technique are explained in the listed discussion below. The zero-knowledge proofs (ZKPs) technique was referenced in a majority of studies. In addition, homomorphic encryption, blind signature, and ring signatures have been subject to a moderate degree of exploration. Several techniques, such as mix networks, time-lock encryption, machine learning, circle shuffle, and multi-signature schemes, were briefly discussed in a few publications.

Table 6. Distribution of security and privacy techniques in blockchain-based e-voting papers (if mentioned).

Technique	No. of Papers	Normalized (%)
ZKP	24	100
HE	24	100
BS	16	66.67
RS	13	54.17
SS	3	12.50
QKD	2	8.33
MN	2	8.33
TLE	2	8.33
ML	2	8.33
CS	1	4.17
RoPO	1	4.17
PMS	1	4.17
BC	1	4.17
DP	1	4.17
PB	1	4.17

$$\textbf{Normalized Percentage} = \frac{\text{Number of Papers in a Category}}{\text{Max Number of Papers in any Category}} \times 100.$$

As for the consensus protocols, we provide an overview of each of the techniques.

1. Zero-Knowledge Proofs (ZKPs): a cryptographic technique that enables one party to prove to another party the truthfulness of a statement or claim without disclosing any extra information [33,105].
2. Homomorphic Encryption (HE): a cryptographic technique that facilitates computations to be executed on encrypted data without the need for decryption [106–108].
3. Blind Signature (BS): a cryptographic method that enables a party to receive a valid signature on a message without disclosing the message's contents to the signer [109].

4. Ring Signatures: A cryptographic technique that offers anonymity and unlinkability to the signer within a group (ring) of potential signers. In the context of cryptographic protocols, a ring signature allows the signer to generate a signature on a specific message, thus convincing the verifier that the message was signed by an entity within a specific group while at the same time obscuring the true identity of the singer [110].
5. Shamir's Secret Sharing Scheme (SS): a cryptographic method that enables the division of a secret into multiple shares that are distributed among participants [92].
6. Quantum Key Distribution (QKD): a method of establishing secure cryptographic keys between two parties that makes use of the concepts of quantum physics [111,112].
7. Mix Network (MN): This technique is used to protect the privacy of voters and the secrecy of votes. Through serving as a channel between voters and the authority responsible for counting the votes [113,114].
8. Time-lock encryption (TLE): in this technique, a time-based delay is added to the encoding of encrypted data [114].
9. Machine Learning (ML): By integrating machine learning and blockchain technology, along with deep learning algorithms, significant enhancements can be achieved in biometric ID authentication. This involves utilizing machine learning methods to analyze facial features and verify the identities of users [84,115].
10. Circle Shuffle (CS): this method relies on a circular arrangement of votes, wherein each vote is assigned to a particular place in the circular structure [92].
11. Reputation-Based PayOff algorithm (RoPO): an incentive mechanism that is used in different decentralized systems to motivate players based on their reputation or performance history [116].
12. Proxy Multi-Signature Scheme (PMS): a variant of the common multi-signature method that includes the idea of a proxy or delegate to make signing on behalf of multiple individuals [117].
13. Bit Commitment (BC): a cryptographic technique in which one party (the committer) makes a commitment to another (the verifier) about a value without initially disclosing that value to the verifiers until the committer decides to reveal the committed value at a later time [118].
14. Differential Privacy (DP): It intends to maintain voters' sensitive data private while still allowing effective aggregate voting data analysis. It provides a structure for protecting voters' anonymity by adding random noise or perturbations to the data in a controlled manner [119].
15. Provenance-Based solution (PB): this solution involves tracking the origin and transformations of data (provenance) within the blockchain [120].

6.4. Authentication and Identity Verification Techniques

In blockchain-based e-voting systems, reliable authentication and identity verification is important to protect the integrity and security of the voting process. Authentication and identity verification in blockchain-based e-voting systems play an essential duty in satisfying various important objectives, such as ensuring voter eligibility, preventing fraud, and maintaining vote secrecy [121,122].

1. Biometric authentication: This method uses an individual's unique characteristics to validate their authenticity. These qualities can include fingerprints, facial recognition, iris or retina patterns, and even voice.
2. OTP (One-Time Password): a password that can only be used for one login session or transaction, often used to give a higher level of protection to sensitive transactions or systems [123,124] .
3. Aadhaar ID verification: the Unique Identification Authority of India (UIDAI) issues Indian residents a 12-digit Aadhaar number based on the resident's self-portrait, ten fingerprints, and two iris scans [125,126].

4. Multifactor authentication: this is the safety mechanism that requires multiple authentication methods from different categories to validate a user's identity for a login or other transaction.
5. Multi-step authentication: a security procedure that requires a user to provide extra evidence of identification when an additional level of assurance is required.
6. PKI-based X.509: PKI-based X.509 is a widely adopted standard that outlines how public key certificates are structured [127,128].
7. Unique IDs based on hash values: this method entails creating a unique identifier by applying a hash function to the biometric data, name, and date of birth of the voters [129].

Table 7 summarizes the distribution of authentication approaches utilized in different research papers. According to the results, the biometric authentication approach is frequently addressed across different studies.

Table 7. Distribution of authentication and identity verification techniques in blockchain-based e-voting papers (if mentioned).

Technique	No. of Papers	Normalized (%)
Biometric Authentication	27	100
Aadhaar ID Verification	7	25.93
OTP (One-Time Password)	6	22.22
Multifactor Authentication	3	11.11
Multi-Step Authentication	3	11.11
PKI-based X.509	2	7.41
Unique Hash IDs	1	3.70

Normalized Percentage $= \frac{\text{Number of Papers in a Category}}{\text{Max Number of Papers in any Category}} \times 100.$

6.5. Other Concepts

We identified several key concepts that deserve further consideration during the development and implementation of blockchain-based e-voting systems. These concepts address areas such as

- Cryptography techniques;
- Choice of development environments for smart contracts;
- Utilization of testing and benchmarking tools.

Table 8 categorizes them and provides relevant tools, environments, and techniques. This table serves as guidance for future research and facilitates exploration in the advancement of blockchain-based e-voting systems.

Table 8. Key concepts in blockchain-based e-voting systems.

Category	Tool	Description
Smart Contract Development and Execution	Solidity	Programming language for writing smart contracts on various blockchain platforms.
	Remix	A popular web-based development environment and IDE (Integrated Development Environment) specifically designed for writing, testing, and deploying smart contracts on the Ethereum blockchain.
	RIDE language	A specific language used for developing decentralized applications (DApps) on the Waves blockchain.

Table 8. *Cont.*

Category	Tool	Description
Smart Contract Development and Execution	Chaincode	Smart contract code written in Hyperledger Fabric for executing transactions.
	Truffle	Development framework for Ethereum smart contracts, providing testing and deployment.
	Hyperledger Composer	Framework for building blockchain applications and smart contracts on Hyperledger.
Blockchain Development and Testing Tools	Ganache	Personal Ethereum blockchain for local development and testing of smart contracts.
	Hyperledger Caliper	Benchmarking tool for measuring the performance of blockchain systems.
Performance Testing	Gatling Performance tool	A load testing tool used to simulate and measure the performance of systems, including blockchain-based applications.
Monitoring and Visualization	Grafana Monitoring tool	A tool used for monitoring and visualizing various metrics and data from systems, including blockchain networks.
Blockchain Interaction	Metamask	A browser extension that allows users to interact with the Ethereum blockchain, manage wallets, and execute transactions.
Cryptography	SHA	A family of cryptographic hash functions used for data integrity verification and password hashing.
	Chameleon hash	A type of hash function that allows for the creation of "trapdoor" information, enabling efficient collision generation.
	Advanced Encryption Standard (AES)	A widely-used symmetric encryption algorithm. It operates on fixed-size blocks of data and supports key lengths of 128, 192, and 256 bits.
	ElGamal cryptosystem	An asymmetric encryption algorithm based on the discrete logarithm problem.
	Paillier cryptosystem	An asymmetric encryption algorithm that allows for homomorphic operations, such as encrypted data manipulation.
	Cryptography over an elliptic curve	Encryption schemes based on elliptic curve mathematics, offering efficient and secure asymmetric encryption.
	RSA-based Public Key	A reference to the RSA encryption algorithm and key generation, which involves the use of a public key and a private key pair.
	RSA digital signature	A signature algorithm that utilizes the RSA encryption scheme for signing and verifying digital signatures.
	ECDSA (Elliptic Curve Digital Signature Algorithm)	A widely-used digital signature algorithm based on elliptic curve cryptography.

Table 8. *Cont.*

Category	Tool	Description
Cryptography	Schnorr signature	A digital signature algorithm known for its simplicity and security, offering efficient signature generation and verification.
	Lattice	A mathematical structure used in lattice-based cryptography, which relies on the hardness of certain lattice problems for security.
	SM2	The Chinese national standard introduced the SM2 algorithm, which utilizes a specific 256-bit elliptic curve for Elliptic Curve Diffie–Hellman key agreement and signature. This version incorporates functionalities for both signature generation and verification [130].
	SM9	It was issued by the Chinese State Cryptographic Authority and utilized for identity-based cryptography. It includes three components: a digital signature algorithm, an identity encryption algorithm, and a key agreement protocol [131].

6.6. Analysis of Results

This study reviewed a variety of blockchain platforms in Section 6.1, including Ethereum, Hyperledger Fabric, Bitcoin, and Multichain, each offering unique capabilities crucial for e-voting systems. Platforms like Ethereum are notable due to their smart contract functionality, which allows the creation of complex voting protocols, thus enhancing security and transparency. The choice of platform plays a critical role in determining the scalability, security, and flexibility of the e-voting system [132].

In Section 6.2, we analyzed the consensus mechanisms employed in the blockchain platforms, which are fundamental to the integrity and reliability of e-voting systems. Algorithms such as Proof of Work and Proof of Stake each bring different strengths and trade-offs in terms of security, energy efficiency, and processing speed. For e-voting systems, particularly on a national scale, selecting an appropriate consensus algorithm is critical, as it directly influences the system's ability to handle plenty of votes securely and efficiently while also preserving voter privacy.

The findings in Section 6.3 indicated the importance of incorporating advanced security and privacy techniques in e-voting systems. Techniques like homomorphic encryption and zero-knowledge proofs play a major role in ensuring that a voter's anonymity is maintained without compromising the transparency and verifiability of their vote. Implementing these techniques is essential for improving public trust in the electoral process. Furthermore, in Section 6.4, this study indicated the significance of methods such as biometric verification and identity management systems in maintaining the integrity of the voting process. These methods are crucial for preventing unauthorized access to the voting system, ensuring that each vote cast is legitimate, and preserving the principle of only one vote for one eligible person.

Lastly, in Section 6.5, the role of additional concepts like cryptographic development and thorough testing methods and tools cannot be neglected. As blockchain technology and cybersecurity threats continue to develop, continuously advancing cryptographic techniques and meticulous monitoring and testing tools are essential for ensuring the security and reliability of e-voting systems.

7. Discussion and Outlook

Many papers provide a discussion of current limitations and suggestions for future research. We summarize both non-functional and functional properties directly extracted from the selected studies, but we also take into account the technology concerns from the previous section.

In the second part of this section, we provide some observations on the different aspects—benefits, challenges, impact, and also identified future research—that we gained by comparing the answers across those aspects, checking them for consistency, and emerging patterns and trends.

7.1. Results—Suggested Roadmap for Blockchain-based E-Voting Systems

Table 9 provides an overview of the importance of suggested study areas for future exploration. Each category is accompanied by the number of research papers related to it as well as the normalized frequency associated with it. We summarize the areas in terms of two categories. The first refers to the properties (P) that e-voting systems need to maintain. The second focuses on the features or functions (F) that such systems should offer.

Properties singled out for further investigation are the following, again in order of frequency:

1. Scalability and Performance Improvements (Scal&Perf): Future work in this matter concentrates on developing more efficient consensus algorithms and investigating how to integrate blockchain technology into large-scale e-voting systems. The primary goal is to improve transaction processing rates, block generation rates, and block sizes while maintaining privacy, security, and energy efficiency [32,133–135].
2. Security and Privacy (Sec&Priv): This requires the development and implementation of advanced cryptographic techniques, such as zero-knowledge proofs, secure multiparty computation, blind signatures, ring signatures, and homomorphic encryption, to safeguard the identities and voting preferences of voters. To ensure a robust, anonymous, and trustworthy e-voting system, research concentrates on enhancing transparency and mitigating various types of attacks, like scalability attacks and transaction malleability [136–138].
3. Implementation, Evaluation, and Testing (Impl&Eval): This involves implementing, evaluating, and testing blockchain-based e-voting systems on a larger scale to measure their performance, scalability, and usability in real-world scenarios. Additionally, efforts will be made to address security evaluations, incorporate privacy-by-design features, explore different blockchain protocols, and conduct user acceptance testing with real voters to validate the system's effectiveness and feasibility for large-scale elections [113,133,139–141].
4. Authentication and Identity Verification (Auth&ID): Future work involves creating a comprehensive and secure authentication system for applications in e-voting using biometric measures and blockchain technology. This should focus on enhancing biometric algorithm accuracy and efficiency, investigating decentralized identifiers, incorporating several biometric recognition technologies, and addressing issues related to user eligibility and trust assumptions throughout the voting process. These schemes intend to improve the overall security and convenience of user authentication and verification in blockchain-based e-voting systems [125,142–144].
5. Coercion-Resistance (Coerc-Res): Future research should examine techniques that allow voters to make choices without the influence of coercers. This can be achieved by enabling voters to modify their votes multiple times, incorporating randomized tokens, leveraging face expression analysis, and employing facial tracking to enhance coercion detection. Additionally, ensuring receipt-free voting can be accomplished using various techniques, including ring signatures, while safeguarding voter privacy and security. The focus should remain on the proper design and execution of these tools to protect the integrity and privacy of the voting process [104,145–147].
6. Accessibility (Access): This involves deploying a voting module on mobile devices that supports offline voting and provides accessibility options for disabled voters. Proper mobility, enhanced design, and increased system availability seek to provide all eligible voters with a user-friendly, accessible, and effective voting experience, with potential solutions proposed for locations where remote voting is not feasible [115,148,149].

7. Legal and Governance Aspects (Leg&Gov): Future work refers to the establishment of regulations and standards for the deployment of blockchain technology, particularly in the context of electoral integrity. It comprises researching the influence of blockchain-based systems on election processes, developing a privacy-compliant framework, and exploring the sociological and psychological variables influencing online voter behavior in order to make blockchain technology more adaptable and suitable in more countries [89,150].

Features or functions that should be developed better in order of frequency:

1. Integration and Interoperability (Int&Inter): The creation and testing of blockchain-based e-voting systems that effectively interact with current voting infrastructures while maintaining compatibility with various legacy systems. The aim is to investigate the growth of blockchain-based voting solutions beyond elections, including agent-based methods and smart city services, as well as support adjusting in other industries like healthcare and auctions [151–153].
2. Consensus Algorithms and Smart Contracts (Cons&SC): Future work for e-voting systems aims to develop self-administering blockchain systems that do not require central authorities while improving scalability and privacy using new consensus algorithms and privacy-preserving approaches such as homomorphic encryption and zero-knowledge proofs. The investigation looks at the use of various consensus techniques, such as PBFT, BFT, and PoW, as well as smart contracts, to automate electoral processes, integrate complex voting rules, and increase security in e-voting systems. Furthermore, improving consensus techniques can also contribute to scalability and energy efficiency [154–157].
3. Usability and User Interface (Usab&UI): future work includes User Interface Enhancement, integrating it with a mobile app [156,158].
4. Machine Learning (ML): future work in Machine Learning for e-voting systems consists of detecting fraudulent behavior and fake voters, predicting voting patterns and identifying anomalies for enhanced security and transparency, and investigating the use of deep learning mechanisms to optimize sidechain parameters [84,159,160].
5. Acceptance (Accept): it involves conducting User Acceptance Testing (UAT) with a diverse group of stakeholders in order to improve system quality, reduce failures, and promise voter satisfaction [161–163].
6. General Concept (Gen): future research includes studying a variety of electoral systems employing blockchain technology.
7. Hybrid Systems (HS): future work should address the integration of paper ballots with electronic or blockchain-based voting mechanisms, studying the possibility of combining online and offline voting methods in different scenarios such as quadratic voting [125,164].
8. Blockchain and IoT (BC&IoT): The future should involve integrating blockchain and IoT technologies in e-voting systems to improve voting process security, transparency, and verifiability. The focus of the research is on developing IoT-based applications to ensure easy data exchange between devices and the blockchain network, checking user authentication through biometrics and other secure methods, and examining the integration of blockchain to revolutionize different industries [38,70,165].

Future work indications were extracted from the evaluation and results, discussion, future work, and conclusion sections of the papers, where 88 of the studies analyzed lacked clear statements regarding future work.

Table 9. Prominence of topics for future research (if mentioned).

Category	Type	No. of Papers	Normalized (%)
Scal&Perf	P	74	100.00
Sec&Priv	P	70	94.59
Impl&Eval	P	59	79.73
Int&Inter	F	34	45.95
Cons&SC	F	24	32.43
Auth&ID	P	23	31.08
Coerc-Res	P	15	20.27
Usab&UI	F	13	17.57
Accept	F	10	13.51
ML	F	7	9.46
Gen	F	7	9.46
Leg&Gov	P	6	8.11
Access	P	5	6.76
HS	F	4	5.41
BC-IoT	F	4	5.41

Normalized Percentage = $\frac{\text{Number of Papers in a Category}}{\text{Max Number of Papers in any Category}} \times 100$.

The "Scalability and Performance" research field emerged as the most prominent, showing its crucial importance. Furthermore, the areas "Security and Privacy", "Implementation, Evaluation, and Testing" and "Interaction and Interoperability" received attention. Figure 5 highlights these critical directions for future study.

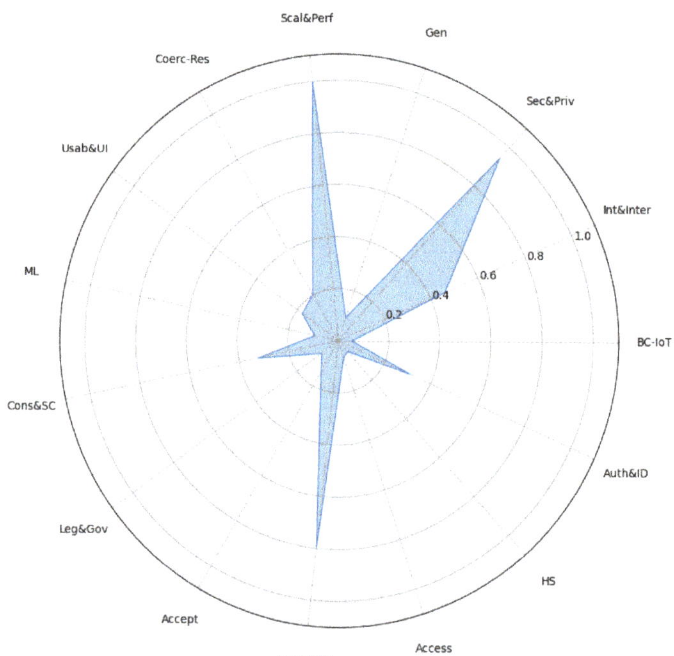

Figure 5. Prospective of research topics for future investigation

7.2. Final Observations

We have defined a number of research questions covering benefits, challenges, impacts, and future research as four perspectives based on system properties related to a list of requirements. A further technology review has helped to make the demonstrated solutions described in the studies, as well as concrete implementation gaps, more clear.

Based on the definitions of the different perspectives, we would expect that the suggested benefits have been demonstrated and shown to positively impact the field and that the challenges have been reiterated as areas for future work.

In order to detect inconsistencies and clarify possible conflicts between, for instance, assumed and demonstrated benefits, we note some observations on these concerns. For this, we mainly refer to the frequency position of a respective property in the frequency lists of the tables above.

- Security: This is the most frequently named property in relation to e-voting systems in general and blockchain-based systems in particular. An initial discrepancy emerges in that security appears at rank 1 or 2 in all lists, showing it as a demonstrated benefit as well as an open challenge. A closer investigation, however, shows that some principle blockchain properties such as integrity, immutability, and durability are acknowledged, but specific concerns relating to attacks on keys or smart contracts still exist, and possible remediation techniques such as zero-knowledge proofs, signature schemes, and homomorphic encryption are proposed.
- Privacy: As a property specifically relevant to the voter and their votes, this is separated from security. Here the picture is consistent by being ranked higher on challenges and future research (ranks 1 and 2, compared to 3 and 4 for benefits and impact), thus clearly showing this as a concern to be better addressed.
- Scalability: not even listed in the benefits, with positions 3 and 1 in challenges and future work, it is clearly seen as a serious open problem of blockchain solutions on a par with security and privacy.
- Usability: Although not a core property associated with blockchain platforms, it is mentioned in the context of a wider e-voting system with front end being integrated. As for privacy, it is consistently discussed across the factors. The ranks (between 8 and 10) are slightly lower, probably showing this as important but not being a core concern of blockchains but of a wider e-voting system.
- Coercion-freeness: this is similar to usability consistently ranked, with ranks 10 and 12 for benefits and impacts and 7 and 10 for impact and future also seen as a property still to be demonstrated, though with potential to improve via blockchains as a transparent and secure ledger mechanism.
- Technical concerns: these appear in the challenges and future work at a relatively high rank (between positions 3 and 4), referring to general implementation and evaluation methods, but also more specifically to interoperability and integration with other platforms and concrete blockchain-specific research needed on consensus protocols and smart contracts.
- Transparency and auditability: these are the only ones that are undisputed as demonstrated benefits of blockchain-based e-voting systems, with no concerns or open problems noted.
- Other properties: properties such as verifiability, accessibility, accuracy/reliability, and acceptability are also consistently referred to as properties of relevance, but not as critical ones.

7.3. Insights and Implications from the Observations

Through this study, convincing evidence for supporting the benefits of blockchain in enhancing security, transparency, decentralization, and privacy suggests that election organizations and governments should consider adopting blockchain technology in their voting systems. The improvement of the mentioned features of blockchain-based systems can increase voter confidence in the voting process and by clearly demonstrating these features to the public, electoral authorities can achieve a more trusting relationship with voters.

Observations of this research indicates the applicability of blockchain technology in e-voting systems. However, it is important to address the challenges highlighted in Section 5.2. These challenges indicate critical areas requiring further investigation and development.

Future research should focus on the challenge areas to enhance the understanding and application of blockchain in e-voting. In addition, the benefits of blockchain, as evidenced in e-voting, can inspire its application in other areas requiring similar levels of security, efficiency, and privacy, including but not limited to digital identity management, healthcare, financial service, supply chain, and education. As well, the success of blockchain in e-voting systems should encourage collaborative efforts between researchers to explore innovative applications of blockchain in public service.

8. Conclusions

We presented a systematic review of the state of research into blockchain-based e-voting systems. This study is motivated by the need to comparatively assess benefits, challenges, and impacts and open future research in comparison to other types of voting systems. Furthermore, a discussion of technology aspects to address the required properties was lacking.

The evolution of blockchain-based e-voting systems from 2017 to 2023 has been marked by significant advancements, as evidenced by research papers from this period. Significant studies emerged, proposing a novel approach to utilizing blockchain technology for recording votes for different voting scenarios. These systems aimed to address common limitations in existing voting systems and involved a critical evaluation of popular blockchain frameworks suitable for e-voting applications. During the years, the primary research emphasis shifted towards enhancing security and developing robust frameworks for blockchain-based e-voting systems. In recent years, the other aspects of e-voting systems, scalability and cost efficiency, have received more attention. Moreover, the importance of privacy-preserving protocols grew significantly, prompting the development of coercion-resistant and privacy-preserving e-voting protocols.

This study followed the PRISMA protocol, resulting in a selection of 252 papers. Five research questions centered on benefits, challenges, impacts, and open future research, as well as technology aspects, guided this study. To provide context, we supplemented this study of the literature with a comprehensive definition of voting system types as a framework, but also technology definitions, also extracted from the literature, in order to make the concerns better understood from an implementation perspective.

The results show that blockchain technology has the potential to successfully implement e-voting systems. Transparency and auditability are seen as undisputed benefits. Security and privacy are, as would be expected for voting processes, the central properties. Here, the potential is seen in blockchain technology over other platform technologies, but whereas some specific aspects are acknowledged, both remain serious open problems, which their top rankings in the frequency lists for challenges and future directions show.

An undisputed limitation of blockchains is their lack of scalability, which is the most serious non-security concern. Beyond core platform concerns, usability, verifiability, accessibility, reliability, and acceptability are properties of concern that in the wider voting systems implementation require more attention. Where evident from the studies considered, we supplemented these observations with concrete solution techniques.

Therefore, this study effectively clarifies both the potential and the limitations of blockchain-based e-voting systems. It achieves this by jointly integrating an analysis of fundamental properties with practical technological implementations and exploring a future roadmap, concluding in a comprehensive discussion that offers a holistic view of the topic.

Author Contributions: All authors have contributed substantially to the work presented. Conceptualization and methodology, M.H.B., H.R.B., N.E.I., and C.P.; data (study) curation, M.H.B.; data (study) analysis, M.H.B.; validation, M.H.B., H.R.B., N.E.I., and C.P.; writing—original draft preparation, M.H.B. and C.P.; writing—review and editing, H.R.B. and N.E.I.; supervision: C.P. All authors have read and agreed to the published version of the manuscript.

Funding: This research was funded by the Südtiroler Informatik AG (SIAG) through the PhD bursary for the years 2021–2025.

Data Availability Statement: Relevant data for the survey has been included in the manuscript. Further data sharing is not applicable to this article.

Conflicts of Interest: The authors declare no conflict of interest.

References

1. Page, M.J.; McKenzie, J.E.; Bossuyt, P.M.; Boutron, I.; Hoffmann, T.C.; Mulrow, C.D.; Shamseer, L.; Tetzlaff, J.M.; Akl, E.A.; Brennan, S.E.; et al. The PRISMA 2020 statement: An updated guideline for reporting systematic reviews. *Int. J. Surg.* **2021**, *88*, 105906. [CrossRef] [PubMed]
2. Voting Technology. Available online: https://electionlab.mit.edu/research/voting-technology (accessed on 22 April 2023).
3. Krimmer, R.; Volkamer, M. Bits or Paper? Comparing Remote Electronic Voting to Postal Voting. In Proceedings of the EGOV (Workshops and Posters), Copenhagen, Denmark, 22–26 August 2005; Citeseer: State College, PA, USA, 2005; pp. 225–232.
4. Jones, D.W. The evaluation of voting technology. In *Secure Electronic Voting*; Springer: New York, NY, USA, 2003; pp. 3–16.
5. Fischer, E.A.; Coleman, K.J. *The Direct Recording Electronic Voting Machine (DRE) Controversy: FAQs and Misperceptions*; Congressional Research Service, Library of Congress: Washington, DC, USA, 2007.
6. Electoral Technology. Available online: https://aceproject.org/ace-en/topics/et/eta/default (accessed on 19 March 2023).
7. Verified Voting–The Verifier. Available online: https://verifiedvoting.org/verifier/#mode/navigate/map/ppEquip/mapType/normal/year/2024 (accessed on 19 March 2023).
8. Oostveen, A.-M.; van den Besselaar, P. E-voting and media effects, an exploratory study. In Proceedings of the Conference on New Media, Technology and Everyday Life in Europe, Amsterdam, The Netherlands, 18–19 September 2003.
9. Buchstein, H. Online democracy, is it viable? Is it desirable? Internet voting and normative democratic theory. In *Electronic Voting and Democracy: A Comparative Analysis*; Palgrave Macmillan UK: London, UK, 2004; pp. 39–58.
10. Akbari, E.; Wu, Q.; Zhao, W.; Arabnia, H.R.; Yang, M.Q. From blockchain to internet-based voting. In Proceedings of the 2017 International Conference on Computational Science and Computational Intelligence (CSCI), Las Vegas, NV, USA, 14–16 December 2017; IEEE: Piscataway, NJ, USA, 2017; pp. 218–221.
11. Kshetri, N.; Voas, J. Blockchain-enabled e-voting. *IEEE Softw.* **2018**, *35*, 95–99. [CrossRef]
12. Tanwar, S.; Gupta, N.; Kumar, P.; Hu, Y.-C. Implementation of blockchain-based e-voting system. *Multimed. Tools Appl.* **2023**, 1–32. [CrossRef]
13. Gritzalis, D.A. Principles and requirements for a secure e-voting system. *Comput. Secur.* **2002**, *21*, 539–556. [CrossRef]
14. Anane, R.; Freeland, R.; Theodoropoulos, G. E-voting requirements and implementation. In Proceedings of the the 9th IEEE International Conference on E-Commerce Technology and the 4th IEEE International Conference on Enterprise Computing, E-Commerce and E-Services (CEC-EEE 2007), Tokyo, Japan, 23–26 July 2007; IEEE: Piscataway, NJ, USA, 2007; pp. 382–392.
15. Volkamer, M. *Evaluation of Electronic Voting: Requirements and Evaluation Procedures to Support Responsible Election Authorities*, 1st ed.; Springer Science & Business Media: Berlin, Germany, 2009; Volume 30.
16. Wolf, P.; Nackerdien, R.; Tuccinardi, D. *Introducing Electronic Voting: Essential Considerations*, 1st ed.; International Institute for Democracy and Electoral Assistance (International IDEA): Stockholm, Sweden, 2011.
17. Neumann, S. Evaluation and Improvement of Internet Voting Schemes Based on Legally-Founded Security Requirements. Ph.D. Thesis, Technische Universität Darmstadt, Darmstadt, Germany, 2016.
18. De Faveri, C.; Moreira, A.; Araújo, J.; Amaral, V. Towards security modeling of e-voting systems. In Proceedings of the 2016 IEEE 24th International Requirements Engineering Conference Workshops (REW), Beijing, China, 12–13 September 2016; IEEE: Piscataway, NJ, USA, 2016; pp. 145–154.
19. Recommendation CM/Rec (2017) 5 of the Committee of Ministers to Member States on Standards for E-Voting. Available online: https://rm.coe.int/0900001680726f6f (accessed on 20 March 2023).
20. Election Assistance Commission. *Voluntary Voting System Guidelines VVSG 2.0.(2021)*; Election Assistance Commission: Washington, DC, USA, 2023.
21. Kong, X.; Wu, Y.; Wang, H.; Xia, F. Edge Computing for Internet of Everything: A Survey. *IEEE Internet Things J.* **2022**, *9*, 23472–23485. [CrossRef]
22. Arbabi, M.S.; Lal, C.; Veeraragavan, N.R.; Marijan, D.; Nygård, J.F.; Vitenberg, R. A Survey on Blockchain for Healthcare: Challenges, Benefits, and Future Directions. *IEEE Commun. Surv. Tutorials* **2023**, *25*, 386–424. [CrossRef]
23. Ali, O.; Ally, M.; Dwivedi, Y. The state of play of blockchain technology in the financial services sector: A systematic literature review. *Int. J. Inf. Manag.* **2020**, *54*, 102199. [CrossRef]

24. Du, M.; Chen, Q.; Xiao, J.; Yang, H.; Ma, X. Supply Chain Finance Innovation Using Blockchain. *IEEE Trans. Eng. Manag.* **2020**, *67*, 1045–1058. [CrossRef]
25. Gai, K.; Guo, J.; Zhu, L.; Yu, S. Blockchain Meets Cloud Computing: A Survey. *IEEE Commun. Surv. Tutorials* **2020**, *22*, 2009–2030. [CrossRef]
26. Steiu, M. Blockchain in education: Opportunities, applications, and challenges. *First Monday* **2020**, *25*. . [CrossRef]
27. Hu, J.; Zhu, P.; Qi, Y.; Zhu, Q.; Li, X. A patent registration and trading system based on blockchain. *Expert Syst. Appl.* **2022**, *201*, 117094. [CrossRef]
28. Zhu, P.; Hu, J.; Li, X.; Zhu, Q. Using blockchain technology to enhance the traceability of original achievements. *IEEE Trans. Eng. Manag.* **2023**, *70*, 1693–1707. [CrossRef]
29. Abdelmaboud, A.; Ahmed, A.I.A.; Abaker, M.; Eisa, T.A.E.; Albasheer, H.; Ghorashi, S.A.; Karim, F.K. Blockchain for IoT applications: Taxonomy, platforms, recent advances, challenges and future research directions. *Electronics* **2022**, *11*, 630. [CrossRef]
30. Taş, R.; Tanrıöver, Ö.Ö. A systematic review of challenges and opportunities of blockchain for E-voting. *Symmetry* **2020**, *12*, 1328. [CrossRef]
31. Jafar, U.; Ab Aziz, M.J.; Shukur, Z. Blockchain for Electronic Voting System—Review and Open Research Challenges. *Sensors* **2021**, *21*, 5874. [CrossRef] [PubMed]
32. Pawlak, M.; Poniszewska-Marańda, A. Trends in blockchain-based electronic voting systems. *Inf. Process. Manag.* **2021**, *58*, 102595. [CrossRef]
33. Huang, J.; He, D.; Obaidat, M.S.; Vijayakumar, P.; Luo, M.; Choo, K.-K.R. The application of the blockchain technology in voting systems: A review. *ACM Comput. Surv. (CSUR)* **2021**, *54*, 1–28. [CrossRef]
34. Jafar, U.; Ab Aziz, M.J. A state of the art survey and research directions on blockchain based electronic voting system. In Proceedings of the Second International Conference, ACeS 2020, Penang, Malaysia, 8–9 December 2020; Revised Selected Papers 2; Springer: Singapore, 2021.
35. Devi, U.; Bansal, S. Secure e-Voting System—A Review. In Proceedings of the Hybrid Intelligent Systems, Olten, Switzerland; Porto, Portugal; Vilnius, Lithuania; Kochi, India, 12–14 December 2023; Springer Nature: Cham, Switzerland, 2023; pp. 1209–1224.
36. Benabdallah, A.; Audras, A.; Coudert, L.; El Madhoun, N.; Badra, M. Analysis of Blockchain Solutions for E-Voting: A Systematic Literature Review. *IEEE Access* **2022**, *10*, 70746–70759. [CrossRef]
37. Jafar, U.; Ab Aziz, M.J.; Shukur, Z.; Hussain, H.A. A Systematic Literature Review and Meta-Analysis on Scalable Blockchain-Based Electronic Voting Systems. *Sensors* **2022**, *22*, 7585. [CrossRef]
38. Vladucu, M.-V.; Dong, Z.; Medina, J.; Rojas-Cessa; R.Vladucu, M.-V.; Dong, Z.; Medina, J.; Rojas-Cessa, R. E-Voting Meets Blockchain: A Survey. *IEEE Access* **2023**, *11*, 23293–23308. [CrossRef]
39. Luxoft. Available online: https://www.luxoft.com/ (accessed on 18 November 2023).
40. Votem. Available online: https://votem.com/ (accessed on 20 November 2023).
41. Voatz. Available online: https://voatz.com/ (accessed on 20 November 2023).
42. Polyas. Available online: https://www.polyas.com/ (accessed on 21 November 2023).
43. Kaspersky Box. Available online: https://box.kaspersky.com/f/e68a161d8e7241909ea3/ (accessed on 21 November 2023).
44. Decentra.Vote. Available online: https://decentra.vote/ (accessed on 25 November 2023).
45. Harley, K.; Cooper, R. Information Integrity: Are We There Yet? *ACM Comput. Surv.* **2021**, *54*, 1–35. [CrossRef]
46. Çabuk, U.C.; Adiguzel, E.; Karaarslan, E. A Survey on Feasibility and Suitability of Blockchain Techniques for the E-Voting Systems. *arXiv* **2020**, arXiv:2002.07175.
47. Kugusheva, A.; Yanovich, Y. Ring Signature-Based Voting on Blockchain. In Proceedings of the 2019 2nd International Conference on Blockchain Technology and Applications, Xi'an, China, 9–11 December 2020; Association for Computing Machinery: New York, NY, USA, 2020; pp. 70–75.
48. Haiyan, X.; Lifang, W.; Yuechuan, W. A New Fair Electronic Contract Signing Protocol. In Proceedings of the Advances in Intelligent Networking and Collaborative Systems (INCoS-2019), Oita, Japan, 5–7 September 2019; Springer International Publishing: Cham, Switzerland, 2020; pp. 289–295.
49. Hjálmarsson, F.Þ.; Hreiðarsson, G.K.; Hamdaqa, M.; Hjálmtýsson, G. Blockchain-Based E-Voting System. In Proceedings of the 2018 IEEE 11th International Conference on Cloud Computing (CLOUD), San Francisco, CA, USA, 2–7 July 2018; IEEE: Piscataway, NJ, USA, 2018; pp. 983–986.
50. Kumar, M.; Katti, C.P.; Saxena, P.C. A secure anonymous e-voting system using identity-based blind signature scheme. In Proceedings of the 13th International Conference, ICISS 2017, Mumbai, India, 16–20 December 2017; Springer International Publishing: Cham, Switzerland, 2017.
51. Russo, A.; Anta, A.F.; Vasco, M.I.G.; Romano, S.P. Chirotonia: A Scalable and Secure e-Voting Framework based on Blockchains and Linkable Ring Signatures. In Proceedings of the 2021 IEEE International Conference on Blockchain (Blockchain), Melbourne, Australia, 6–8 December 2021; IEEE: Piscataway, NJ, USA, 2021; pp. 417–424.
52. Ikundi, O.; Nwosu, K.C.; Abdulgader, M. LegitVote: A Blockchain-Based System to Facilitate E-Voting Process. In Proceedings of the 2022 International Conference on Computer and Applications (ICCA), Cairo, Egypt, 20–22 December 2022; IEEE: Piscataway, NJ, USA, 2022; pp. 1–7.

53. Fusco, F.; Lunesu, M.; Pani, F.; Pinna, A. Crypto-voting, a Blockchain based e-Voting System. In Proceedings of the 10th International Joint Conference on Knowledge Discovery, Knowledge Engineering and Knowledge Management (IC3K 2018)—Volume 3: KMIS, Seville, Spain, 18–20 September 2018; pp. 223–227.
54. Vivek, S.K.; Yashank, R.S.; Prashanth, Y.; Yashas, N.; Namratha, M. E-Voting Systems using Blockchain: An Exploratory Literature Survey. In Proceedings of the 2020 Second International Conference on Inventive Research in Computing Applications (ICIRCA), Coimbatore, India, 15–17 July 2020; IEEE: Piscataway, NJ, USA, 2020; pp. 890–895.
55. Mello-Stark, S.; Lamagna, E.A. The Need for Audit-Capable E-Voting Systems. In Proceedings of the 2017 31st International Conference on Advanced Information Networking and Applications Workshops (WAINA), Taipei, Taiwan, 27–29 March 2017; IEEE: Piscataway, NJ, USA, 2017; pp. 535–540.
56. Hsu, J.; Bronson, G. E-Voting Technologies Usability: A Critical Element for Enabling Successful Elections. In *Emerging Challenges in Business, Optimization, Technology, and Industry: Proceedings of the Third International Conference on Business Management and Technology, Vancouver, BC, Canada, 13–17 March 2017*; Springer International Publishing: Cham, Switzerland, 2018.
57. Shahzad, B.; Crowcroft, J. Trustworthy Electronic Voting Using Adjusted Blockchain Technology. *IEEE Access* **2019**, *7*, 24477–24488. [CrossRef]
58. Sheer Hardwick, F.; Gioulis, A.; Naeem Akram, R.; Markantonakis, K. E-Voting With Blockchain: An E-Voting Protocol with Decentralisation and Voter Privacy. In Proceedings of the 2018 IEEE International Conference on Internet of Things (iThings) and IEEE Green Computing and Communications (GreenCom) and IEEE Cyber, Physical and Social Computing (CPSCom) and IEEE Smart Data (SmartData), Halifax, NS, Canada, 30 July–3 August 2018; IEEE: Piscataway, NJ, USA, 2018; pp. 1561–1567.
59. Küsters, R.; Müller, J. Cryptographic security analysis of e-voting systems: Achievements, misconceptions, and limitations. In Proceedings of the Second International Joint Conference, E-Vote-ID 2017, Bregenz, Austria, 24–27 October 2017, Springer International Publishing: Cham, Switzerland, 2017.
60. Conti, V.; Taş, R.; Tanrıöver, Ö.Ö. A Manipulation Prevention Model for Blockchain-Based E-Voting Systems. *Secur. Commun. Networks* **2021**, *2021*, 6673691.
61. Borras, J. *Overview of the Work on E-Voting Technical Standards*; Cabinet Office, UK Government: London, UK, 2002.
62. Prajapati, P.; Dave, K.; Shah, P. A review of recent blockchain applications. *Int. J. Sci. Technol. Res.* **2020**, *9*, 897–903.
63. Kho, Y.-X.; Heng, S.-H.; Chin, J.-J. A Review of Cryptographic Electronic Voting. *Symmetry* **2022**, *14*, 858. [CrossRef]
64. buidris, Y.; Kumar, R.; Wenyong, W. A Survey of Blockchain Based on E-Voting Systems. In Proceedings of the 2019 2nd International Conference on Blockchain Technology and Applications, Xi'an, China, 25–30 June 2019; Association for Computing Machinery: New York, NY, USA, 2020; pp. 99–104.
65. Nguyen, T.; Thai, M.T. zVote: A Blockchain-based Privacy-preserving Platform for Remote E-voting. In Proceedings of the ICC 2022—IEEE International Conference on Communications, Seoul, Republic of Korea, 16–20 May 2022; IEEE: Piscataway, NJ, USA, 2022; pp. 4745–4750.
66. Anita, N.; Vijayalakshmi, M. Blockchain Security Attack: A Brief Survey. In Proceedings of the 2019 10th International Conference on Computing, Communication and Networking Technologies (ICCCNT), Kanpur, India, 6–8 July 2019; IEEE: Piscataway, NJ, USA, 2019; pp. 1–6.
67. Alamleh, H.; AlQahtani, A.A.S. Analysis of the Design Requirements for Remote Internet-Based E-Voting Systems. In Proceedings of the 2021 IEEE World AI IoT Congress (AIIoT), Seattle, WA, USA, 10–13 May 2021; IEEE: Piscataway, NJ, USA, 2021; pp. 386–390.
68. Chaeikar, S.S.; Jolfaei, A.; Mohammad, N.; Ostovari, P. Security Principles and Challenges in Electronic Voting. In Proceedings of the 2021 IEEE 25th International Enterprise Distributed Object Computing Workshop (EDOCW), Gold Coast, Australia, 25–29 October 2021; IEEE: Piscataway, NJ, USA, 2021; pp. 38–45.
69. Hajian Berenjestanaki, M.; Barzegar, H.R.; El Ioini, N.; Pahl, C. An Investigation of Scalability for Blockchain-Based E-Voting Applications. In Proceedings of the Blockchain and Applications, 5th International Congress, Guimarães, Portugal, 12–14 July 2023; Springer Nature Switzerland: Cham, Switzerland, 2023; pp. 134–143.
70. Geng, T.; Njilla, L.; Huang, C.T. A Survey of Blockchain-Based Electronic Voting Mechanisms in Sensor Networks. In Proceedings of the 20th ACM Conference on Embedded Networked Sensor Systems, Boston, MA, USA, 6–9 November 2023; Association for Computing Machinery: New York, NY, USA, 2023; pp. 1222–1228.
71. Hapsara, M.; Imran, A.; Turner, T. E-Voting in Developing Countries. In Proceedings of the Electronic Voting, Bregenz, Austria, 24–26 October 2017; Springer International Publishing: Cham, Switzerland, 2017; pp. 36–55.
72. Goel, A.K.; Rai, A.; Narain, A., Richard, A.; Kumar, K. Trusted Vote: Reorienting eVoting using Blockchain. In Proceedings of the 2022 Sixth International Conference on I-SMAC (IoT in Social, Mobile, Analytics and Cloud) (I-SMAC), Coimbatore, India, 15–17 July 2022; IEEE: Piscataway, NJ, USA, 2022; pp. 129–138.
73. Majumder, S.; Ray, S. Usage of Blockchain Technology in e-Voting System Using Private Blockchain. In *Intelligent Data Engineering and Analytics*; Springer Nature: Singapore, 2022; pp. 51–61.
74. Pawlak, M.; Poniszewska-Marańda, A.; Kryvinska, N. Towards the Intelligent Agents for Blockchain E-Voting System. *Procedia Comput. Sci.* **2018**, *141*, 239–246. [CrossRef]
75. Gong, B.; Lu, X.; Fat, L.W.; Au, M.H. Blockchain-Based Threshold Electronic Voting System. In *Security and Privacy in Social Networks and Big Data, Proceedings of the 5th International Symposium, SocialSec 2019, Copenhagen, Denmark, 14–17 July 2019; Revised Selected Papers 5*; Springer: Singapore, 2019.

76. Tirodkar, V.; Patil, S. Proposed Infrastructure for Census Enumeration and Internet Voting Application in Digital India with Multichain Blockchain. In *Advanced Computing Technologies and Applications, Proceedings of the 2nd International Conference on Advanced Computing Technologies and Applications—ICACTA 2020, Mumbai, India, 28–29 Feburary 2020*; Springer: Singapore, 2020.
77. Yang, Z.; Hu, H.; Ou, J.; Qian, B.; Luo, Y.; He, P.; Zhou, M.; Chen, Z. A Practical Anonymous Voting Scheme Based on Blockchain for Internet of Energy. *Secur. Commun. Netw.* **2022**, *2022*, 4436824.
78. Daramola, O.; Thebus, D. Architecture-centric evaluation of blockchain-based smart contract e-voting for national elections. *Informatics* **2020**, *7*, 16. [CrossRef]
79. Wang, Z.; Luo, X.; Li, M.; Sun, W.; Xue, K. WeVoting: Blockchain-based Weighted E-Voting with Voter Anonymity and Usability. In Proceedings of the GLOBECOM 2022—2022 IEEE Global Communications Conference, Rio de Janeiro, Brazil, 4–8 December 2022; IEEE: Piscataway, NJ, USA, 2022; pp. 2585–2590.
80. Spadafora, C.; Longo, R.; Sala, M. *A Coercion-Resistant Blockchain-Based E-Voting Protocol with Receipts*; Department of Mathematics, University Of Trento: Trento, Italy, 2020. Available online: https://eprint.iacr.org/2020/674 (accessed on 18 March 2023).
81. Isirova, K.; Potii, O. Development Principles for Electronic Voting System Using Distributed Ledger Technology. In Proceedings of the 2020 IEEE 11th International Conference on Dependable Systems, Services and Technologies (DESSERT), Kyiv, Ukraine, 14–18 May 2020; IEEE: Piscataway, NJ, USA, 2020; pp. 446–450.
82. Lijuan, Z.; Dunyue, L.; Rui, Z.; Yongbin, Z.; Rouxin, F.; Ziyang, C. Electronic Voting Scheme Based on Blockchain and SM2 Cryptographic Algorithm Zero-Knowledge Proof. In Proceedings of the 2022 IEEE International Conference on Web Services (ICWS 2022), Barcelona, Spain, 11–15 July 2022; Springer Nature: Cham, Switzerland, 2022; pp. 88–103.
83. Gao, S.; Zheng, D.; Guo, R.; Jing, C.; Hu, C. An Anti-Quantum E-Voting Protocol in Blockchain With Audit Function. *IEEE Access* **2019**, *7*, 115304–115316. [CrossRef]
84. Cheema, M.A.; Ashraf, N.; Aftab, A.; Qureshi, H.K.; Kazim, M.; Azar, A.T. Machine Learning with Blockchain for Secure E-voting System. In Proceedings of the 2020 First International Conference of Smart Systems and Emerging Technologies (SMARTTECH), Riyadh, Saudi Arabia, 3–5 November 2020; IEEE: Piscataway, NJ, USA, 2020. pp. 177–182.
85. Denis González, C.; Frias Mena, D.; Massó Muñoz, A.; Rojas, O.; Sosa-Gómez, G. Electronic Voting System Using an Enterprise Blockchain. *Appl. Sci.* **2022**, *12*, 531. [CrossRef]
86. Churi, M.; Bajaj, A.; Pannu, G.; Patil, M. Blockchain Based E-Voting System. In *Intelligent Computing and Networking: Proceedings of IC-ICN 2022*; Springer Nature: Singapore, 2023; pp. 123–142.
87. Oprea, S.-V.; Bâra, A.; Andreescu, A.-I.; Cristescu, M.P. Conceptual Architecture of a Blockchain Solution for E-Voting in Elections at the University Level. *IEEE Access* **2023**, *11*, 18461–18474. [CrossRef]
88. Pawlak, M.; Poniszewska-Marańda, A. Blockchain E-Voting System with the Use of Intelligent Agent Approach. In Proceedings of the 17th International Conference on Advances in Mobile Computing & Multimedia (MoMM2019), Munich, Germany, 2–4 December 2019; Association for Computing Machinery: New York, NY, USA, 2020; pp. 145–154.
89. Ohammah, K.L.; Thomas, S.; Obadiah, A.; Mohammed, S.; Lolo, Y.S. A Survey on Electronic Voting On Blockchain. In Proceedings of the 2022 IEEE Nigeria 4th International Conference on Disruptive Technologies for Sustainable Development (NIGERCON), Lagos, Nigeria, 17–19 May 2022; IEEE: Piscataway, NJ, USA, 2022; pp. 1–4.
90. Neziri, V.; Dervishi, R.; Rexha, B. Survey on Using Blockchain Technologies in Electronic Voting Systems. In Proceedings of the 2021 25th International Conference on Circuits, Systems, Communications and Computers (CSCC), Crete Island, Greece, 19–22 July 2021; IEEE: Piscataway, NJ, USA, 2021; pp. 61–65.
91. Werth, J.; Hajian Berenjestanaki, M.; Barzegar, H.; El Ioini, N.; Pahl, C. A Review of Blockchain Platforms Based on the Scalability, Security and Decentralization Trilemma. In Proceedings of the 25th International Conference on Enterprise Information Systems (ICEIS 2023), Prague, Czech Republic, 24–26 April 2023; Volume 1, pp. 146–155.
92. Bartolucci, S.; Bernat, P.; Joseph, D. SHARVOT: Secret SHARe-Based VOTing on the Blockchain. In Proceedings of the 1st International Workshop on Emerging Trends in Software Engineering for Blockchain (WETSEB '18), Gothenburg, Sweden, 27 May 2018; Association for Computing Machinery: New York, NY, USA, 2018; pp. 30–34.
93. Jafar, U.; Aziz, M.J.A.; Shukur, Z.; Hussain, H.A. A Cost-efficient and Scalable Framework for E-Voting System based on Ethereum Blockchain. In Proceedings of the 2022 International Conference on Cyber Resilience (ICCR), Dubai, United Arab Emirates, 6–7 October 2022; IEEE: Piscataway, NJ, USA, 2022.
94. Nakamoto, S. Bitcoin: A Peer-to-Peer Electronic Cash System. Available online: https://bitcoin.org/bitcoin.pdf (accessed on 14 May 2023).
95. King, S.; Nadal, S. PPCoin: Peer-to-Peer Crypto-Currency with Proof-of-Stake. 2012. Available online: https://peercoin.net/assets/paper/peercoin-paper.pdf (accessed on 19 August 2012).
96. Kovan—Stable Ethereum Public Testnet. Available online: https://github.com/kovan-testnet/proposal/blob/master/README.md (accessed on 14 May 2023).
97. Buchman, E. Tendermint: Byzantine Fault Tolerance in the Age of Blockchains. Ph.D. Thesis, University of Guelph, Guelph, ON, Canada, 2016.
98. Castro, M.; Liskov, B. Practical Byzantine Fault Tolerance. In Proceedings of the 3rd Symposium on Operating Systems Design and Implementation (OSDI 99), New Orleans, LA, USA, 22–25 February 1999.
99. Ongaro, D.; Ousterhout, J. In Search of an Understandable Consensus Algorithm (Extended Version). In Proceedings of the USENIX Annual Technical Conference, USENIX ATC, Philadelphia, PA, USA, 19–20 June 2014; pp. 19–20.

100. Chaisawat, S.; Vorakulpipat, C. Towards Achieving Personal Privacy Protection and Data Security on Integrated E-Voting Model of Blockchain and Message Queue. *Secur. Commun. Netw.* **2021**, *2021*, 1–14. [CrossRef]
101. Delegated Proof of Stake (DPOS). Available online: https://how.bitshares.works/en/master/technology/dpos.html (accessed on 15 May 2023).
102. Li, W.; Meese, C.; Nejad, M.; Li, W.; Meese, C.; Nejad, M.; Guo, H. P-CFT: A Privacy-preserving and Crash Fault Tolerant Consensus Algorithm for Permissioned Blockchains. In Proceedings of the 2021 4th International Conference on Hot Information-Centric Networking (HotICN), Nanjing, China, 25–27 November 2021; IEEE: Piscataway, NJ, USA, 2021; pp. 26–31.
103. Stellar Consensus Protocol (SCP). Stellar Documentation. Available online: https://developers.stellar.org/docs/fundamentals-and-concepts/stellar-consensus-protocol (accessed on 1 September 2023).
104. Abuidris, Y.; Kumar, R.; Yang, T.; Onginjo, J. Secure Large-Scale E-Voting System Based on Blockchain Contract Using a Hybrid Consensus Model Combined with Sharding. *ETRI J.* **2021**, *43*, 357–370. [CrossRef]
105. Fatrah, A.; El Kafhali, S.; Haqiq, A.; Salah, K. Proof of Concept Blockchain-Based Voting System. In Proceedings of the 4th International Conference on Big Data and Internet of Things (BDIoT '19), Rabat, Morocco, 23–24 October 2019; Association for Computing Machinery: New York, NY, USA, 2020; pp. 1–5.
106. Zhang, S.; Wang, L.; Xiong, H. Chaintegrity: Blockchain-Enabled Large-Scale E-Voting System with Robustness and Universal Verifiability. *Int. J. Inf. Secur.* **2020**, *19*, 323–341. [CrossRef]
107. Gupta, S.P.; Tripathi, A.M. E-Voting using Blockchain. *J. Physics Conf. Ser.* **2021**, *1911*, 1–14. [CrossRef]
108. Qu, W.; Wu, L.; Wang, W.; Liu, Z.; Wang, H. A Electronic Voting Protocol Based on Blockchain and Homomorphic Signcryption. *Concurr. Comput. Pract. Exp.* **2022**, *34*, e5817. [CrossRef]
109. Carcia, J.C.P.; Benslimane, A.; Boutalbi, S. Blockchain-based system for e-voting using Blind Signature Protocol. In Proceedings of the 2021 IEEE Global Communications Conference (GLOBECOM), Madrid, Spain, 7–11 December 2021; IEEE: Piscataway, NJ, USA, 2021; pp. 01–06.
110. Kurbatov, O.; Kravchenko, P.; Poluyanenko, N.; Shapoval, O.; Kuznetsova, T. Using Ring Signatures For An Anonymous E-Voting System. In Proceedings of the 2019 IEEE International Conference on Advanced Trends in Information Theory (ATIT), Kyiv, Ukraine, 18–20 December 2019; IEEE: Piscataway, NJ, USA, 2022; pp. 187–190.
111. Verma, G. A Secure Framework for E-Voting Using Blockchain. In Proceedings of the 2022 Second International Conference on Computer Science, Engineering and Applications (ICCSEA), Gunupur, India, 8 September 2022; IEEE: Piscataway, NJ, USA, 2022; pp. 1–5.
112. Gupta, S.; Gupta, A.; Pandya, I.Y.; Bhatt, A.; Mehta, K. End to End Secure E-Voting Using Blockchain & Quantum Key Distribution. *Mater. Today Proc.* **2023**, *80*, 3363–3370.
113. Chaieb, M.; Yousfi, S. LOKI Vote: A Blockchain-Based Coercion Resistant E-Voting Protocol. In Proceedings of the Information Systems: 17th European, Mediterranean, and Middle Eastern Conference, EMCIS 2020, Dubai, United Arab Emirates, 25–26 November 2020; Springer International Publishing: Cham, Switzerland, 2020.
114. Golnarian, D.; Saedi, K.; Bahrak, B. A decentralized and trustless e-voting system based on blockchain technology. In Proceedings of the 2022 27th International Computer Conference, Computer Society of Iran (CSICC), Tehran, Islamic Republic of Iran, 23–24 February 2022; IEEE: Piscataway, NJ, USA, 2022; pp. 1–7.
115. Parmar, A.; Gada, S.; Loke, T.; Jain, Y.; Pathak, S.; Patil, S. Secure E-Voting System using Blockchain technology and authentication via Face recognition and Mobile OTP. In Proceedings of the 2021 12th International Conference on Computing Communication and Networking Technologies (ICCCNT), Kharagpur, India, 6–8 July 2021; IEEE: Piscataway, NJ, USA, 2021; pp. 1–5.
116. Li, M.; Luo, X.; Sun, W.; Li, J.; Xue, K. AvecVoting: Anonymous and verifiable E-voting with untrustworthy counters on blockchain. In Proceedings of the ICC 2022-IEEE International Conference on Communications, Seoul, Republic of Korea, 16–20 May 2022; IEEE: Piscataway, NJ, USA, 2022; pp. 4751–4756.
117. Luo, T. An Efficient Blockchain Based Electronic Voting System Using Proxy Multi-signature. In Proceedings of the 2021 3rd International Academic Exchange Conference on Science and Technology Innovation (IAECST), Guangzhou, China, 10–12 December 2021; IEEE: Piscataway, NJ, USA, 2021; pp. 513–516.
118. Doost, M.; Kavousi, A.; Mohajeri, J.; Salmasizadeh, M. Analysis and Improvement of an E-voting System Based on Blockchain. In Proceedings of the 2020 28th Iranian Conference on Electrical Engineering (ICEE), Tabriz, Iran, 4–6 August 2020; IEEE: Piscataway, NJ, USA, 2020; pp. 1–4.
119. Xu, Z.; Cao, S. Efficient Privacy-Preserving Electronic Voting Scheme Based on Blockchain. In Proceedings of the 2020 IEEE International Conference on Smart Internet of Things (SmartIoT), Beijing, China, 14–16 August 2020; IEEE: Piscataway, NJ, USA, 2020; pp. 190–196.
120. Khan, K.M.; Arshad, J.; Khan, M.M. Empirical Analysis of Transaction Malleability within Blockchain-Based E-Voting. *Comput. Secur.* **2021**, *100*, 102081. [CrossRef]
121. Panja, S.; Roy, B. A Secure End-to-End Verifiable E-Voting System Using Blockchain and Cloud Server. *J. Inf. Secur. Appl.* **2021**, *59*, 102815. [CrossRef]
122. Ch, R.; Kumari D, J.; Gadekallu, T.R.; Iwendi, C. Distributed-Ledger-Based Blockchain Technology for Reliable Electronic Voting System with Statistical Analysis. *Electronics* **2022**, *11*, 3308. [CrossRef]

123. Abegunde, J.; Spring, J.; Xiao, H. SEVA: A Smart Electronic Voting Application Using Blockchain Technology. In Proceedings of the 2021 IEEE International Conference on Blockchain (Blockchain), Melbourne, Australia, 6–8 December 2021; IEEE: Piscataway, NJ, USA, 2021; pp. 353–360.
124. Kumar, R.; Badwal, L.; Avasthi, S.; Prakash, A. A Secure Decentralized E-Voting with Blockchain & Smart Contracts. In Proceedings of the 2023 13th International Conference on Cloud Computing, Data Science & Engineering (Confluence), Noida, India, 19–20 January 2023; IEEE: Piscataway, NJ, USA, 2023; pp. 419–424.
125. Jain, A.K.; Kalra, S.; Kapoor, K.; Jangra, V. Blockchain-Based Secure E-Voting System Using Aadhaar Authentication. In *Predictive Data Security Using AI: Insights and Issues of Blockchain, IoT, and DevOps*; Springer Nature: Singapore, 2022; pp. 89–103.
126. Sudha, N.; Reddy, A.B. E-Voting System Using U-Net Architecture with Blockchain Technology. In *Intelligent Computing and Applications: Proceedings of ICDIC 2020*; Springer Nature: Singapore, 2022; pp. 69–79.
127. Díaz-Santiso, J.; Fraga-Lamas, P. E-Voting System Using Hyperledger Fabric Blockchain and Smart Contracts. *Eng. Proc.* **2021**, *7*, 11.
128. Saeed, S.H.; Hadi, S.M.; Hamad, A.H. Iraqi Paradigm E-Voting System Based on Hyperledger Fabric Blockchain Platform. *Ing. Syst. Inf.* **2022**, *27*, 737–745. [CrossRef]
129. Awalu, I.L.; Kook, P.H.; Lim, J.S. Development of a Distributed Blockchain EVoting System. In Proceedings of the 2019 10th International Conference on E-Business, Management and Economics (ICEME), Beijing, China, 15–17 July 2019; Association for Computing Machinery: New York, NY, USA, 2019; pp. 207–216.
130. OpenSSL Foundation, Inc. OpenSSL. Available online: https://www.openssl.org/docs/man1.1.1/man7/SM2.html (accessed on 24 May 2023).
131. Cheng, Z. The SM9 Cryptographic Schemes. Cryptology ePrint Archive, Paper 2017/117. 2017. Available online: https://eprint.iacr.org/2017/117 (accessed on 6 March 2023).
132. Werth, J.; El Ioini, N.; Hajian Berenjestanaki, M.; Barzegar, H.R.; Pahl, C. A Platform Selection Framework for Blockchain-Based Software Systems Based on the Blockchain Trilemma. In Proceedings of the ENASE, Prague, Czech Republic, 24–25 April 2023; pp. 362–371.
133. Mustafa, M.K.; Waheed, S. An E-Voting Framework with Enterprise Blockchain. In *Advances in Distributed Computing and Machine Learning: Proceedings of ICADCML 2020*; Springer: Singapore, 2021.
134. Anwar ul Hassan, C.; Hammad, M.; Iqbal, J.; Hussain, S.; Ullah, S.S.; AlSalman, H.; Mosleh, M.A.A.; Arif, M. A Liquid Democracy Enabled Blockchain-Based Electronic Voting System. *Sci. Program.* **2022**, *2022*, 1–10. [CrossRef]
135. Olaniyi, O.M.; Dogo, E.M.; Nuhu, B.K.; Treiblmaier, H.; Abdulsalam, Y.S.; Folawiyo, Z. A Secure Electronic Voting System Using Multifactor Authentication and Blockchain Technologies. In *Blockchain Applications in the Smart Era*; Springer International Publishing: Cham, Switzerland, 2022; pp. 41–63.
136. Madhani, N.; Gajria, V.; Kanani, P. Distributed and Anonymous E-Voting Using Blockchain and Ring Signatures. In *Communication and Intelligent Systems: Proceedings of ICCIS 2020*; Springer: Singapore, 2021.
137. Subah, Z.; Rozario, S.; Islam, N.; Amir, S.A.B. Blockchain Technology Integrated Electronic Vote Casting System. In Proceedings of the 2nd International Conference on Computing Advancements, Dhaka, Bangladesh, 10–12 March 2022; ACM: New York, NY, USA, 2022; pp. 133–137.
138. Neziri, V.; Shabani, I.; Dervishi, R.; Rexha, B. Assuring Anonymity and Privacy in Electronic Voting with Distributed Technologies Based on Blockchain. *Appl. Sci.* **2022**, *12*, 5477. [CrossRef]
139. Verwer, M.B.; Dionysiou, I.; Gjermundrød, H. TrustedEVoting (TeV) a Secure, Anonymous and Verifiable Blockchain-Based e-Voting Framework. In Proceedings of the E-Democracy—Safeguarding Democracy and Human Rights in the Digital Age, Athens, Greece, 12–13 December 2019; Springer International Publishing: Cham, Switzerland, 2020; pp. 129–143.
140. Khan, K.M.; Arshad, J.; Khan, M.M. Simulation of Transaction Malleability Attack for Blockchain-Based E-Voting. *Comput. Electr. Eng.* **2020**, *83*, 106583. [CrossRef]
141. Indrason, N.; Khongbuh, W.; Saha, G. Blockchain-Based Boothless E-Voting System. In *International Conference on Innovative Computing and Communications: Proceedings of ICICC 2020*; Springer: Singapore, 2021; Volume 1, p. 1.
142. Pooja, S.; Raju, L.K.; Chhapekar, U. Face Detection Using Deep Learning to Ensure a Coercion Resistant Blockchain-Based Electronic Voting. *Eng. Sci.* **2021**, *16*, 341–353.
143. Tandon, S.; Singh, N.; Porwal, S.; Satiram; Maurya, A.K. E-Matdaan: A Blockchain based Decentralized E-Voting System. In Proceedings of the 2022 IEEE Students Conference on Engineering and Systems (SCES), Prayagraj, India, 1–3 July 2022; IEEE: Piscataway, NJ, USA, 2022; pp. 1–6.
144. S. A, S.; Kumar, K.T.G. E-voting System using Public Blockchain. In Proceedings of the 2022 IEEE 2nd Mysore Sub Section International Conference (MysuruCon), Mysuru, India, 16–17 October 2022; IEEE: Piscataway, NJ, USA, 2022; pp. 1–6.
145. Adiputra, C.K.; Hjort, R.; Sato, H. A Proposal of Blockchain-Based Electronic Voting System. In Proceedings of the 2018 Second World Conference on Smart Trends in Systems, Security and Sustainability (WorldS4), London, UK, 30–31 October 2018; IEEE: Piscataway, NJ, USA, 2018; pp. 22–27.
146. Killer, C.; Rodrigues, B.; Matile, R.; Scheid, E.; Stiller, B. Design and Implementation of Cast-as-Intended Verifiability for a Blockchain-Based Voting System. In Proceedings of the 35th Annual ACM Symposium on Applied Computing (SAC 2020), Brno, Czech Republic, 30 March–3 April 2020; pp. 286–293.

147. Kyazhin, S.; Popov, V. Yet Another E-Voting Scheme Implemented Using Hyperledger Fabric Blockchain. In Proceedings of the Computational Science and Its Applications—ICCSA 2020, Cagliari, Italy, 1–4 July 2020; Springer International Publishing: Cham, Switzerland, 2020; pp. 37–47.
148. Ouyang, J.; Deng, Y.; Tang, H. Blockchain Electronic Voting System for Preventing One Vote and Multiple Investment. In Proceedings of the Blockchain and Trustworthy Systems: First International Conference, BlockSys 2019, Guangzhou, China, 7–8 December 2019; Springer: Singapore, 2020; Volume 1.
149. APEH, J.; Ayo, C.K.; Adebiyi, A. Implementing a Secured Offline Blockchain Based Electronic Voting System. *J. Theor. Appl. Inf. Technol.* **2022**, *100*, 18.
150. Malhotra, M.; Kumar, A.; Kumar, S.; Yadav, V. Untangling E-Voting Platform for Secure and Enhanced Voting Using Blockchain Technology. In *Transforming Management with AI, Big-Data, and IoT*; Springer International Publishing: Cham, Switzerland, 2022; pp. 51–72.
151. Tyagi, A.K.; Fernandez, T.F.; Aswathy, S.U. Blockchain and Aadhaar based Electronic Voting System. In Proceedings of the 2020 4th International Conference on Electronics, Communication and Aerospace Technology (ICECA), Coimbatore, India, 5–7 November 2020; IEEE: Piscataway, NJ, USA, 2020; pp. 498–504.
152. Kumar, A.V.; Sarvani, G.V.; Satya, D. Blockchain Based Public Cloud Security for E-Voting System on IoT Environment. In Proceedings of the IOP Conference Series: Materials Science and Engineering, Warangal, India, 9–10 October 2020; IOP Publishing: Bristol, UK, 2020; p. 042013.
153. Barański, S.; Szymański, J.; Sobecki, A.; Gil, D.; Mora, H. Practical I-voting on stellar blockchain. *Appl. Sci.* **2020**, *10*, 7606. [CrossRef]
154. Pandey, A.; Bhasi, M.; Chandrasekaran, K. VoteChain: A Blockchain Based E-Voting System. In Proceedings of the 2019 Global Conference for Advancement in Technology (GCAT), Bangalore, India, 18–20 October 2019; IEEE: Piscataway, NJ, USA, 2019; pp. 1–4.
155. Kumar, M. Securing the E-voting system through blockchain using the concept of proof of work. In Proceedings of the 2021 International Conference on Technological Advancements and Innovations (ICTAI), Tashkent, Uzbekistan, 10–12 November 2021; IEEE: Piscataway, NJ, USA, 2021; pp. 423–427.
156. Echchaoui, H.; Roumaissa, B.; Boudour, R. A Proposal of Blockchain and NFC-Based Electronic Voting System. In Proceedings of the Advanced Computational Techniques for Renewable Energy Systems, Tamanghasset, Algeria, 20–22 November 2021; Springer International Publishing: Cham, Switzerland, 2023; pp. 66–75.
157. Kumar, D.; Dwivedi, R.K. Designing a Secure E Voting System Using Blockchain with Efficient Smart Contract and Consensus Mechanism. In Proceedings of the International Conference on Advanced Network Technologies and Intelligent Computing, Varanasi, India, 22–24 December 2022; Springer Nature Switzerland: Cham, Switzerland, 2023; pp. 452–469.
158. Rosasooria, Y.; Mahamad, A.K.; Saon, S.; Isa, M.A.M.; Yamaguchi, S.; Ahmadon, M.A. E-Voting on Blockchain using Solidity Language. In Proceedings of the 2020 Third International Conference on Vocational Education and Electrical Engineering (ICVEE), Surabaya, Indonesia, 3–4 October 2020; IEEE: Piscataway, NJ, USA, 2020; pp. 1–6.
159. VasanthaKumar, C.; Kabilan, V.; Kathiravan, M.; Ragashanmugam, R.G. A Study on Decentralized Electronic-Voting Using Blockchain. In Proceedings of the 2022 Second International Conference on Advanced Technologies in Intelligent Control, Environment, Computing & Communication Engineering (ICATIECE), Bangalore, India, 16–17 December 2022; IEEE: Piscataway, NJ, USA, 2022; pp. 1–7.
160. Kohad, H.; Kumar, S.; Ambhaikar, A. Scalability of Blockchain based E-voting system using Multiobjective Genetic Algorithm with Sharding. In Proceedings of the 2022 IEEE Delhi Section Conference (DELCON), New Delhi, India, 11–13 February 2022; IEEE: Piscataway, NJ, USA, 2022; pp. 1–4.
161. Baudier, P.; Kondrateva, G.; Ammi, C.; Seulliet, E. Peace engineering: The contribution of blockchain systems to the e-voting process. *Technol. Forecast. Soc. Chang.* **2021**, *162*, 120397. [CrossRef]
162. Khudoykulov, Z.; Tojiakbarova, U.; Bozorov, S.; Ourbonalieva, D. Blockchain based e-voting system: Open issues and challenges. In Proceedings of the 2021 International Conference on Information Science and Communications Technologies (ICISCT), Tashkent, Uzbekistan, 3–5 November 2021; IEEE: Piscataway, NJ, USA, 2021; pp. 1–5.
163. Wahab, Y.; Ghazi, A.; Al-Dawoodi, A.; Alisawi, M.; Abdullah, S.; Hammood, L.; Nawaf, A. A Framework for Blockchain Based E-Voting System for Iraq. *Int. J. Interact. Mob. Technol.* **2022**, *16*, 210–222. [CrossRef]
164. Sudharsan, B.; Tharun, V.R.; Nidhish, K.M.P.; Raj, J.B.; Surya, A.M.; Alagappan, M. Secured Electronic Voting System Using the Concepts of Blockchain. In Proceedings of the 2019 IEEE 10th Annual Information Technology, Electronics and Mobile Communication Conference (IEMCON), Vancouver, Canada, 17–19 October 2019; IEEE: Piscataway, NJ, USA, 2019; pp. 675–681.
165. Dhulavvagol, P.M.; Bhajantri, V.H.; Totad, S.G. Blockchain Ethereum Clients Performance Analysis Considering E-Voting Application. *Procedia Comput. Sci.* **2020**, *167*, 2506–2515. [CrossRef]

electronics

MDPI

Review

A Survey on Parameters Affecting MANET Performance

Ahmed M. Eltahlawy [1], Heba K. Aslan [1], Eslam G. Abdallah [2], Mahmoud Said Elsayed [3], Anca D. Jurcut [3,*] and Marianne A. Azer [1,4]

[1] Faculty of Information Technology and Computer Science, Nile University, Cairo 12677, Egypt
[2] Information Systems Security Management, Concordia University of Edmonton, Edmonton, AB T5B 4E4, Canada
[3] School of Computer Science, University College Dublin, Belfield, D04 V1W8 Dublin, Ireland
[4] National Telecommunication Institute, Nile University, Cairo 12677, Egypt
* Correspondence: anca.jurcut@ucd.ie

Abstract: A mobile ad hoc network (MANET) is an infrastructure-less network where mobile nodes can share information through wireless links without dedicated hardware that handles the network routing. MANETs' nodes create on-the-fly connections with each other to share information, and they frequently join and leave MANET during run time. Therefore, flexibility in MANETs is needed to be able to handle variations in the number of existing network nodes. An effective routing protocol should be used to be able to route data packets within this dynamic network. Lacking centralized infrastructure in MANETs makes it harder to secure communication between network nodes, and this lack of infrastructure makes network nodes vulnerable to harmful attacks. Testbeds might be used to test MANETs under specific conditions, but researchers prefer to use simulators to obtain more flexibility and less cost during MANETs' environment setup and testing. A MANET's environment is dependent on the required scenario, and an appropriate choice of the used simulator that fulfills the researcher's needs is very important. Furthermore, researchers need to define the simulation parameters and the other parameters required by the used routing protocol. In addition, if the MANET's environment handles some conditions where malicious nodes perform network attacks, the parameters affecting the MANET from the attack perspective need to be understood. This paper collects environmental parameters that might be needed to be able to set up the required environment. To be able to evaluate the network's performance under attack, different environmental parameters that evaluate the overall performance are also collected. A survey of the literature contribution is performed based on 50 recent papers. Comparison tables and statistical charts are created to show the literature contribution and the used parameters within the scope of the collected papers of our survey. Results show that the NS-2 simulator is the most popular simulator used in MANETs.

Keywords: AODV; DSR; MANET attacks; MANET configuration parameters; MANET evaluation; MANET simulators; NS-2; OLSR

Citation: Eltahlawy, A.M.; Aslan, H.K.; Abdallah, E.G.; Elsayed, M.S.; Jurcut, A.D.; Azer, M.A. A Survey on Parameters Affecting MANET Performance. *Electronics* **2023**, *12*, 1956. https://doi.org/10.3390/electronics12091956

Academic Editor: Hamed Taherdoost

Received: 6 February 2023
Revised: 29 March 2023
Accepted: 17 April 2023
Published: 22 April 2023

1. Introduction

MANETs' nodes create on-the-fly connections with other network nodes without a need for existing infrastructure. These established connections allow all nodes to exchange information and forward packets between each other [1]. Each node contributes to the network by acting as a router that forwards data packets between the source node and the destination node [2].

Before researchers proceed with the setup and testing of a MANET environment, they should be able to select a suitable simulator. Researchers need to know the simulator's key features, and the points of strength and weakness of each simulator to select the simulator which fits the required MANET environment. In this paper, a comparison between the universally used simulators in the MANET is covered.

After selecting the simulation tool, researchers need to understand the different parameters that affect the behavior of MANETs. The efficiency of the network's performance is dependent on the defined environment parameter sets. In this paper, three main categories of parameter sets are defined as follows: (1) simulation parameters are the list of parameters related to the simulation tool where these parameters control the overall network definition, for example, simulation area, simulation time, and the mobility speed of nodes; (2) routing parameters control the routing protocol mechanism; and (3) attack parameters control the effect of malicious nodes on network performance. The performance measurements of a MANET are achieved using evaluation metrics used to evaluate the network's efficiency. In this paper, evaluation metric terms are also described. Figure 1 depicts the MANET simulation environment.

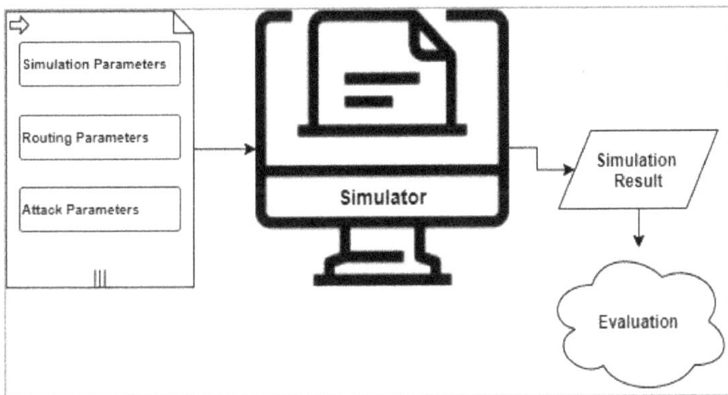

Figure 1. MANETs' simulation environment.

An abundance of the literature covered the effect of changing different environmental parameters on MANETs' performance. In this paper, a survey of 50 recent papers that cover the literature contribution is collected. The main contributions of this paper are summarized as follows:

1. The commonly used simulation tools in a MANET are described, covering the advantages and disadvantages of each. Additionally, statistics on the percentage of usage of these simulation tools are collected against 50 recent papers.
2. The list of routing protocol parameters that control the routing behavior is provided for three routing protocols. Comprehensive flowcharts for the covered routing protocols are provided. Additionally, the routing parameters' usage statistics against 50 recent papers are presented.
3. The simulation parameters used to define a MANET environment are collected, illustrating the usage of each, and statistics on the literature usage percentage of the simulation parameters are covered. The literature range of values used for each simulation parameter is also provided.
4. The main parameters that influence the MANET performance under attack are covered, a list of common attack types on a MANET is collected, and the percentage of usage is shown.
5. The evaluation metric terms used for a performance analysis of MANETs are described. Additionally, statistical tables are collected to show the used environment parameters in our survey papers.

The remainder of the paper is organized as follows. Section 2 is an introduction to different routing protocols in MANETs and their related routing parameters. In Section 3, the list of simulation tools that support MANETs is covered, as well as the simulation parameters and attack parameters that affect MANETs' performance when under attack. Section 4 covers different evaluation metrics used to analyze the network's performance.

In Section 5, the literature contribution is presented. Finally, conclusions and future work suggestions are presented in Section 6.

Table 1 summarizes the abbreviations of terminologies used in this paper.

Table 1. The list of abbreviations.

Notation	Meaning
MANETs	Mobile ad hoc networks
AODV	Ad hoc on-demand distance vector
DSR	Dynamic source routing
OLSR	Optimized link state routing protocol
RREQ	Route request
RREP	Route reply
DPC	Delete period constant
RERR	Route error
TC	Topology control
MPR	Multipoint relays
mph	Mile per hour
DB	Decibel
DDoS	Distributed denial of service
THPT	Average throughput
AETED	Average end-to-end delay time
PDR	Packet delivery ratio
PLR	Packet loss ratio
ROR	Routing overhead ratio
NRL	Normalized routing load
NL	Network load

2. Routing in MANETs

In MANETs, each node is responsible for packet forwarding on behalf of the source node, and it also initiates routing discovery mechanisms to discover its neighbors in the network, then find the best route to reach a destination node [3]. When a new node joins the network, it announces itself by broadcasting a hello message to all neighbors and starts learning about the network [4]. In addition, each node holds a routing table database to maintain a record of the current network nodes as well as the number of hops to reach each node inside the network [5]. There are a multitude of routing protocols related to MANETs' discovery and data forwarding. The three main categories for routing protocols in MANETs are as follows:

- Proactive routing protocols: For example, OLSR, each node maintains its routing table by periodically updating its information [6]; this increases network overhead. On the other hand, routes will always be available with a minimum delay. Proactive protocols provide better performance than reactive protocols as each node continuously updates its awareness of network changes. When a request is received, the packet forwarding procedure is directly handled.
- Reactive routing protocols: For example, AODV and DSR, when a source node tries to perform a packet transmission, it initiates a route discovery mechanism to know how to reach the destination. After the route is determined and updated in the routing table, the packet is forwarded [7]. Reactive protocols have minimal network overhead, but there is a delay time consumed in the route discovery.
- Hybrid routing protocols: For example, ZRP, the close local neighbors to a node are periodically updated, and the global nodes that are not direct neighbors will be updated on demand such as in reactive routing protocols [8].

This paper describes the AODV, DSR, and OLSR routing protocols and the related routing parameters for each. Figure 2 shows a simple classification of the MANETs' routing protocols.

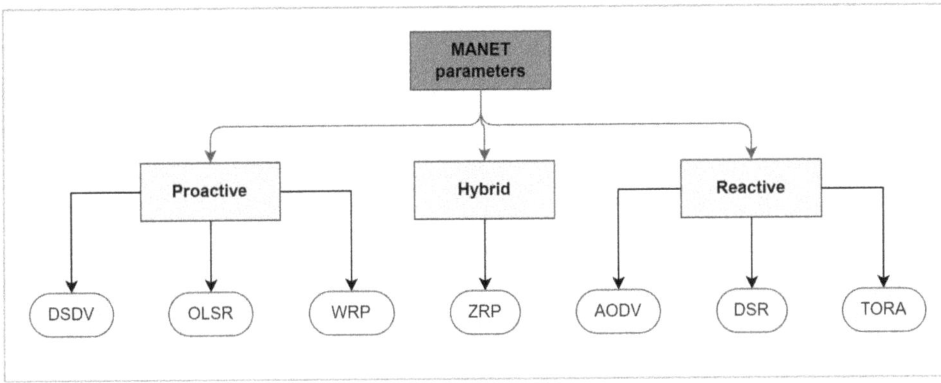

Figure 2. MANETs' routing protocols classification.

2.1. AODV Routing Protocol

AODV is a reactive routing protocol used for MANETs where mobile hosts provide a packet forwarding service acting as an intermediate node between source and destination. In AODV, each node acts as a router and their local routing tables are updated on demand when a request to forward a packet is received or the node is the packet originator [9].

To maintain connectivity between a node and its neighbors, a discovery mechanism is used. AODV discovery mechanism is used to increase the response time for new requests. The route discovery mechanism is initiated by transmitting a RREQ packet to neighbors, asking them to search for the shortest path to the destination. This mechanism increases node awareness with the smallest number of hops needed to reach the destination node. When an intermediate node receives a RREQ, it rebroadcasts the RREQ to all neighbor nodes only in case it does not have a direct connectivity link with the destination node [10].

When an intermediate node has a fresh route to the destination node and the RREQ conditions are fulfilled, the intermediate node sends a RREP in the backward direction to the source. During the forward and reverse path of RREQ and RREP packet forwarding, all intermediate nodes update their local routing table with the latest information contained in the forwarded packet [11].

Each routing table entry contains the following information fields [12]:

1. Destination node address;
2. Number of hop counts to reach the destination;
3. Intermediate nodes address;
4. Route entry expiry time;
5. Destination node sequence number.

When the source node receives the RREP packet, it can begin sending the data needed. If the source node is out of a MANET's range during the active route request, it can initiate another route discovery request with a different request identification.

To ensure that connectivity is present between neighbors, each node periodically sends a hello message. A hello message is a type of RREP packet that is used to announce the node's existence inside the network. If a node has not participated in any packet forwarding or has not sent a hello message for a specific period, the link toward this node will be considered broken. The broken node neighbors send RERR packets to their active neighbors in the network to invalidate any existing route that uses this broken node 'as an intermediate node' in data forwarding [13]. The AODV routing protocol flow chart is illustrated in Figure 3.

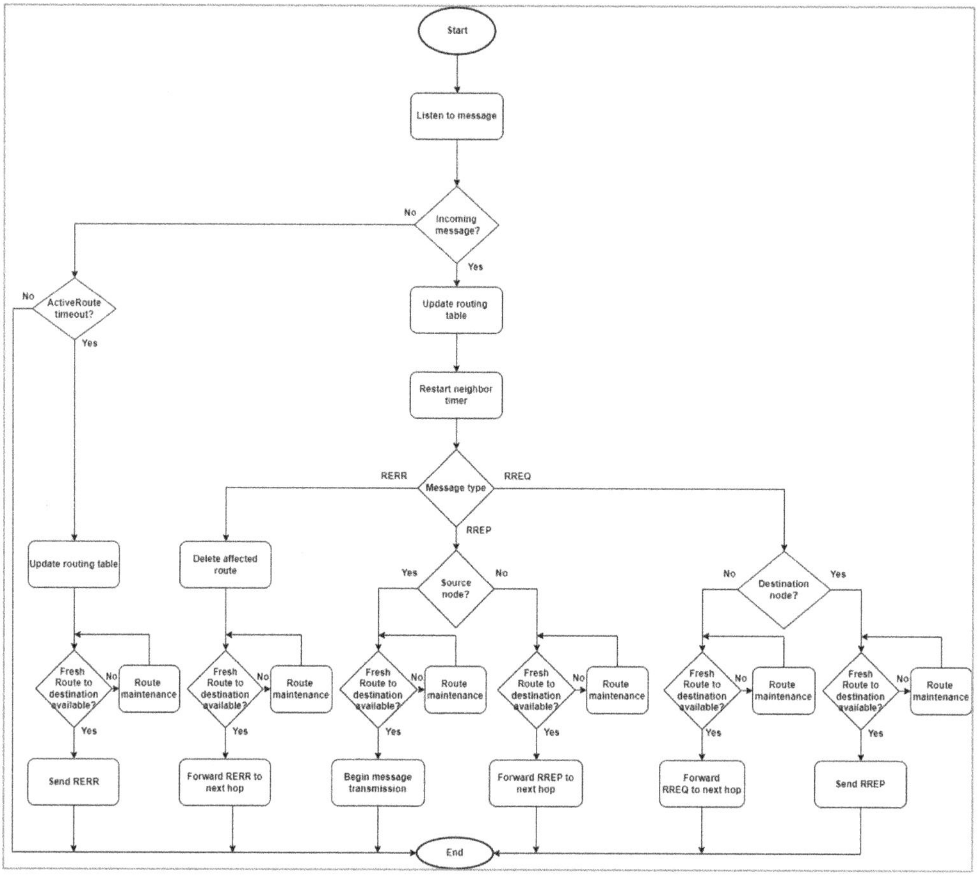

Figure 3. The AODV routing protocol flowchart.

A mobile node holds AODV configuration parameters with default values to control routing protocol operations. The main configuration parameters that affect the AODV protocol are as follows [14]:

- Network diameter: The network diameter value sets the maximum number of hop counts between two nodes in MANETs. The network diameter default value is up to thirty-five hops at most as per RFC 3561 standard.
- Node transversal time: The node transversal time is the estimation of packet transversal time between two neighbor nodes; this estimation should consider the network, processing, and transfer delay time. The default configuration time is 40 ms.
- Network transversal time: The network transversal time is the expected time between sending the RREQ packet and the reception of the RREP packet as per the equation [14]:

$$\text{NetworkTransversalTime} = 2 \times \text{NetworkDiameter} \times \text{NodeTransversalTime} \quad (1)$$

- Route request retry: If a route reply is not received by the source node within the maximum network transversal time, the source node can retry to request the route discovery again for a maximum route request retry times. If the route discovery exceeds the route request retry times, the destination node should be considered unreachable. The default value for the route request retry parameter is equal to 2 retries.
- Blacklist timeout: When the RREP transmission from node A to node B fails, node A records node B in its blacklist buffer. During this blocking time, node A discards any

RREQ from neighbor node B until the blacklist timeout is reached. After the blacklist timeout expires, node B is removed from the blacklist [14].

$$\text{BlackListTimeout} = \text{RouteRequestRetry} \times \text{NetworkTransversalTime} \quad (2)$$

- Route request rate limits: The route request rate limit is the maximum number of RREQ packets for the source node to originate per second. The route request rate limit's default value is ten packets per second.
- Active route timeout: The neighbor node is recorded in the routing table and considered an active node when the active route timeout is not exceeded. When a neighbor node is active, the recorded route to this neighbor should be used [15]. The active route timeout default value is 3000 ms.
- Hello interval: All MANET nodes should reveal their existence in the network within a hello interval time [16]. If a node does not contribute to the routing activities for a hello interval time, it should broadcast a hello message with TTL = 1. Hello interval default value is set to be 1000 ms.
- Allowed hello loss: If a node does not receive any contribution to routing activities from its direct neighbor node for more than (HelloInterval × AllowedHelloLoss), the node should assume a link failure to this neighbor [17]. The allowed hello loss default value is two link failures.
- DPC: After the delete period constant time is expired, the expired route will be deleted from the routing table [18]. The default value for DPC is 5 s.

Table 2 summarizes all AODV configuration parameters and their default values.

Table 2. AODV parameters' default values.

AODV Parameter	Default Value
NetworkDiameter	35 hops
NodeTransversalTime	40 ms
NetworkTransversalTime	1400 ms
RouteRequestRetry	2 retries
BlackListTimeout	2800 ms
RouteRequestRateLimits	10 packets/s
ActiveRouteTimeout	3000 ms
HelloInterval	1000 ms
AllowedHelloLoss	2 times
Delete Period Constant	5 s

2.2. DSR Routing Protocol

DSR is an efficient reactive routing protocol for MANETs. Each data packet contains a header that carries the IP address of all intermediate nodes between a source node and a destination node. The DSR header holds the sequence of hops to reach the destination [19].

In DSR, each node holds a cache memory to store the routing information needed for all MANET nodes; a source node can also cache multiple routes to the same destination. This mechanism allows the routing of data packets to be much more rapid in comparison to other MANETs' routing protocols. There is no need for periodic packets in DSR to minimize network overhead [20]. The DSR protocol is divided into two mechanisms: route discovery and route maintenance [21].

The route discovery mechanism is initiated when a source node does not hold the needed routing information to reach the destination node. The source node broadcasts a RREQ message to all neighbors within the source's wireless range to initiate a route discovery. The RREQ message contains the following information:

1. source node identifier;
2. destination node identifier;
3. route request identifier;
4. record listing the address of all intermediate nodes.

A route maintenance mechanism is issued when the cached route to a destination is no longer valid. When a link to the destination node is broken, the source node can try using another cached route to this destination or it can initiate a route discovery mechanism to find new routes and update the cache. Figure 4 depicts the DSR routing protocol flowchart.

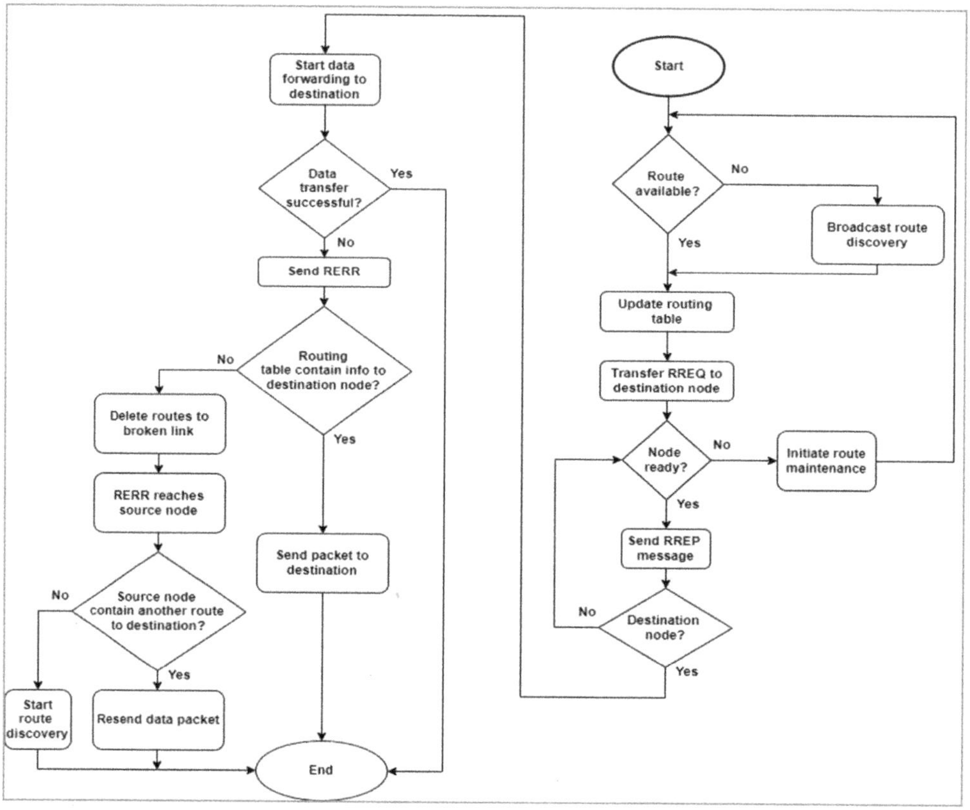

Figure 4. The DSR routing protocol flowchart.

When the destination node receives a RREQ, it examines the route back again to the source node, then it returns a RREP message that holds the accumulated record list back again to the initiator. If the examination of the reverse path to reach the source node fails, the destination node should broadcast a route discovery and then send the RREP message after updating the cached route. The DSR protocol contains a set of configuration parameters that could affect routing in MANETs as follows [22]:

- Discovery hop limit: The discovery hop limit value is defined as the limit to the route request re-broadcast. If the first attempt of RREQ does not reach the destination node, the default value of the discovery hop limit is 255 hops, and the minimum value is one hop.
- Broadcast jitter: The destination node should delay the RREP message by a random value that does not exceed the broadcast jitter's maximum delay time. The broadcast jitter default value is ten milliseconds.
- Route cache timeout: The route cache timeout is associated with each route entry in the cache [23]. When the timeout is reached, this means that the related route is not used and needs to be deleted from the node's cache. Route cache timeout default value is three hundred milliseconds.

- Send buffer timeout: When a packet cannot be transmitted to the next-hop node, this packet is queued inside a buffer to try sending it when possible. Send buffer timeout is the maximum time associated with a packet to be sent before being removed from the send buffer. The default value for send buffer timeout is 30 s.
- Max request period: After a route discovery attempt fails to find a route to the destination node, the time between successive route discovery attempts doubles until the maximum request period is reached. The default value for the maximum request period time is 10 s.
- Re-transmit buffer size: Re-transmit buffer holds the maximum number of packets waiting for the next-hop reachability confirmation. If the buffer is not sufficient to keep the new packet, this packet is discarded without notification. The re-transmit buffer size defines the buffer size with a default value of 50 packets.
- Max maintenance re-transmission: The maximum number of re-transmissions for a packet waiting for a confirmation from the next hop should be limited by the configuration value of the max maintenance re-transmission parameter. The default value is only two transmissions.

Table 3 summarizes the DSR configuration parameters with their default values.

Table 3. DSR parameters' default values.

DSR Parameter	Default Value
DiscoveryHopLimit	1 hop
BroadcastJitter	10 ms
RouteCacheTimeout	300 ms
SendBufferTimeout	30 s
MaxRequestPeriod	10 s
RetransmitBufferSize	50 packet
MaxMaintenanceRetransmit	2 times

2.3. OLSR Routing Protocol

OLSR is a proactive routing protocol that is based on the periodic exchange of control packages to maintain the network topology [24]. Routes to neighbor nodes should be available when needed. OLSR reduces the control packet data rate by only declaring a subset of the neighbors [25]. MPR nodes in most cases are neighbor nodes that are only two hops away with bidirectional links. Multipoint relays can only re-transmit the received broadcast messages, and this technique reduces the useless broadcast messages' re-transmission. Nodes that are not MPR normally process the received messages but do not re-transmit the broadcast messages in MANETs.

In OLSR, a node periodically broadcasts a hello message with all information about the node's neighbors. This hello message allows the neighbors to know the one-hop neighbors and their link state to create the neighbor's table [26]. Additionally, using the information contained in the hello messages, they learn the two hops' neighbors to form the MPR selector table.

To be able to identify the whole network topology and have better scalability, each node periodically transmits another control message (TC) along with the periodic hello messages. A TC message contains the MPR selector list of the transmitter, and this allows network nodes to create their topology table. TC messages are only sent when a node senses a change in its MPR table that needs to be advertised to other nodes with constraints on time between two consecutive TC message transmissions. After receiving a TC message, the receiver should maintain its topology table, either by creating a new entry record or by maintaining an existing node record. Figure 5 covers the OLSR routing protocol mechanism.

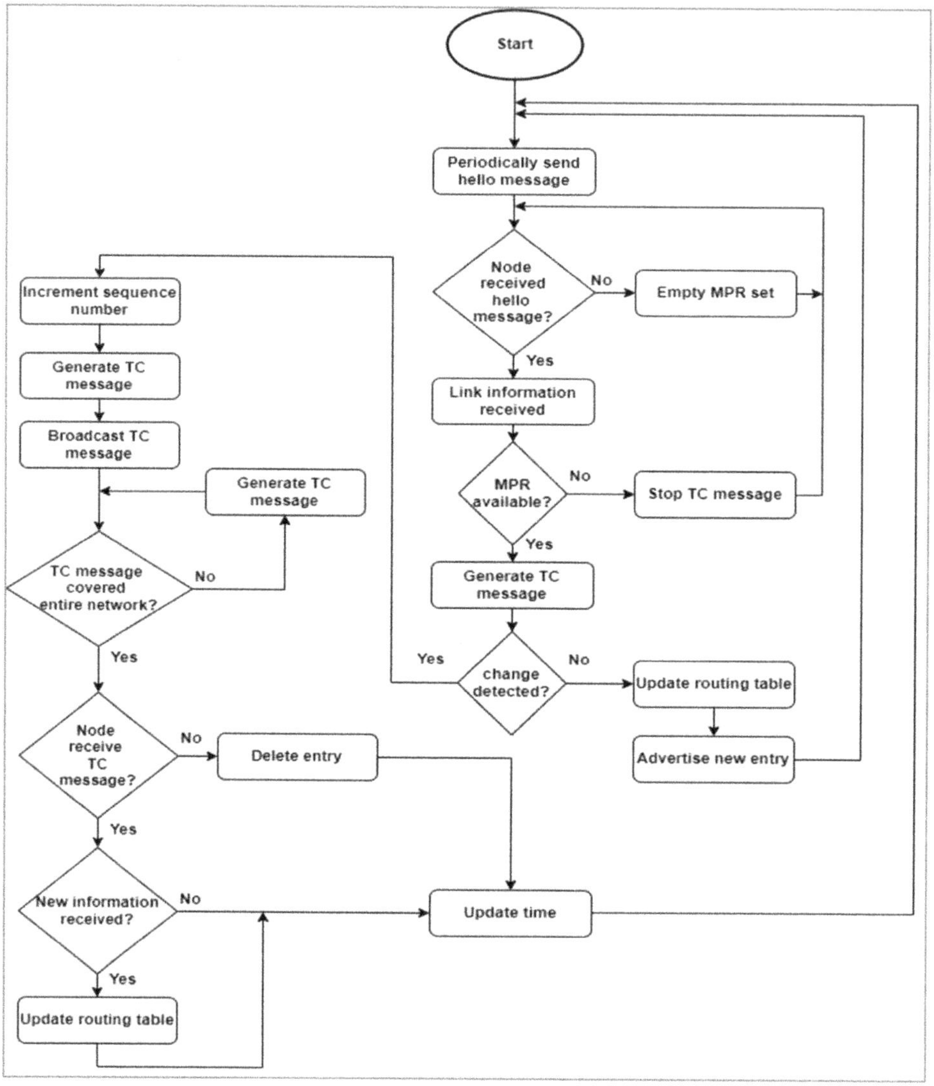

Figure 5. The OLSR routing protocol flowchart.

To be able to control the OLSR performance, some configuration parameters are used below [27]:

- Willingness: Willingness is a configuration parameter that specifies the node's willingness to forward traffic packets to other network nodes [28]. A node may change the willingness during run-time based on conditions such as resource constraints and power limitations. Willingness is an integer value with a range between 0 and 7. 'WILL_NEVER = 0' is the lowest willingness value where this node must not be selected as a MPR for any node. 'WILL_ALWAYS = 7' is the highest willingness for a node to advertise its willingness to forward traffic on behalf of other network nodes. The default willingness value for a node is 'WILL_DEFAULT = 3'.
- Hello interval: Hello interval is the set periodic time between two consecutive hello messages in seconds. The default value is 2 s.

- TC interval: This is the interval time in seconds between two consecutive topology control messages that carry the connectivity information. The TC interval default value is 5 s.
- Refresh interval: Each node must cooperate in the network by sending a periodic hello message before the refresh interval period reaches a timeout. A hello interval must be smaller than or equal to the refresh interval. The default value for the refresh interval parameter is 2 s.
- Neighbor hold time: Defines the link expiry time before declaring it as a broken link [29]. The neighbor hold time default value is 6 s.
- Topology hold time: This is the timeout for the entries in the topology table before being deleted [29]. The topology hold time default value is 15 s.

Table 4 summarizes the default values of OLSR configuration parameters.

Table 4. OLSR parameters' default values.

OLSR Parameter	Default Value
Willingness	WILL_DEFAULT (3)
TCInterval	5 s
RefreshInterval	2 s
NeighbHoldTime	6 s
TopHoldTime	15 s
HelloInterval	2 s

3. Simulation in MANETs

MANET technology is rapidly changing, and new protocols and mechanisms are continuously proposed by researchers. Evaluating a network's performance under different attacks is important to be able to propose protection mechanisms. Therefore, a cost-effective method that empowers researchers to set up and test MANETs plays an important role in research.

3.1. MANETs Simulators

Simulators are software tools used to create a virtual environment that supports researchers to set up and test a network's performance under different conditions. Simulators are GUI-driven tools used to set up a network environment and then perform different attacks on the defined network, or make comparisons between a standard routing protocol and a newly proposed protocol. Using the defined evaluation metrics, a simulator is also capable of collecting the network's results and evaluating the overall performance [30].

There is another method for developers to define and test MANETs using testbeds. Testbeds are experimentation in-lab networks that researchers can set up using dedicated hardware sets for this purpose. Testbeds lack the flexibility to define a MANET network, as MANETs are dynamic networks where nodes continuously join and leave the network. Additionally, the cost is much higher than software simulations to define a MANET using testbeds.

To be able to select a suitable simulator, the researchers need to know the simulator's key features [31]. Table 5 is a comparison between the widely used simulators in MANETs.

Table 5. Comparison between simulation tools in MANETs

Simulator Name	Languages Supported	Platform Support	License	Advantages	Disadvantages
OPNET	C, C++	Windows, Sun Solaris, RedHat Linux	Commercial, Free Educational License	- User-friendly and easy to use. - Provides additional supportive tools.	- Limited wireless mobility. - Not open source and supported protocols are limited. - Expensive. - Lack of energy model.

Table 5. *Cont.*

Simulator Name	Languages Supported	Platform Support	License	Advantages	Disadvantages
OMNeT++	C++, NED	Windows, MacOS, and any Unix-like systems	Open source	- Used by a wide number of users. - Extensive GUI interface. - Intelligence support. - Rich C++ libraries. - Parallelly distributed simulation is supported.	- Documentation is poor. - Performance measures are weak. - Does not cover all protocols.
NS-2	C++, OTCL	Windows, MacOS, Ubuntu, Sun Solaris, Fedora Linux, and any Unix-like systems	Open source	- The most used simulator for research. - Good with complex systems' evaluation. - Provides energy model. - Supports wired and wireless networks.	- Documentation is poor. - Simulation is not real-time. - Lack of supporting tools. - Not suitable for large systems. - Difficult to use and poor GUI. - High computational overhead and memory usage.
NS-3	C++, Python	MacOS, FreeBSD, Linux	Open source	- Very fast simulator where parallel simulation is supported with real-time scheduling. - Supports emulation. - Provides debugging traces. - Organized source code with low-level abstraction. - Good documentation.	- Lacks backward compatibility with NS-2. - Virtualization support is limited. - Difficult to use.
GloMoSim	C, PERSEC	Windows XP/7, FreeBSD, Sun Solaris, Fedora Linux	Free	- Scalable and can handle very large systems with thousands of nodes. - Parallel simulation environment. - Scalable simulation library.	- Documentation is poor. - The simulator is outdated. - Does not support end devices such as simulators.
QualNet and EXATA/cyber	JAVA	Windows NT/2000/XP/Professional, macOS, Sun Solaris, and most Unix-like systems	Commercial	- Provides animation tools. - Scalable and can handle very large systems with thousands of nodes. - Support wired and wireless networks. - Realtime simulator	- Slow interfaces. - Difficult to install. - Expensive.
JIST/SWANS	JAVA, Tcl	Windows, macOS, Sun Solaris Linux	Commercial	- Powerful simulator and suitable for simulating real-world systems. - Less memory usage.	- Features not competing with other simulators.
J-SIM	JAVA	Windows, Sun Solaris Linux	Open source	- Supports wired and wireless networks. - Reusable models with good flexibility.	- Worst execution time.

3.2. Attacks on MANETs' Routing Protocols

The MANET's environment is dynamic and nodes continuously join and leave. An attacker could easily take a critical location in the network to block data packets from being delivered to the destination node. Moreover, a malicious node might produce a high-power signal that covers a wide range of network nodes to introduce itself as the best routing path to forward the packet between the source node and the destination node [32]. This malicious node would then block the data packets from being forwarded to the destination node. Such malicious activity leads to increasing the loss of important data packets, and it is reducing the network's overall throughput.

MANETs suffer from malicious activities where malicious nodes tend to impact the routing protocol mechanism. The direct impact of the attacks on routing protocols is to degrade the MANET's performance. To disrupt the MANET routing protocol, attackers tend to use several techniques such as follows:

1. Routing table overflow attack: In this attack, the attacking node tends to crowd the network by advertising several non-existing nodes to overflow the routing table [33].

This prevents legitimate nodes from being aware of network nodes and routing their packets normally.

2. Flooding attack: In a flooding attack, malicious nodes tend to waste network resources such as memory, bandwidth, and battery by flooding the network with bogus packets [34]. For example, flooding RREQ packets prevents the MANET from functioning normally.
3. DDoS attack: In a DDoS attack, attackers tend to keep the targeted legitimate node busy by continuously requesting RREQ messages from collaborative attackers at the same time without respecting the TTL time [35].
4. False removal of working route: In this attack, the malicious node advertises a false state of the link with the destination node as if the link is broken. This enforces the source node to re-initiate route discovery protocol to find another path to reach the destination. Additionally, it slows down packet transmission. False removal of working route attack could be used with another collaborative attack to isolate the targeted legitimate node from MANET.
5. Node isolation attack: Attackers isolate an innocent node by blocking routing information about this targeted node from the entire network [36]. This leads to an ignorance of the presence of this innocent node.
6. Routing table poisoning: In this attack, the attacker sends false RREQ packets with a higher sequence number to force all nodes to delete the old genuine route to a destination and update this route with a corrupted one.
7. Blackhole attack: The attacker tends to change the routing protocol packets to be the best route known for a targeted destination, and when it is requested to forward data packets to the destination node, it starts discarding the received packets to slow down the network performance [37].
8. Grayhole attack: Grayhole attack is an instance of a blackhole attack where an attacker selectively drops some data packets and normally forwards others [38], or drops all packets but only at a certain time. This makes the attack difficult to detect.
9. Wormhole attack: In a wormhole attack, two attacking nodes cooperate where one attacker at a specific location encapsulates some packets and tunnels them to the second attacker, bypassing all intermediate nodes to introduce itself as the fastest route to a destination and then drop the data packets later [39]. It can also be used to replay the received data packets in the other side of the network to disrupt the routing protocol.
10. Rushing attack: In a rushing attack, the malicious node sends RREQ messages with high-power transmission to introduce itself as the shortest path to any destination with only one hop count [40], this manipulates all network nodes to use this routing path. The rushing attack is most likely used alongside another attack such as dropping the network packets that need forwarding.

3.3. Simulation and Attack Parameters

Researchers need to understand the different parameters used to control the MANET simulation environment, as well as the parameters that affect the network's behavior under attack. The list of simulation and attack parameters is described as follows:

- Maximum simulation time (s): While running any simulator, a simulation time parameter is set to stop the simulation after this timeout is reached [41]; for more accurate results it is preferred to increase the simulation time.
- Medium packet rate (packet/s): To avoid interference and packet loss between nodes due to the wireless medium limitation, a packet rate ratio should be pre-set between all MANET nodes. This parameter depends on the road capacity (number of nodes/mile), the available frequency used for packet transfer, and the used wireless protocol (ex. IEEE 802.11) [42].
- Mobility speed of nodes (m/s): MANET nodes do not have a fixed location, which means that they are moving from one place to another at varying speeds. The speed of nodes affects the result of the simulation.

- Nodes' mobility movement pattern: The mobility pattern of mobile nodes in a MANET comprises one of the following patterns: (1) random way mobility, (2) linear mobility in a straight line, (3) circle mobility, and (4) stationary mobility for fixed nodes across the network.
- Number of intermediate nodes: Increasing the number of intermediate nodes that forward packets between source nodes and destination decreases the routing protocol performance.
- Number of source nodes: In MANETs, source nodes initiate packet transmission procedures; increasing the number of source nodes in MANET will overload the channel with more packets overhead.
- Number of malicious nodes: Increasing the number of malicious nodes in MANETs decreases overall network performance.
- Position of intermediate nodes: The position of intermediate nodes inside MANETs affects the performance. As the number of intermediate nodes between the source node and the destination node increases, network performance increases.
- Position of malicious nodes: Attackers tend to take a good physical position between source and destination nodes to be able to perform the planned attack and drop the network packets.
- Data packet payload (byte/packet): Data packet payload is the percentage of real data bytes (excluding the control and header data bytes) divided by the overall packet size in bytes. The data packet payload is an indication of the actual gain from packet transmission.
- Simulation area: Simulates the MANETs' network coverage area in m^2. The simulation area reflects on the density of nodes inside the network, which impacts the routing protocol mechanisms.
- Antenna type: The following are the antenna types and properties used for wireless communication: (1) the isotropic antenna transmits equal signal power in all directions; (2) the omnidirectional antenna transmits equal power in all horizontal directions, decreasing to zero along the vertical axis; and (3) the directional antenna transmits only in one direction at a specified angle.
- Transportation protocol type: Transport protocol is based on two types: (1) the TCP protocol is a connection-oriented protocol that requires a connection establishment between the sender and the receiver first before sending data packets. This leads to a more secure and guaranteed delivery of data packets. On the other hand, the TCP protocol slows down packet delivery due to the needed overhead of handshaking. (2) UDP is a connectionless protocol that needs no connection establishment, which is faster but less reliable for packet delivery.
- Transmission power: Each node needs to configure the transmission power that defines the range that this node could reach in one hop. Increasing transmission power leads to more coverage, but also means more energy consumption and quick battery drain.
- Mobility speed of malicious nodes: MANETs have a dynamic network structure, which means that at certain times the network consists of some nodes that could leave the network after a while. Malicious node mobility speed is a key factor in affecting network performance. The attacker could use its speed to target an innocent node and isolate it from the network by simply taking a position between this innocent node and the destination node while traveling.
- Transmission power of malicious nodes: The power of transmission for a malicious node could be valuable when the attacker aims to introduce itself as the shortest path between source and destination nodes. This malicious node can then drop the network packets later.

Figure 6 summarizes the different types of MANETs' parameters.

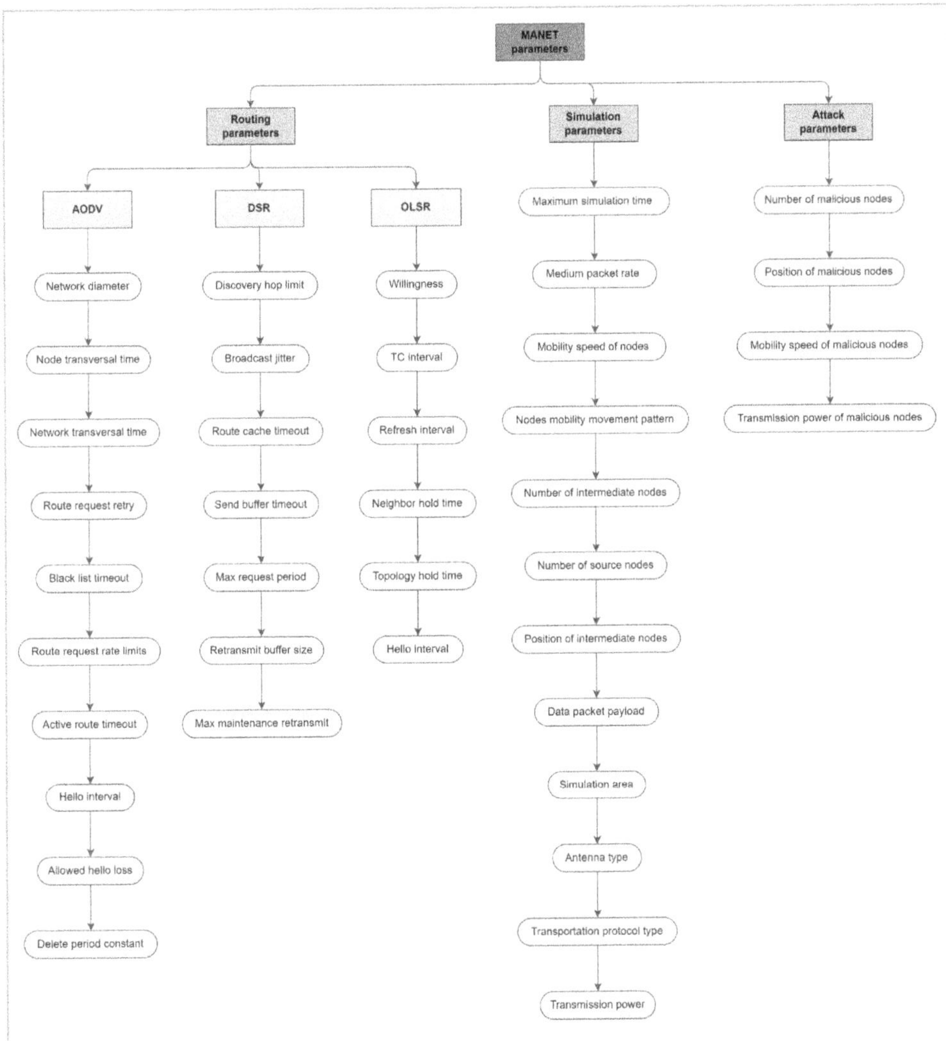

Figure 6. The different types of MANETs' parameters.

4. Evaluation Metrics and Performance Analysis in MANETs

Different evaluation metrics are used to define the characteristics of the MANET performance under certain conditions. After researchers set up the simulation environment and define the parameters needed to control the MANET environment, the results of the simulation tool need to be evaluated. To analyze the network performance, some metrics are used as follows:

- THPT: Throughput is the rate of successfully delivered packets that reached the receiver node per time slot [43]. Throughput is affected by topology changes, noise on communication links, the power of transmission from the source node, and the existence of malicious nodes affecting the throughput ratio.
- AETED: Average end-to-end delay is the average time taken to send a packet to the destination node [44]. This delay is due to many reasons such as route discovery queuing and process latency, delays caused by wireless links, and processing delays at both the sender and the receiver sides.

- PDR: Packet delivery ratio is the ratio of packets that are received by the destination across the overall transmitted packets from the source node [45]. The packet delivery ratio represents the maximum throughput that can be achieved by the MANET network.
- PLR: Packet loss ratio is the opposite of PDR; PLR measures the total lost packets that did not reach the destination node across the overall transmitted packets [46].
- ROR: Routing overhead ratio is the size of control and header packets needed by the protocol for route discovery and maintenance over the total data packets received by the destination node [47].
- NRL: Normalized routing load is the ratio between the total number of control packets sent by a source node over the total number of data packets received by a destination node [48]. An increase in normalized routing load metric indicates the efficiency of the used routing protocol.
- NL: The network load is the average amount of data packets that are being carried by the entire network over time [49]. Increasing the network load ratio increases the possibility of data collision in the wireless medium.

Figure 7 is a conclusion of the evaluation metric terms used in MANETs.

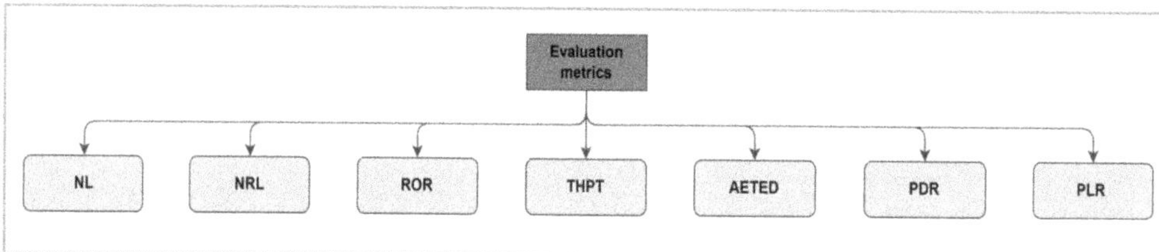

Figure 7. The different evaluation metrics used in MANETs.

5. Related Work

An abundance of the literature covered the effect of changing different environmental parameters on MANETs' performance. Statistical analyses regarding the topics covered by the researchers and the areas which require more attention in the future are performed. All selected references share in common the AODV routing protocol. AODV protocol is one of the widely used routing protocols in MANETs [49] as it has a wide range of advantages compared with other protocols. AODV is loop-free and scales to a large number of nodes, is adaptable to topology changes and responds to changes quickly, supports both unicast and multicast transmissions, has a minimal routing overhead, and has lower setup delay [50]. Some researchers conduct a performance analysis comparison between the AODV routing protocol and other routing protocols such as OLSR and DSR protocols, while other researchers focus on the performance of the AODV protocol under attack. A part of the literature contribution focuses on analyzing the effect of changing some parameters, such as mobility speed or network density, to analyze the effect of changing such parameters on the AODV protocol. Furthermore, other researchers propose enhancements to existing protocols while others propose new mechanisms for routing.

The current survey is based on 50 recent papers that share in common the AODV routing protocol. Table 6 summarizes the used routing protocol parameters within the scope of collected papers.

Out of 50 papers, only five references covered the effect of changing routing protocol parameters on the overall performance. As shown in Table 7, the percentage of the usage of routing protocols in MANETs does not exceed 6% of the literature contribution. Other routing parameters that are not mentioned in Table 6 were not used in the current survey papers. More focus and contributions are needed from the literature to address the effect of changing the routing parameters.

Table 6. Survey on routing parameter usage in MANETs.

Reference Name	Routing Protocol	Network Diameter	Node Transversal Time	RREQ Retries	Max RREQ Timeout	Active Route Timeout	Delete Period Timeout
Observation of AODV Routing Protocol's Performance at Variation in ART Value for Various Node's Mobility [15]	AODV	x	x	x	-	x	x
Impact of Active Route Time Out and Delete Period Constant on AODV Performance [18]	AODV, DSR	-	-	-	-	x	x
Comparative Performance Analysis of AODV for CBR and VBR Traffic under Influence of ART and DPC [23]	AODV	-	-	-	-	x	x
Performance Optimization of MANET Networks through Routing Protocol Analysis [51]	AODV, OLSR	-	-	x	x	-	-

(x) parameter is used, (-) parameter is not used.

Table 7. Percentage of routing parameter usage in MANETs.

Routing Parameter	Papers	Percentage of Usage
Network diameter	1 of 50	2%
Node transversal time	1 of 50	2%
RREQ retries	2 of 50	4%
Max RREQ timeout	1 of 50	2%
Active route timeout	3 of 50	6%
Delete period	3 of 50	6%
Other parameters	0 of 50	0%

Some researchers analyzed the effect of attacking the MANET routing protocol under different environments and attack scenarios. As shown in Table 8, the literature has placed more focus on blackhole and grayhole attacks. Based on a study of the most common attacks on the MANET network layer [52], the study shows that blackhole and grayhole attacks are globally introduced to affect MANETs and they also have a high impact on MANET performance. Table 8 compares the simulation and attack parameters used in the collected papers.

Table 8. Survey on simulation and attack parameters usage in MANETs.

Reference Name	Simulator	Network Area	Simulation Time	Mobility Speed (m/s)	Number of Network Nodes	Number of Malicious Nodes	Attack Type	Packet Rate (Packet/s)	Mobility Model
Performance Analysis of MANET under Grayhole Attack Using AODV Protocol [1]	NS-2	1000 m × 850 m	1200 s	-	10	1	Grayhole	-	Random waypoint
A Comparative Study of Reactive, Proactive, and Hybrid Routing Protocol in Wireless Sensor Network Under Wormhole Attack [7]	QualNet 5.0	400 m × 100 m	17 min	10	50	1, 8	Wormhole	-	Random waypoint
Securing Blackhole Attacks in MANETs using Modified Sequence Number in AODV Routing Protocol [10]	NS-2	500 m × 500 m	100 s	-	[10–50]	[1–10]	Blackhole	-	Random waypoint
Simulation-Based Study of Blackhole Attack under AODV Protocol [12]	NS-2	500 m × 500 m	[20–100] s	0, 50	[20–100]	0–1	Blackhole	[5–25]	Random waypoint

Table 8. *Cont.*

Reference Name	Simulator	Network Area	Simulation Time	Mobility Speed (m/s)	Number of Network Nodes	Number of Malicious Nodes	Attack Type	Packet Rate (Packet/s)	Mobility Model
Blackhole Attack Detection in Vehicular Ad Hoc Network Using Secure AODV Routing Algorithm [32]	NS-2	650 m × 1000 m	100 s	-	100	1	Blackhole	-	-
Identifying the Impacts of Active and Passive Attacks on Network Layer in a Mobile Ad Hoc Network: A Simulation Perspective [33]	NS-2	-	5 s	-	10, 15, 20, 25, 30	1, 2	Blackhole, Wormhole, Grayhole	-	-
An Effective Approach to Detect and Prevent Collaborative Grayhole Attack by Malicious Node in MANET [38]	NS-3	300 m × 1500 m	200 s	-	50	0, 10	Grayhole	-	Random waypoint
Comparative Analysis of Blackhole and Rushing Attack in MANET [40]	NS-2	1000 m × 1000 m	200 s	-	50	5, 10, 15, 20	Blackhole, Rushing	-	-
VRA-AODV: Routing Protocol Detects Blackhole and Grayhole Attacks in Mobile Ad Hoc Network [43]	NS-2	3200 m × 1000 m	200 s	-	100	1	Blackhole, Grayhole	2 packets/s	Random waypoint
A Dynamic Threshold-based Algorithm for Improving Security and Performance of AODV Under Black-hole Attack in MANET [45]	NS-2	750 m × 750 m	500 s	20	10, 60	0, 1	Blackhole, Grayhole	-	Random waypoint
Defending Against Smart Grayhole Attack Within MANETs: A Reputation Based Ant Colony Optimization Approach for Secure Route Discovery in DSR Protocol [46]	NS-2	200 m × 200 m	300 s	-	-	1	Grayhole	-	Random waypoint
A Novel Approach for Mitigating Gray hole Attack in MANET [47]	NS-2	750 m × 750 m	500 s	5, 15, 25, 35	48	0, 1, 2	Grayhole	-	Random waypoint
Evaluation of Blackhole Attack with Avoidance Scheme using AODV Protocol in VANET [53]	NS-2	650 m × 650 m	1000 s	-	20	0, 1	Blackhole	-	Random waypoint, Highway, City
Entity-Centric Combined Trust (ECT) Algorithm to Detect Packet Dropping Attack in Vehicular Ad Hoc Networks (VANETs) [54]	NS-2	3000 m × 3000 m	500 s	30	[100–600]	10, 20, 30, 40, 50, 60	Blackhole	-	Highway
Blackhole Attack Prevention in MANET Using Enhanced AODV Protocol [55]	GloMoSim 2.03	1600 m × 1600 m	1 h	1, 5, 10, 20, 50	20	1	Blackhole	1, 2, 4, 6, 8 packet/s	Random waypoint
Design and Analysis of an Improved AODV Protocol for Black hole and Flooding Attack in Vehicular Ad Hoc Network (VANET) [56]	NS-2	-	-	-	3, 5, 10	1	Blackhole, Flooding	-	-
Detection and Prevention of Black Hole Attacks in Mobile Ad Hoc Networks [57]	NS-2	1000 m × 1000 m	500 s	[0–20]	50	0, 1, 2	Blackhole	-	Random waypoint
Gray Hole Attack Analysis in AODV Based Mobile Adhoc Network with Reliability Metric [58]	NS-2	7000 m × 500 m	100 s	5, 10, 15, 20, 25	50, 100, 150, 500	0, 5, 10	Grayhole	-	Random waypoint
Effect of Wormhole Attacks on MANET [59]	NS-2	1000 m × 850 m	1200 s		5, 30	0, 2	Wormhole	-	Random waypoint
An Approach to Detect Wormhole Attack in AODV based MANET [60]	NS-2	750 m × 750 m	-	-	10, 20, 50	0,1	Wormhole	-	Random waypoint
An Approach to Prevent Gray-hole Attacks on Mobile Ad Hoc Networks [61]	NS-2	750 m × 550 m	500 s	-	20, 30, 40	-	Grayhole	-	-

Table 8. *Cont.*

Reference Name	Simulator	Network Area	Simulation Time	Mobility Speed (m/s)	Number of Network Nodes	Number of Malicious Nodes	Attack Type	Packet Rate (Packet/s)	Mobility Model
A Novel Solution for Grayhole Attack in AODV Based MANETs [62]	NS-2	800 m × 800 m	50 s	20	5, 30	1, 7	Grayhole	-	-
BP-AODV: Blackhole Protected AODV Routing Protocol for MANETs Based on Chaotic Map [63]	NS-2	1000 m × 500 m	200 s	20, 25	25	1	Blackhole	-	-
Intelligent Detection of Black Hole Attacks for Secure Communication in Autonomous and Connected Vehicles [64]	NS-2	1000 m × 1000 m	500 s	30	50, 60, 70, 80	0, 4	Blackhole	-	-
Impact Analysis of Blackhole, Flooding, and Grayhole Attacks and Security Enhancements in Mobile Ad Hoc Networks Using SHA3 Algorithm [65]	NS-2	1200 m × 1200 m	-	30	10, 100	1, 5	Blackhole, Grayhole, Flooding	-	-
Comparative Performance Analysis of AODV and DSR Routing Protocols under Wormhole Attack in Mobile Ad Hoc Network on Different Node's Speeds [66]	QualNet 5.0	1500 m × 1500 m	300 s	10, 15, 20, 25, 30	20	2	Wormhole	-	Random waypoint
Performance Evaluation of AODV and AOMDV Routing Protocols under Collaborative Blackhole and Wormhole Attacks [67]	NS-2	1200 m × 800 m	-	-	50, 80, 100, 120	0, 1, 2	Blackhole, Wormhole	-	-
Black Hole Attacks Analysis for AODV and AOMDV Routing Performance in VANETs [68]	NS-2	1000 m × 1000 m	100 s	11, 16, 22	10	1	Blackhole	-	-
Performance Analysis of AODV and DSR Routing Protocols of MANET under Wormhole Attack and a Suggested Trust-Based Routing Algorithm for DSR [69]	EXata/Cyber 1.2	2500 m × 2500 m	300 s	-	20, 40, 60, 80, 100, 120, 140, 160, 180, 200	2, 3, 4	Wormhole	-	Random waypoint

(-) parameter is not used.

Table 8 shows a wide variety in the simulation and attack parameters used to set up the MANET environment. All research papers share in common the random waypoint mobility model. The network area for small networks was found to be 200 m × 200 m, while for extensive networks, the network area does not exceed 2500 m × 2500 m. The range of simulation time was found to be from 5 s up to 1 h, and the mobility speed range is between 0 for static nodes up to 50 m per second. Additionally, from Table 8, the number of network nodes for small networks is between 3 and 50 nodes, and for very large networks, the number of nodes reaches 600 nodes with a varying number of malicious nodes inside—the number of malicious nodes varies between 0 and 60 malicious nodes. Table 9 presents a conclusion of the ranges used in the literature for simulation and attack parameters.

Table 9. Simulation and attack parameters' range of used values in MANETs.

Parameter	Range of Used Values
Network area (m^2)	[200 × 200, 2500 × 2500]
Simulation time (s)	[5, 3600]
Mobility speed (m/s)	[0, 50]
Number of network nodes	[3, 600]
Number of malicious nodes	[0, 60]
Packet rate (packet/s)	[1, 25]

In Table 8, only 29 papers out of the 50 collected papers cover scenarios where MANETs are under attack where each researcher uses a different simulation tool and parameters to deploy the MANET environment. From Table 8, the NS-2 simulation tool is the most used tool, followed by both NS-3 and OPNET simulators. Figure 8 shows the percentage of simulation tool usage in MANETs.

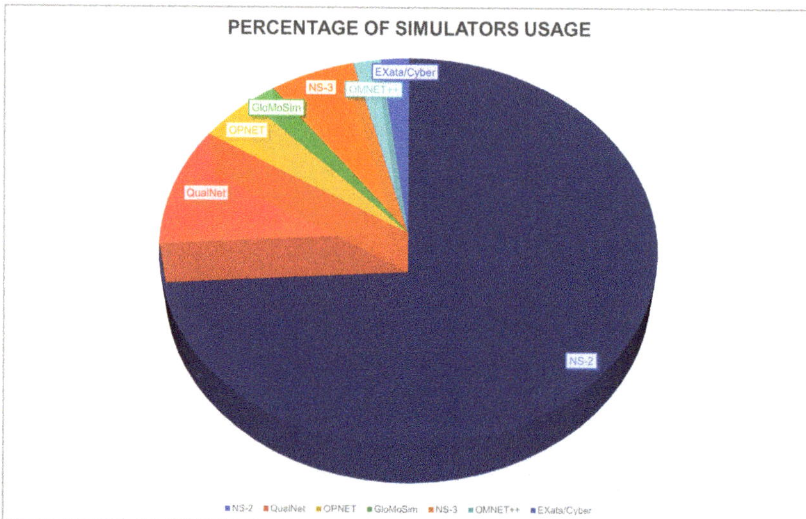

Figure 8. Percentage of simulation tool usage in MANETs.

Table 10 shows the use of evaluation metrics according to the collected papers of this survey.

Table 10. Survey on evaluation metrics use in MANETs.

Reference Name	Throughput	Average End-to-End Delay	Packet Delivery Ratio	Packet Loss Ratio	Routing Overhead Ratio	Normalized Routing Load	Network Load
Performance Analysis of MANET under Grayhole Attack Using AODV Protocol [1]	x	-	-	-	-	-	-
Performance Evaluation of AODV, OLSR, and GRP for Transmitting Video Conferencing over MANETs [2]	x	x	x	-	-	-	x
Performance Analysis of Routing Protocols AODV, OLSR, and DSDV on MANET using NS3 [3]	x	x	x	x	-	-	-
Performance Evaluation and Analysis of Proactive and Reactive MANET Protocols at Varied Speeds [4]	-	x	x	-	-	-	-
A Comparative Study of Reactive, Proactive, and Hybrid Routing Protocol in Wireless Sensor Network Under Wormhole Attack [6]	x	x	-	-	-	-	-
Performance Comparison and Evaluation of the Proactive and Reactive Routing Protocols for MANETs [7]	x	x	x	-	-	-	-
Securing Blackhole Attacks in MANETs using Modified Sequence Number in AODV Routing Protocol [10]	x		x	-	-	-	-

Table 10. *Cont.*

Reference Name	Throughput	Average End-to-End Delay	Packet Delivery Ratio	Packet Loss Ratio	Routing Overhead Ratio	Normalized Routing Load	Network Load
Simulation-Based Study of Blackhole Attack under AODV Protocol [12]	x	x	x	x	x	-	-
Observation of AODV Routing Protocol's Performance at Variation in ART Value for Various Node's Mobility [15]	x	x	-	-	-	-	-
Impact of Active Route Time Out and Delete Period Constant on AODV Performance [18]	x	x	x	-	-	-	-
Survey on Performance Analysis of AODV, DSR, and DSDV in MANET [19]	x	x	x	x	-	-	-
Analysis of Routing Protocols for Ad Hoc Networks [20]	x	x	x	-	x	-	-
Comparative Performance Analysis of AODV for CBR and VBR Traffic under Influence of ART and DPC [23]	x	x	-	-	-	-	-
Performance Evaluation of OLSR and AODV Routing Protocols over Mobile Ad Hoc Networks [24]	x	x	x	x	x	-	-
Investigating the Impact of Mobility Models on MANET Routing Protocols [25]	x	x	-	-	-	-	x
Blackhole Attack Detection in Vehicular Ad Hoc Network Using Secure AODV Routing Algorithm [32]	x	-	x	-	-	-	-
Identifying the Impacts of Active and Passive Attacks on Network Layer in a Mobile Ad Hoc Network: A Simulation Perspective [33]	x	x	x	x	-	-	-
Performance Analysis of Black Hole Attack and Flooding Attack AODV Routing Protocol on VANET (Vehicular Ad Hoc Network) [34]	x	x	-	-	-	-	-
An Effective Approach to Detect and Prevent Collaborative Grayhole Attack by Malicious Node in MANET [38]	x	-	x	x	x	-	-
Comparative Analysis of Blackhole and Rushing Attack in MANET [40]	x	x	x	x	-	-	-
VRA-AODV: Routing Protocol Detects Blackhole and Grayhole Attacks in Mobile Ad Hoc Network [43]	x	-	x	-	x	-	-
A Dynamic Threshold-based Algorithm for Improving Security and Performance of AODV Under Black-hole Attack in MANET [45]	x	-	x	x	x	x	-
Defending Against Smart Grayhole Attack Within MANETs: A Reputation Based Ant Colony Optimization Approach for Secure Route Discovery in DSR Protocol [46]	x	x	x	x	-	-	-
A Novel Approach for Mitigating Grayhole Attack in MANET [47]	x	x	x	x	x	x	-
Comparative Study of Routing Protocols for Mobile Ad Hoc Networks [49]	x	x	x	x	x	x	-

Table 10. *Cont.*

Reference Name	Throughput	Average End-to-End Delay	Packet Delivery Ratio	Packet Loss Ratio	Routing Overhead Ratio	Normalized Routing Load	Network Load
Performance Optimization of MANET Networks through Routing Protocol Analysis [51]	x	x	x	x	x	-	-
Evaluation of Black Hole Attack with Avoidance Scheme Using AODV Protocol in VANET [53]	x	x	x	x	-	-	-
Entity-Centric Combined Trust (ECT) Algorithm to Detect Packet Dropping Attack in Vehicular Ad Hoc Networks (VANETs) [54]	x	x	x	x	x	-	-
Blackhole Attack Prevention in MANET Using Enhanced AODV Protocol [55]	-	x	x	-	x	-	-
Design and Analysis of an Improved AODV Protocol for Black Hole and Flooding Attack in Vehicular Ad Hoc Network (VANET) [56]	-	x	x	x	x	-	-
Detection and Prevention of Black Hole Attacks in Mobile Ad Hoc Networks [57]	x	-	-	x	x	-	-
Grayhole Attack Analysis in AODV Based Mobile Adhoc Network with Reliability Metric [58]	x	-	-	-	-	-	-
Effect of Wormhole Attacks on MANET [59]	x	-	-	-	-	-	-
An Approach to Detect Wormhole Attack in AODV based MANET [60]	-	-	x	-	-	-	-
An Approach to Prevent Gray-hole Attacks on Mobile Ad Hoc Networks [61]	x	x	x	-	-	-	-
A Novel Solution for Grayhole Attack in AODV Based MANETs [62]	-	x	x	-	-	x	-
BP-AODV: Blackhole Protected AODV Routing Protocol for MANETs Based on Chaotic Map [63]	x	x	x	-	-	-	-
Intelligent Detection of Black Hole Attacks for Secure Communication in Autonomous and Connected Vehicles [64]	x	x	x	x	x	-	-
Impact Analysis of Blackhole, Flooding, and Grayhole Attacks and Security Enhancements in Mobile Ad Hoc Networks Using SHA3 Algorithm [65]	x	x	x	-	-	-	-
Comparative Performance Analysis of AODV and DSR Routing Protocols under Wormhole Attack in Mobile Ad Hoc Network on Different Node's Speeds [66]	x	x	-	-	-	-	-
Performance Evaluation of AODV and AOMDV Routing Protocols under Collaborative Blackhole and Wormhole Attacks [67]	x	x	x	-	-	-	-
Black Hole Attacks Analysis for AODV and AOMDV Routing Performance in VANETs [68]	x	-	-	x	-	-	-
Performance Analysis of AODV and DSR Routing Protocols of MANET Under Wormhole Attack and a Suggested Trust-Based Routing Algorithm for DSR [69]	x	x	-	-	-	-	-

Table 10. *Cont.*

Reference Name	Throughput	Average End-to-End Delay	Packet Delivery Ratio	Packet Loss Ratio	Routing Overhead Ratio	Normalized Routing Load	Network Load
Analyzing the Impact of the Number of Nodes on the Performance of the Routing Protocols in a MANET Environment [70]	x	x	x	-	-	-	-
A Performance Study of Various Mobility Speed on AODV Routing Protocol in Homogeneous and Heterogeneous MANET [71]	x	-	x	-	-	-	-
Logistic Regression Based Reliability Analysis for Mobile Ad Hoc Network with Fixed Maximum Speed and Varying Pause Times [72]	x	-	-	-	-	-	-
A Performance Review of Intra and Inter-Group MANET Routing Protocols under Varying Speed of Nodes [73]	x	x	x	-	x	-	-
Energy Analysis of AODV Routing Protocol in MANET [74]	x	-	x	-	-	-	-
Performance Comparison of Modified AODV-ETX with AODV and AODV-ETX Routing Protocol in a MANET [75]	x	x	x	-	-	-	-

(x) parameter is used, (-) parameter is not used.

For the 50 surveyed references, throughput is the most used evaluation metric, and average end-to-end delay and packet delivery ratio are also widely used in the evaluation. Figure 9 shows the usage statistics of evaluation metrics.

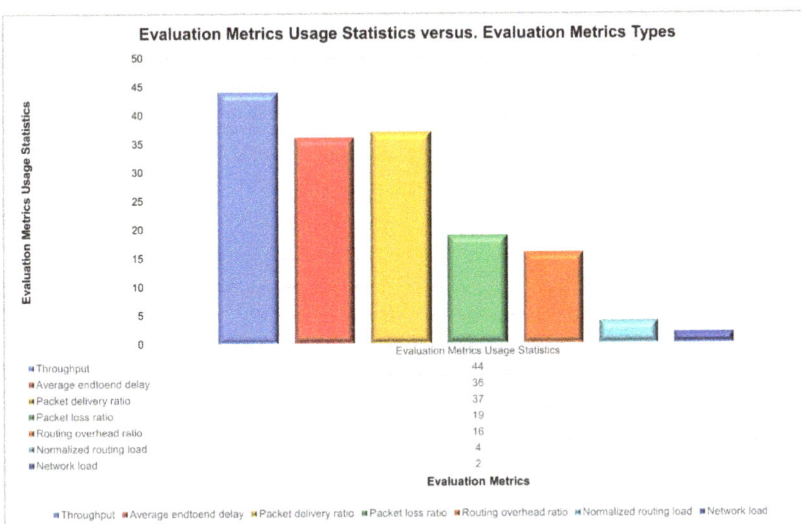

Figure 9. Percentage of evaluation metrics usage in MANETs.

Based on Table 11, a comparison is performed to show the progress made in the current survey compared to another recent survey paper [76]. After reviewing the literature contribution, in [76] is the only paper we found that covers the area of interest of the current paper. Another one of the literature contribution was found to be either outdated or partially covered the current area of interest.

Table 11. Comparison of current contribution with the literature.

	Our Survey Paper	**A Review on Parameters of Internet Gateway Discovery in MANETS [76]**
Number of papers used in the survey	50	72
Covered simulation tools	Provide simulation tool key features + Cover statistics of the following simulators: - NS-2 - NS-3 - OMNET++ - OPNET - GloMoSim - QualNet and EXATA/cyber - JIST/SWANS - J-SIM	Cover statistics of the following simulators: - NS-2 - MATLAB - OMNET++ - OPNET
Covered simulation parameters	Cover statistics of the following simulation parameters: - Simulation time - Packet rate - Mobility speed - Movement pattern/model - Number of intermediate nodes - Number of source nodes - Position of nodes - Packet payload/size - Simulation area - Antenna type - Transport protocol - Transmission power	Cover statistics of the following simulation parameters: - Mobility model - Speed of the nodes - Pause time - Packet size - Packet rate - Topology size - Number of nodes - Transmission range - Simulation time - Traffic type
Covered routing parameters	Cover the following routing protocols + All related routing parameters: - AODV - DSR - OLSR	-
Covered attack parameters	Cover attack types in MANETs + Cover the following attack parameters: - Number of malicious nodes - Position of malicious nodes - Speed of malicious nodes - Transmission power of malicious nodes	-
Covered evaluation metrics	Cover statistics of the following evaluation metrics: - THPT - PDR - PLR - AE2ED - ROR - NRL - NL	Cover statistics of the following evaluation metrics: - THPT - PDR - PDF - AE2ED - ROR - NRL

(-) parameter is not covered.

6. Conclusions and Future Work

The efficiency of packet forwarding between nodes depends on the network environment. To set up the MANET environment, researchers need to select a suitable simulator that fits the needed environment. Researchers use MANET simulation tools for different purposes, some of them conduct a performance analysis comparison between different routing protocols, whereas others check the performance of specific protocols under attack. Moreover, a part of the literature contribution analyzes the effect of changing the environment parameters on performance, and others use simulation tools to evaluate the performance of a newly introduced protocol. To be able to control MANET behavior and set up the needed environment for evaluation, researchers should be familiar with different parameters that affect the MANET environment. The efficiency of the MANET's performance is controlled by different parameters that are clustered into three group sets: (1) simulation parameters, (2) routing parameters, and (3) attack parameters.

In this paper, the key features of different simulation tools in MANETs are provided. A survey is performed against 50 recent papers to summarize the literature contribution. The list of simulation parameter values used in the surveyed papers is mentioned. Additionally, the performed statistics show that NS-2 is the most popular simulator used in the MANET. In addition, the results of this survey show that the minimum defined network area for small networks was found to be 200 m × 200 m, and for extensive networks, the network area does not exceed 2500 m × 2500 m. The range of simulation time was found to be from 5 s up to 1 h, and the mobility speed range is between 0 for static nodes up to 50 m per second. Furthermore, the number of network nodes for small networks is between 3 and 50 nodes, and for extremely large networks, the number of nodes reaches 600 nodes with a varying number of malicious nodes inside. Additionally, the statistics show that the number of malicious nodes varies between 0 and 60 malicious nodes. All parameters that control the MANET behavior are described along with a list of commonly used evaluation metrics that are used to evaluate network performance. Furthermore, the literature contribution is collected for all parameters. It is noticed that checking the effect of changing routing parameters on the network's performance is not particularly focused on in the literature.

Future work is recommended to focus on evaluating the effect of changing routing parameters on a MANET's performance. Additionally, an analysis of malicious activities on MANETs under different environments is needed. Finally, the detection and prevention of MANET attacks is an active research area of great interest to many researchers that warrants further exploration.

Author Contributions: Conceptualization, A.M.E., H.K.A., E.G.A. and M.A.A.; methodology, A.M.E., H.K.A., E.G.A. and M.A.A.; software, A.M.E.; validation, H.K.A., E.G.A. and M.A.A.; formal analysis, A.M.E.; investigation, A.M.E.; resources, A.M.E.; data curation, A.M.E., H.K.A., E.G.A., M.S.E., A.D.J. and M.A.A.; writing—original draft preparation, A.M.E.; writing—review and editing, A.M.E., H.K.A., E.G.A., M.S.E., A.D.J. and M.A.A.; visualization, A.M.E. and M.A.A.; supervision, H.K.A., E.G.A., M.S.E., A.D.J. and M.A.A.; project administration, H.K.A. and M.A.A.; funding acquisition, A.D.J. All authors have read and agreed to the published version of the manuscript.

Funding: This research received no external funding.

Data Availability Statement: Not Applicable.

Conflicts of Interest: The authors declare that they have no known competing financial interest or personal relationships that could have appeared to influence the work reported in this paper.

References

1. Reddy, B.; Dhananjaya, B. The AODV routing protocol with built-in security to counter blackhole attack in MANET. *Mater. Today Proc.* **2022**, *50*, 1152–1158. [CrossRef]
2. Ahmed, D.E.; Ibrahim, H.; Khalifa, O. Performance Evaluation of AODV, OLSR, and GRP for Transmitting Video Conferencing over MANETs. *Int. J. Comput. Sci. Inf. Secur.* **2020**, *18*, 45–51.

3. Kurniawan, A.; Kristalina, P.; Hadi, M.Z.S. Performance Analysis of Routing Protocols AODV, OLSR and DSDV on MANET using NS3. In Proceedings of the 2020 International Electronics Symposium (IES), Surabaya, Indonesia, 29–30 September 2020; pp. 199–206. [CrossRef]
4. Skaggs-Schellenberg, R.; Wang, N.; Wright, D. Performance Evaluation and Analysis of Proactive and Reactive MANET Protocols at Varied Speeds. In Proceedings of the 10th Annual Computing and Communication Workshop and Conference (CCWC), Las Vegas, NV, USA, 6–8 January 2020; pp. 0981–0985. [CrossRef]
5. Ferdous, R.; Muthukkumarasamy, V. A Comparative Performance Analysis of MANETs Routing Protocols in Trust-Based Models. In Proceedings of the International Conference on Computational Science and Computational Intelligence (CSCI), Las Vegas, NV, USA, 15–17 December 2016; pp. 880–885. [CrossRef]
6. Govindasamy, J.; Punniakody, S. A comparative study of reactive, proactive and hybrid routing protocol in wireless sensor network under wormhole attack. *J. Electr. Syst. Inf. Technol.* **2018**, *5*, 735–744. [CrossRef]
7. Bai, Y.; Mai, Y.; Wang, N. Performance comparison and evaluation of the proactive and reactive routing protocols for MANETs. In Proceedings of the Wireless Telecommunications Symposium (WTS), Chicago, IL, USA, 26–28 April 2017; pp. 1–5. [CrossRef]
8. Panda, N.; Patra, B.; Hota, S. Manet Routing Attacks and Their Countermeasures: A Survey. *J. Crit. Rev.* **2020**, *7*, 2777–2792. [CrossRef]
9. Saudi, N.A.M.; Arshad, M.A.; Buja, A.G.; Fadzil, A.F.A.; Saidi, R. Mobile Ad-Hoc Network (MANET) Routing Protocols: A Performance Assessment. In Proceedings of the Third International Conference on Computing, Mathematics and Statistics, Singapore, 27 March 2019; pp. 53–59. [CrossRef]
10. Shrestha, S.; Baidya, R.; Giri, B.; Thapa, A. Securing Blackhole Attacks in MANETs using Modified Sequence Number in AODV Routing Protocol. In Proceedings of the 8th International Electrical Engineering Congress (iEECON), Chiang Mai, Thailand, 4–6 March 2020; pp. 1–4. [CrossRef]
11. Kumari, A.; Krishnan, S. Simulation-Based Study of Blackhole Attack Under AODV Protocol. In Proceedings of the Fourth International Conference on Computing Communication Control and Automation (ICCUBEA), Pune, India, 16–18 August 2018; pp. 1–6. [CrossRef]
12. Jubair, M.A.; Muniyandi, R.C. NS2 Simulator to Evaluate the Effective of Nodes Number and Simulation Time on the Reactive Routing Protocols in MANET. *Int. J. Appl. Eng. Res.* **2016**, *11*, 11394–11399.
13. Agrawal, R.; Tripathi, R.; Tiwari, S. Performance Evaluation and Comparison of AODV and DSR Under Adversarial Envi-ronment. In Proceedings of the International Conference on Computational Intelligence and Communication Networks, Gwalior, India, 7–9 October 2011; pp. 596–600. [CrossRef]
14. Perkins, C.; Belding-Royer, E.; Das, S. *RFC3561: Ad Hoc On-Demand Distance Vector (AODV) Routing*; IETF: Santa Barbara, CA, USA, 2003. [CrossRef]
15. Gupta, S.K.; Saket, R.K. Observation of AODV Routing Protocol's Performance at Variation in ART Value for Various Node's Mobility. In Proceedings of the First International Conference on Information and Communication Technology for Intelligent Systems: Volume 1. Smart Innovation, Systems and Technologies, Maghreb, Tunisia, 18–20 December 2018; Springer: Cham, Switzerland, 2016; Volume 50. [CrossRef]
16. Sharma, Y.; Sharma, A.; Sengupta, J. Performance evaluation of Mobile Ad hoc Network routing protocols under various se-curity attacks. In Proceedings of the International Conference on Methods and Models in Computer Science, New Delhi, India, 13–14 December 2010; pp. 117–124. [CrossRef]
17. Gupta, S.K.; Alsamhi, S.; Saket, R.K. Optimal Relation between ART and Mobility & Transmission Range at Default QualNet & Calculated Transmission Powers. In Proceedings of the 6th International Conference on Advances in Engineering Sciences and Applied Mathematics (ICAESAM-2016), Kuala Lumpur, Malaysia, 21–22 December 2016. [CrossRef]
18. Agrawal, N.; Fatima, M. Impact of Active Route Time Out and Delete Period Constant on AODV Performance. *Int. J. Comput. Appl.* **2016**, *147*, 19–25. [CrossRef]
19. Aggarwal, A.; Gandhi, S.; Chaubey, N. Performance analysis of aodv, dsdv and dsr in manets. *Int. J. Distrib. Parallel Syst.* **2014**, *2*, 167–177. [CrossRef]
20. Desai, R.; Patil, B.P. Analysis of routing protocols for Ad Hoc Networks. In Proceedings of the 2014 International Conference on Circuits, Systems, Communication and Information Technology Applications (CSCITA), Mumbai, India, 4–5 April 2014; pp. 111–115. [CrossRef]
21. Bobade, N.P.; Mhala, N.N. Performance Evaluation of AODV and DSR On-Demand Routing Protocols with Varying MANET Size. *Int. J. Wirel. Mob. Netw.* **2012**, *4*, 183–196. [CrossRef]
22. Johnson, D.; Hu, Y. *The Dynamic Source Routing Protocol (DSR) for Mobile Ad Hoc Networks for IPv4*; Rice University: Houston, TX, USA, 2007; Available online: https://tools.ietf.org/html/rfc4728 (accessed on 5 February 2023).
23. Gupta, S.K.; Alsamhi, S.H.; Saket, R.K. Comparative performance analysis of AODV for CBR & VBR traffic under influence of ART & DPC. In Proceedings of the 11th International Conference on Industrial and Information Systems, Roorkee, India, 3–4 December 2016; pp. 112–117. [CrossRef]
24. Hashim, A.-A.; Farhan, M.M.; Alshybani, S. Performance Evaluation of OLSR and AODV Routing Protocols over Mobile Ad-hoc Networks. In Proceedings of the First International Conference of Intelligent Computing and Engineering, Hadhramout, Yemen, 15–16 December 2019; pp. 1–8. [CrossRef]

25. Abdullah, A.M.; Ozen, E.; Bayramoglu, H. Investigating the Impact of Mobility Models on MANET Routing Protocols. *Int. J. Adv. Comput. Sci. Appl.* **2019**, *10*, 25–35. [CrossRef]
26. Jacquet, P.; Muhlethaler, P.; Clausen, T.; Laouiti, A.; Qayyum, A.; Viennot, L. Optimized link state routing protocol for ad hoc networks. In Proceedings of the IEEE International Multi Topic Conference: Technology for the 21st Century, IEEE INMIC 2001, Lahore, Pakistan, 30 December 2001; pp. 62–68. [CrossRef]
27. Clausen, T.; Jacquet, P.; Adjih, C.; Laouiti, A.; Minet, P.; Muhlethaler, P.; Qayyum, A.; Viennot, L. Optimized link state routing protocol (OLSR). RFC3626. *Technol. Rep.* **2003**, *1*, 1–75.
28. Boushaba, A.; Benabbou, A.; Benabbou, R.; Zahi, A.; Oumsis, M. Optimization on OLSR protocol for reducing topology control packets. In Proceedings of the International Conference on Multimedia Computing and Systems, Tangiers, Morocco, 10–12 May 2012; pp. 539–544. [CrossRef]
29. Kumar, M.; Sharma, C.; Dhiman, A.; Rangra, A.K. Performance Variation of Routing Protocols with Mobility and Scalability in MANET. In *Advances in Intelligent Systems and Computing*; Springer: Singapore, 2018; Volume 638, pp. 9–21. [CrossRef]
30. Malhotra, J. A survey on MANET simulation tools. In Proceedings of the 2014 Innovative Applications of Computational In-telligence on Power, Energy and Controls with their impact on Humanity (CIPECH), Ghaziabad, India, 28–29 November 2014; pp. 495–498. [CrossRef]
31. Dorathy, I.; Chandrasekaran, M. Simulation tools for mobile ad hoc networks: A survey. *J. Appl. Res. Technol.* **2018**, *16*, 437–445. [CrossRef]
32. Kumar, A.; Varadarajan, V.; Kumar, A.; Dadheech, P.; Choudhary, S.S.; Kumar, V.A.; Panigrahi, B.; Veluvolu, K.C. Black hole attack detection in vehicular ad-hoc network using secure AODV routing algorithm. *Microprocess. Microsyst.* **2020**, *80*, 103352. [CrossRef]
33. Ahamed, U.; Fernando, S. Identifying the Impacts of Active and Passive Attacks on Network Layer in a Mobile Ad-hoc Network: A Simulation Perspective. *Int. J. Adv. Comput. Sci. Appl.* **2020**, *11*, 600–605. [CrossRef]
34. Fiade, A.; Triadi, A.Y.; Sulhi, A.; Masruroh, S.U.; Handayani, V.; Suseno, H.B. Performance Analysis of Black Hole Attack and Flooding Attack AODV Routing Protocol on VANET (Vehicular Ad-Hoc Network). In Proceedings of the 8th International Conference on Cyber and IT Service Management, Pangkal Pinang, Indonesia, 23–24 October 2020; pp. 1–5. [CrossRef]
35. Kaur, T.; Kumar, R. Mitigation of Blackhole Attacks and Wormhole Attacks in Wireless Sensor Networks Using AODV Protocol. In Proceedings of the IEEE International Conference on Smart Energy Grid Engineering, Oshawa, ON, Canada, 12–15 August 2018; pp. 288–292. [CrossRef]
36. Poongodi, T.; Khan, M.S.; Patan, R.; Gandomi, A.H.; Balusamy, B. Robust Defense Scheme Against Selective Drop Attack in Wireless Ad Hoc Networks. *IEEE Access* **2019**, *7*, 18409–18419. [CrossRef]
37. Hameed, A.; Al-Omary, A. Survey of Blackhole attack on MANET. In Proceedings of the 2nd Smart Cities Symposium, Bahrain, Bahrain, 24–26 March 2019; pp. 1–4. [CrossRef]
38. Yadav, S.; Kumar, R.; Tiwari, N.; Bajpai, A. An Effective Approach to Detect and Prevent Collaborative Grayhole Attack by Malicious Node in MANET. In *Intelligent Systems Design and Applications, Advances in Intelligent Systems and Computing*; Springer: Cham, Switzerland, 2021; Volume 1181. [CrossRef]
39. Dash, S.P. Study of Blackhole Attack and Wormhole Attack in Vanet Environment and Their Countermeasure. Master's Thesis, Michigan Technological University, Houghton, MI, USA, 2019.
40. Sivanesh, S.; Dhulipala, V.S. Comparaitive Analysis of Blackhole and Rushing Attack in MANET. In Proceedings of the In-ternational Conference on Microwave Integrated Circuits, Photonics and Wireless Networks, Tiruchirappalli, India, 22–24 May 2019; pp. 495–499. [CrossRef]
41. Chavan, A.; Kurule, D.; Dere, P. Performance Analysis of AODV and DSDV Routing Protocol in MANET and Modifications in AODV against Black Hole Attack. *Procedia Comput. Sci.* **2016**, *79*, 835–844. [CrossRef]
42. Cali, F.; Conti, M.; Gregori, E. IEEE 802.11 protocol: Design and performance evaluation of an adaptive backoff mechanism. *IEEE J. Sel. Areas Commun.* **2000**, *18*, 1774–1786. [CrossRef]
43. Vo, T.T.; Luong, T.N. Vra-Aodv: Routing Protocol Detects Blackhole and Grayhole Attacks in Mobile Ad hoc Network. *J. Comput.* **2018**, *13*, 222–235. [CrossRef]
44. Mai, Y.; Bai, Y.; Wang, N. Performance Comparison and Evaluation of the Routing Protocols for MANETs Using NS3. *J. Electr. Eng.* **2017**, *5*, 187–195. [CrossRef]
45. Gurung, S.; Chauhan, S. A dynamic threshold based algorithm for improving security and performance of AODV under black-hole attack in MANET. *Wirel. Netw.* **2019**, *25*, 1685–1695. [CrossRef]
46. Ourouss, K.; Naja, N.; Jamali, A. Defending Against Smart Grayhole Attack Within MANETs: A Reputation-Based Ant Colony Optimization Approach for Secure Route Discovery in DSR Protocol. *Wirel. Pers. Commun.* **2021**, *116*, 207–222. [CrossRef]
47. Gurung, S.; Chauhan, S. A novel approach for mitigating gray hole attack in MANET. *Wirel. Netw.* **2018**, *24*, 565–579. [CrossRef]
48. Husieen, N.A.; Kadhum, A.N. The Effect of Pause Time on the Performance of Mobile Ad-hoc Network Routing Protocols. In Proceedings of the 4th International Conference on Intelligent Information Technology Application, Qinghuangdao, China, 5–7 November 2010.
49. El-Kabbany, A.F.; Ali, H.M.; Hussein, A.; Tawfeek, B. Comparative study of routing protocols for mobile ad hoc networks. *Int. J. Intell. Comput. Inf. Sci.* **2017**, *17*, 31–43. [CrossRef]

50. Hassnawi, L.A.; Ahmad, R.B.; Yahya, A.; Aljunid, S.A.; Elshaikh, M. Performance Analysis of Various Routing Protocols for Motorway Surveillance System Cameras' Network. *Int. J. Comput. Sci. Issues (IJCSI)* **2012**, *9*, 7.
51. Priyambodo, T.K.; Wijayanto, D.; Gitakarma, M.S. Performance Optimization of MANET Networks through Routing Protocol Analysis. *Computers* **2021**, *10*, 2. [CrossRef]
52. Mohammad, S.N. Security Attacks in MANETS (Survey Prospective). *Int. J. Eng. Adv. Technol.* **2017**, *6*, 93–96.
53. Kumar, M.; Jain, V.; Jain, A.; Bisht, U.S.; Gupta, N. Evaluation of black hole attack with avoidance scheme using AODV protocol in VANET. *J. Discret. Math. Sci. Cryptogr.* **2019**, *22*, 277–291. [CrossRef]
54. Tripathi, K.N.; Jain, G.; Yadav, A.M.; Sharma, S.C. Entity-Centric Combined Trust (ECT) Algorithm to Detect Packet Dropping Attack in Vehicular Ad Hoc Networks (VANETs). In *Next Generation Information Processing System. Advances in Intelligent Systems and Computing*; Springer: Singapore, 2021; Volume 1162. [CrossRef]
55. Alsmady, A.; Alazzam, H.; Al-Shorman, A. Blackhole attack prevention in MANET using enhanced AODV protocol. In Proceedings of the Second International Conference on Data Science, E-Learning and Information Systems. Association for Computing Machinery, Seoul, Republic of Korea, 19–21 July 2019; pp. 1–5. [CrossRef]
56. Kumar, A.; Sinha, M. Design and analysis of an improved AODV protocol for black hole and flooding attack in vehicular ad-hoc network (VANET). *J. Discret. Math. Sci. Cryptogr.* **2019**, *22*, 453–463. [CrossRef]
57. Imran, M.; Khan, F.A.; Abbas, H.; Iftikhar, M. Detection and Prevention of Black Hole Attacks in Mobile Ad hoc Networks. In *Ad-Hoc Networks and Wireless, Lecture Notes in Computer Science*; Springer: Berlin/Heidelberg, Germany, 2015; Volume 8629. [CrossRef]
58. Singh, M.M.; Mandal, J.K. Gray Hole Attack Analysis in AODV Based Mobile Adhoc Network with Reliability Metric. In Proceedings of the IEEE 4th International Conference on Computer and Communication Systems, Singapore, 23–25 February 2019; pp. 565–569. [CrossRef]
59. Jha, H.N.; Gupta, S.; Maity, D. Effect of Wormhole Attacks on MANET. In *Design Frameworks for Wireless Networks*; Lecture Notes in Networks and Systems; Springer: Singapore, 2020; Volume 82. [CrossRef]
60. Dubey, N.; Joshi, K.K. An Approach to Detect Wormhole Attack in AODV based MANET. *Int. J. Comput. Appl.* **2015**, *114*, 32–39. [CrossRef]
61. Sachan, K.; Lokhande, M. An approach to prevent Gray-hole attacks on Mobile Ad-Hoc Networks. In Proceedings of the International Conference on ICT in Business Industry & Government, Qingdao, China, 10–11 July 2017; pp. 1–6. [CrossRef]
62. Jhaveri, R.H.; Patel, S.J.; Jinwala, D.C. A Novel Solution for Grayhole Attack in AODV Based MANETs. In *Advances in Communication, Network, and Computing*; Lecture Notes of the Institute for Computer Sciences, Social Informatics and Telecommunications Engineering; Springer: Berlin, Heidelberg, 2012; Volume 108. [CrossRef]
63. El-Semary, A.M.; Diab, H. BP-AODV: Blackhole Protected AODV Routing Protocol for MANETs Based on Chaotic Map. *IEEE Access* **2019**, *7*, 95197–95211. [CrossRef]
64. Hassan, Z.; Mehmood, A.; Maple, C.; Khan, M.A.; Aldegheishem, A. Intelligent Detection of Black Hole Attacks for Secure Communication in Autonomous and Connected Vehicles. *IEEE Access* **2020**, *8*, 199618–199628. [CrossRef]
65. Ramya, P.; SairamVamsi, T. Impact Analysis of Blackhole, Flooding, and Grayhole Attacks and Security Enhancements in Mobile Ad Hoc Networks Using SHA3 Algorithm. In *Microelectronics, Electromagnetics, and Telecommunications*; Lecture Notes in Electrical Engineering; Springer: Singapore, 2018; Volume 471. [CrossRef]
66. Ali, S.; Nand, P. Comparative performance analysis of AODV and DSR routing protocols under wormhole attack in mobile ad hoc network on different node's speeds. In Proceedings of the International Conference on Computing, Communication, and Automation, Greater Noida, India, 29–30 April 2016; pp. 641–644. [CrossRef]
67. Hai, T.H.; Toi, N.D.; Huh, E.N. Performance Evaluation of AODV and AOMDV Routing Protocols Under Collaborative Blackhole and Wormhole Attacks. In *Advances in Computer Science and Ubiquitous Computing*; Lecture Notes in Electrical Engineering; Springer: Singapore, 2021; Volume 715. [CrossRef]
68. Afdhal, A.; Muchallil, S.; Walidainy, H.; Yuhardian, Q. Black hole attacks analysis for AODV and AOMDV routing performance in VANETs. In Proceedings of the International Conference on Electrical Engineering and Informatics, Banda Aceh, Indonesia, 18–20 October 2017; pp. 29–34. [CrossRef]
69. Tripathi, S. Performance Analysis of AODV and DSR Routing Protocols of MANET under Wormhole Attack and a Suggested Trust Based Routing Algorithm for DSR. In Proceedings of the IEEE International WIE Conference on Electrical and Computer Engineering, Bangalore, India, 15–16 November 2019; pp. 1–5. [CrossRef]
70. Haglan, H.M.; Mostafa, S.A.; Safar, N.Z.M.; Mustapha, A.; Saringatb, M.Z.; AlHakami, H.; AlHakami, W. Analyzing the impact of the number of nodes on the performance of the routing protocols in manet environment. *Bull. Electr. Eng. Inform.* **2021**, *10*, 434–440. [CrossRef]
71. Ismail, Z.; Hassan, R. A performance study of various mobility speed on AODV routing protocol in homogeneous and heterogeneous MANET. In Proceedings of the 17th Asia Pacific Conference on Communications, Sabah, Malaysia, 2–5 October 2011; pp. 637–642. [CrossRef]
72. Singh, M.M.; Mandal, J.K. Logistic Regression Based Reliability Analysis for Mobile Ad Hoc Network with Fixed Maximum Speed and Varying Pause Times. *J. Sci. Ind. Res.* **2017**, *76*, 81–84.
73. Sisodia, D.S.; Singhal, R.; Khandal, V. A Performance Review of Intra and Inter-Group MANET Routing Protocols under Varying Speed of Nodes. *Int. J. Electr. Comput. Eng.* **2017**, *7*, 2721–2730. [CrossRef]

74. Mafirabadza, C.; Khatri, P. Energy analysis of AODV routing protocol in MANET. In Proceedings of the International Con-ference on Communication and Signal Processing, Tamilnadu, India, 6–8 April 2016; pp. 1125–1129. [CrossRef]
75. Purnomo, A.; Najib, W.; Hartono, R. Performance Comparison of Modified AODV-ETX with AODV and AODV-ETX Routing Protocol in an MANET. *IOP Conf. Ser. Mater. Sci. Eng.* **2019**, *578*, 012082. [CrossRef]
76. Osman, H.; Ebedon, M.M.; Saad, A. A Review on Parameters of Internet Gateway Discovery in MANETS. *Int. J. Online Biomed. Eng. (iJOE)* **2021**, *17*, 38–59. [CrossRef]

electronics

MDPI

Review

Blockchain-Based New Business Models: A Systematic Review

Hamed Taherdoost [1,*] and Mitra Madanchian [1,2]

1 Department of Arts, Communications and Social Sciences, University Canada West, Vancouver, BC V6Z 0E5, Canada; mitra@hamta.ca
2 Research and Development Department, Hamta Business Corporation, Vancouver, BC V6E 1C9, Canada
* Correspondence: hamed.taherdoost@gmail.com; Tel.: +1-236-889-5359

Abstract: The role of blockchain in new business model development requires greater focus because the technology is still in its infancy. Thus, there has been little research on the effects of the various blockchain networks (such as public, private, and consortium). This finding prompted a thorough investigation of new blockchain-based business models created between 2012 and 2022 to close this gap. This review's focus is on journals, and duplicate articles have been removed. Works based on interviews, articles in press, non-English articles, reviews, conferences, book chapters, dissertations, and monographs are also not included. Seventy-five papers from the past ten years are included in this evaluation. This study examines the current state of new blockchain-based business models. Additionally, the implications and applications in the related literature have been investigated. These findings highlight numerous open research questions and promising new directions for investigation, which will likely be helpful to academics and professionals. The business strategies built on blockchain are currently on a path with a rapid upward trajectory. Blockchain technology offers businesses numerous chances to modify and develop new company models. By changing the conventional framework, blockchain innovation leads to the development of new methods for developing company models. The supportive potential of blockchain technologies such as NFT and P2E is increasingly being coupled with the development of new corporate projects and the modification of current business models. Since this field of study is still fairly new, researchers will have fresh opportunities to analyze its characteristics.

Keywords: blockchain; business; business models; business development; new business models

Citation: Taherdoost, H.; Madanchian, M. Blockchain-Based New Business Models: A Systematic Review. *Electronics* **2023**, *12*, 1479. https://doi.org/10.3390/electronics12061479

Academic Editor: Mehdi Sookhak

Received: 10 March 2023
Accepted: 15 March 2023
Published: 21 March 2023

1. Introduction

Blockchain is a disruptive technology that can transform businesses and have a wide range of applications [1–4]. Blockchain is a peer-to-peer transaction ledger system that is trusted, immutable, transparent, permanent, decentralized [5,6], and supported by algorithmic trust and distributed consensus mechanisms. It allows for secure information sharing, the long-term preservation of digital records, and the validation and verification of digital transactions. Numerous industries, including insurance, the supply chain industry, banking, real estate, renewable energy, and healthcare, have initiated blockchain initiatives [7]. Since blockchain is decentralized and eliminates intermediaries, commercial transactions can occur anywhere. Blockchain is an excellent means to (a) prove ownership; (b) trade; (c) establish peer-to-peer trust for real-time transactions; (d) increase dependability; and (e) withstand external attacks [6].

A business model constitutes a company's way of conducting business and the measures it takes to gain a competitive edge and improve its goods and services [8]. Firms digitalize their business models to increase their competitiveness in a world characterized by a fluctuating market, new technology, and diverse client demands [9,10]. Blockchain has changed the way business is conducted [11,12]. Initially used in the banking industry, today, blockchain is employed to transfer digital data across businesses. By altering how participants engage in digital transactions, a blockchain may provide new capabilities for businesses [13]. When employed in corporate operations, blockchains have far-reaching implications. For instance, transactions may be validated, disintermediation can be facilitated, and the efficiency and trust among members of an organization's ecosystem can be enhanced [14–16]. These advantages may dramatically transform a company's operations. However, the technology is still in its infancy, and research studies have expressed concern regarding the technological obstacles, ethical difficulties, and implementation security hazards with which it is threatened [17–19].

The current amount of research on business model innovation and blockchains is insufficient. Most research has been conducted from a technical standpoint [20–22]. Such studies have provided conceptual models of blockchain-based information systems and discussed the technological architecture that enables value creation. Scholars have examined the technology from a design perspective and concentrated on the various ways in which blockchain may be applied in business processes such as supply chain management [23]. Focusing on technical aspects complicates the determination of the commercial use of an invention. Several studies [24,25] have investigated the impact of blockchains on the development of strategic skills. In addition, they have examined how technology influences the introduction of new activities and how it impacts the governance and structure of current activities [25]. However, the studies have not examined the significance of technological features in developing, delivering, and collecting value across various types of companies [24]. To determine how blockchain technology produces, distributes, and collects value, as well as how technical circumstances may alter business models, an investigation is required.

Aside from the fact that theorists have not paid sufficient attention to how blockchain technology impacts business models, blockchain startups also fail to provide their promised commercial benefits. Companies do not comprehend how blockchain technology may enhance their corporate practices [19]. In addition, it is yet unclear whether business model patterns have performed effectively with this new technology base. Using these issues as a launch point, the following research topics will be the subject of this paper.

Blockchain can significantly contribute to disruptive developments in management and business. The lack of knowledge and comprehension of blockchain technologies prevents academic study and real-world use. To obtain and maintain a competitive edge, business managers must comprehend the potential effects and threats of blockchain applications. Applications based on blockchain seem to have significant prospects for performance enhancement and revenue generation [4]. The three main ways in which blockchain might influence and disrupt business models are through the disintermediation of middlemen, reducing transaction costs, and the authentication of traded commodities [26]. Earlier studies investigated the connections between business models and blockchain [26–30]. Yet, some of them alter the current conventional business models, while others develop a model for a particular industry or simply concentrate on digital transformation in general. A dynamic capabilities framework with blockchain properties and an awareness of business models was conceived in the research conducted by Aydiner [31]. By exploring the technological factors that can affect business models and probing the function of technological advantages in boosting company value, Marikyan et al. [32] presented a conceptual insight into the use of blockchain in organizations with varied value configurations. Lee [33] examined how various business models are used to create a token economy as a result of how blockchain and cryptocurrencies are still developing and interconnected. Chen and Bellavitis [34] evaluated the advantages of decentralized finance, listed current business models, and considered potential drawbacks and restrictions. The state-of-the-art practices are outlined in an article by Viriyasitavat et al. [35] to pinpoint new areas of study, difficulties, and potentially useful applications when incorporating blockchain into the growth of business process management. The purpose of the paper written by Bürer et al. [36] is to identify key topics to be further researched by focusing on applications for blockchain systems. To comprehend the challenges and opportunities precipitated by blockchain in different business operations, Kimani et al. [37] undertook a literature review.

The objective of this paper is to review the current status of the literature on blockchain-based new business models in a way that will help emerging researchers catch up on the development of the field and provide recommendations for improving the caliber of subsequent studies. To be more precise, this study conducts a thorough literature review of earlier research on new business models based on blockchain. In addition, this paper aims to pinpoint knowledge gaps and promising research directions. The specifics of the research questions (RQs) are as follows:

RQ1 : What is the present state of research in this field?

1. RQ2: What implications will blockchain have for new business models?
2. RQ3: What are the applications of blockchain-based business models?
3. RQ4: What new business models based on blockchain will emerge in the coming years?

The structure of this study is as follows: The research methodology used to find, screen, and select the included studies is comprehensively addressed in the second section. The third section examines the literature on blockchain-based new business models, showcasing the most popular papers, examining their applicability, and highlighting some of the most challenging problems in this field. Future developments are discussed as the section comes to a close. The final section of the report discusses the conclusions.

2. Background of the Study

The development of the internet in the 1990s prompted the serious study of company strategies regarding the frameworks of business models [38,39]. The reason for this new interest was that during this period, businesses were prompted to reevaluate their operations due to the impact of market globalization and the introduction of new communication technologies [40]. This shift hastened the hunt for novel organizational structures, thereby paving the way for the replacement of traditional business models with e-commerce-based ones that maximally exploit the possibilities provided by the Internet Age [41]. In this view, the business models were initially classified as methods for reforming company operations in association with environmental analyses, as stated in the study by Simmons et al. [42]. Given the frequency with which new opportunities and challenges appear in the market, it is clear that business models are crucial for determining how organizational structures can be optimally shaped [43,44].

2.1. Business Models

A business model is a conceptual instrument that aids strategic decision making and directs managers through the implementation process. It emphasizes a system-level, comprehensive explanation of how organizations operate [45,46]. Chesbrough [47] found that all companies operating in a competitive setting have a business model, regardless of what it has been termed.

The conventional business model is centralized and consists of shareholders or owners, an organization, its staff, and its customers. In this approach, the company profits by selling things or services. It expects consumers to buy their services or products at a certain price after they have been produced. The rate will have been determined so that it addresses crucial details such as wages and any other expenses required by the company to deliver the goods or services. Companies that adopt this model use centralized models, which vary by industry but often include franchisees, retailers, distributors, and manufacturers. During the last two decades, studies on business models have increased [46] and taken varied directions [48], with many academics and managers focusing on innovation in business models [42]. This focus on new business models is a direct response to increased competitiveness and the ongoing changes in consumer and market needs [21]. In this regard, market success in recent years has not been reflected in the launch of new services or products on their own but rather in the reinvention of business models [48].

2.2. How Does Blockchain Transform Business Models?

A new type of business strategy needs to be created due to changing consumer standards, needs, technologies, and laws. Additionally, not all components incorporated in the current models are compatible with the various available technologies and blockchain characteristics. Model-based methods help people comprehend general business strategies by directing them toward relevant elements that have an impact on how people conduct business across a range of industries. One of the traits of business-model-based thinking is the ability to recognize and address operational problems. The way in which values are produced and captured for consumers determines business logic, and there are various models for various reasons [49]. Business models are theoretical viewpoints that outline the structure of corporate operations to capture values and demonstrate how these values can be turned into profits [50]. A business model uses a system-level perspective to describe how businesses operate. There are established techniques for defining business strategies for organizations. Digital tools, on the other hand, are changing the current practices governing business structures into new kinds of strategies. Blockchain will disrupt established business strategies as well as value streams that are collected and produced [29].

Osterwalder and Pigneur presented the well-known traditional business model CANVAS, which has nine principles and incorporates the idea of straightforward, pertinent, and comprehensible methods of describing businesses' features [51]. When building a model with factors, the firm level of the business idea is taken into account and the query "what of doing business" is posed [52]. These nine components—cost structures; key partnerships, activities, and resources; income streams; customer relationships; channels; value offerings; and customer groups—analyze capacities for effectiveness and value for stakeholders [29,53]. This model is deficient in that it does not include data and confidence as components of its worth [53].

The literature on business models lacks a collection of universal elements that specify how the models ought to be. Therefore, the St. Gallen Business model navigator created the questions required to be able to describe a business model. To understand the value of a business, the model poses the following questions: "Who is the customer?" "why is the business model financially viable?" "How to build and disseminate the value proposition?", and "What is offered to the target customer?" [53,54]. The value design model, one of the other models that have been proposed, consists of extracts, exchanges, nodes, and value drivers that work together equally. The ecosystem is the primary force behind the value design approach, which uses building blocks to create an integrated view to find values [53]. The business DNA (design, needs, and aspirations) paradigm operates within three value-based structural components that engage with specific system components. Defining each of the DNA segments through "How?", "What?", and "Why?" queries leads to interaction. The D blocks are made up of important collaborators, assets, and tasks. Channels, client connections, and parts are contained in the N block. The value offer, income, and expenses are all covered in the A block [55]. These models always view the value through extra intermediaries to clarify the business model when defining the building block components. However, the blockchain pledges to eliminate the middlemen in commercial arrangements. Due to their static methods, these frequently used business models appear to be unable to describe blockchain-based business properties and values.

2.3. Blockchain

In today's world, as nearly every action, service, or communication involves some sort of data transfer, information has risen to prominence as the most valuable asset in every exchange. Information quality and availability must be guaranteed while working with large data sets. For data-driven decisions to be trusted by stakeholders, data must be transparent, accountable, and verifiable. The agri-food industry is one of many physical and service-providing markets [56] that could benefit from the deployment of blockchain technology, which is now popular in financial applications and transactions [57]. Numerous industries have been affected by blockchain, which has altered how organizations create, function, and interact with customers. Blockchain-based business models have aided businesses in modifying their strategies and identifying new methods by which to thrive in the digital age.

A blockchain-based business model is characterized by the three primary properties of blockchain technology: transparency, immutability, and decentralization [58]. The essence of commerce consists of peer-to-peer interactions within a dependable and trustworthy network. Decentralization profoundly impacts how businesses function. Profit production and the flow of entities and transactions are designed to maximize the benefits for end-users and businesses. The current paradigm comprises decentralized applications that can conduct peer-to-peer transactions without requiring intermediaries or a central authority [59]. Incorporating decentralized applications into this paradigm eliminates the need for shareholders and employees. Even though blockchain technology drastically modifies the impacts and responsibilities of users, a business model that incorporates its use is viable since users become both owners and employees. Eliminating intermediaries reduces the costs and time associated with third parties, thereby enhancing the ecosystem, boosting returns for investors, and lowering prices for consumers [60].

The blockchain business model offers genuine benefits that any organization can exploit. With this business model, both firm owners and end consumers benefit from the value provided. In addition to removing intermediaries and other security investments that unnecessarily raise the price of their products and services, business owners can benefit from recruiting investors or receiving payments from across the globe. On the other hand, consumers may rely on trustworthy agreements brought about by self-executing smart contracts and feel confident knowing that their personal information is secure. Although the blockchain sector is still in its infancy, it continues to demonstrate its immense potential that is still untapped.

3. Research Methodology

3.1. Planning the Review

This study sought to evaluate the current state of new business models based on blockchain technology. With the utmost seriousness, this investigation reviews all the recent, pertinent literature. The review approach applied herein uses organized RQs, databases, and methods for locating and assessing material. To provide a precise, quantitative, and in-depth evaluation of blockchain-based new business models, specific components of the specified reporting items for systematic reviews were chosen. The entire strategy includes the following crucial actions [61]:

I. Examining the current state of the field.
II. Recognizing the study's evolutional trends.
III. Analyzing the field's challenges and potential future directions
IV. Providing a breakdown of the investigation's findings.

3.2. Research Strategy

An inclusive viewpoint is necessary for a comprehensive review of the literature. Several databases were chosen before the search was conducted to increase the chance of finding highly relevant articles. This study uses sources from Scopus.

3.3. Search Criteria

For various reasons, not all outstanding studies have been included in the search criteria. A total of 553 Scopus results have been analyzed (8 December 2022). A total of 75 studies have been included in this review (Figure 1). The search strings' development was influenced by the study domain and research topics. Relevant information was found and gathered through searches for "Blockchain" AND "Business Model"; OR "Blockchain" AND "Business Development"; OR "Blockchain" AND "Business Management"; OR "Blockchain" AND "Business Framework"; OR "Blockchain" AND "Digital Business"; OR

"Blockchain" AND "Online Business"; OR "Blockchain" AND "Electronic Business"; OR "Blockchain" AND "E-Business"; OR "Blockchain" AND "Business".

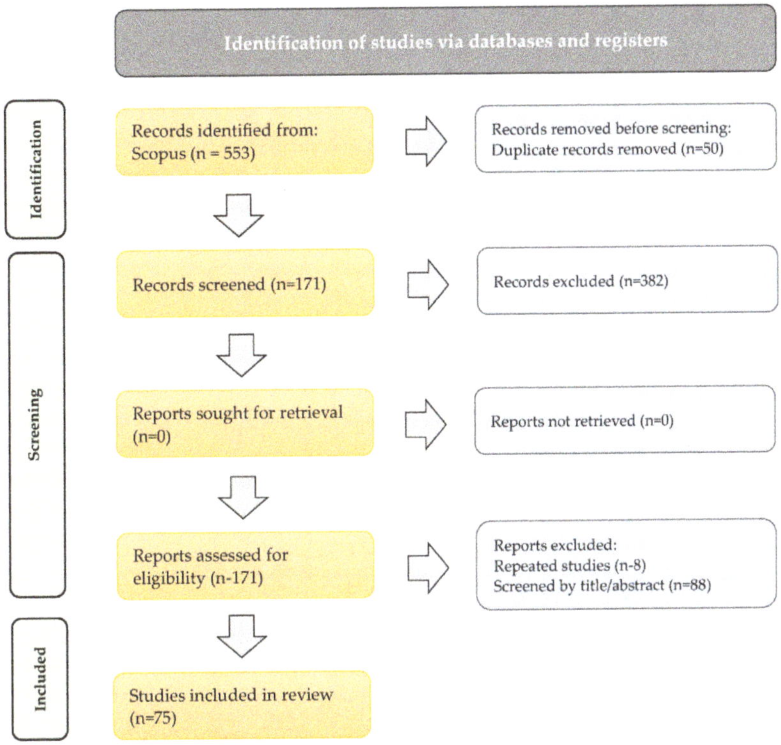

Figure 1. PRISMA flowchart showing how studies were chosen for systematic reviews.

I. Inclusion criteria (IC):
   1. The research could be published at any time between the years 2012 and 2022.
   2. The study is constrained to journals.
   3. The study corresponds to the following type of document: "article".

II. Exclusion criteria (EC):
   1. Studies could not be in-press articles.
   2. Studies could not be duplicated articles.
   3. Studies could not be written work in languages other than English.

4. Results and Discussion

The following is a list of responses to the RQs from the systematic review. The use of new business models based on blockchain appears to have significantly advanced thanks to this study. New business models built on blockchains are described in this section. The future sections will detail the importance of using new business models based on blockchain technology.

4.1. Selection Results

A total of 553 items were obtained in this search, of which 478 were screened. There are 75 articles included in this systematic review. The selected works are listed below, along with an explanation of the general classification results. The review process is constrained by the review database used, which was primarily Scopus.

RQ1: What is the present state of research in this field?

This systematic review examines research papers on blockchain-based new business models published between 2012 and 2022. This systematic review examines descriptive data on annually published papers, topic areas, author nationality, top keywords, and most-cited publications.

Figure 2 displays the number of articles created in each subject area from 2012 to 2022. Business, management, and accounting (35 articles) and computer science (34 articles) are the main topics. Other subject areas covered in the collection include engineering (twenty articles); economics, econometrics, and finance (nine articles); decision sciences (eight articles); energy (eight articles); environmental science (eight articles); social sciences (eight articles); mathematics (five articles); psychology (three articles); materials science (two articles); pharmacology, toxicology, and pharmaceutics (two articles); biochemistry, genetics, and molecular biology (one article); chemistry (one article); medicine (one article); and multidisciplinary studies (one article).

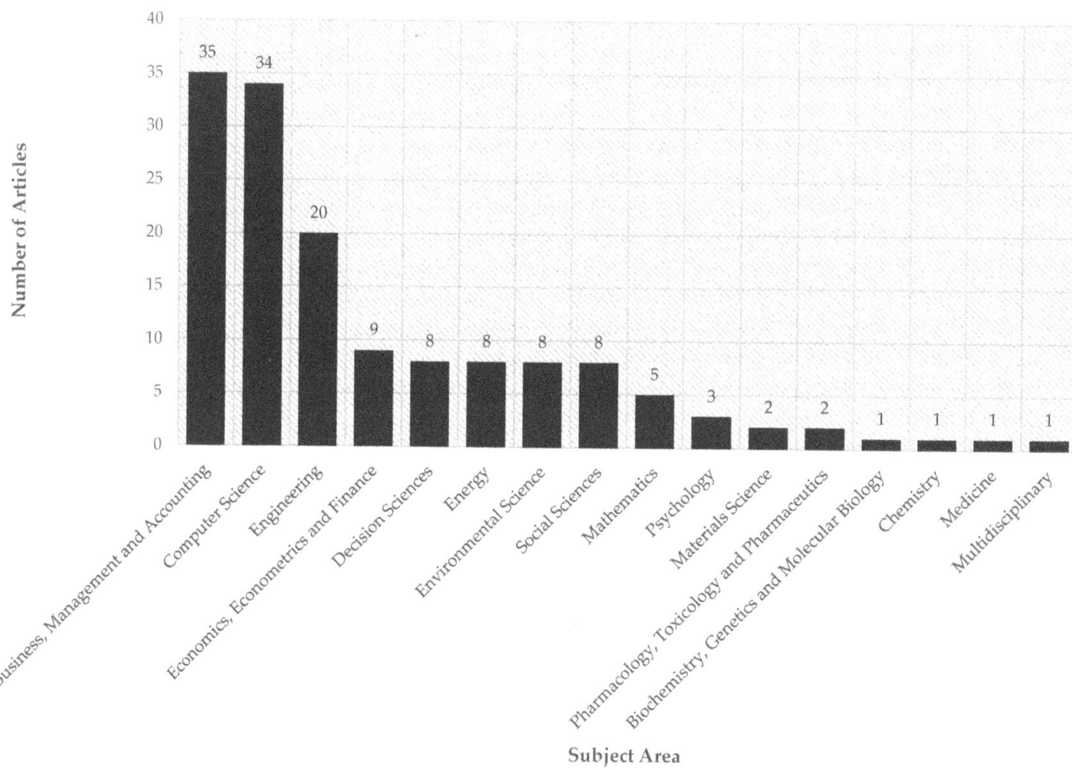

Figure 2. The number of papers published between 2012 and 2022 on the subject.

Figure 3 depicts the number of articles published each year between 2012 and 2022. The first paper was published in 2016, demonstrating how recently blockchain has emerged as a research topic in the literature on new business models. The distribution of publication dates over time is as follows: one paper (nearly 1.33%) was published in 2016, four papers (nearly 5.33%) were published in 2018, eight papers (nearly 10.67%) were published in 2019, fifteen (20%) were published in 2020, nineteen (nearly 25.33%) were published in 2021, and twenty-five papers (nearly 33.33%) were published in 2022. Given the lag-time of academic research and publishing with respect to a compelling new technology that was only presented publicly in 2009, this rapid, upward trajectory is an expected trend.

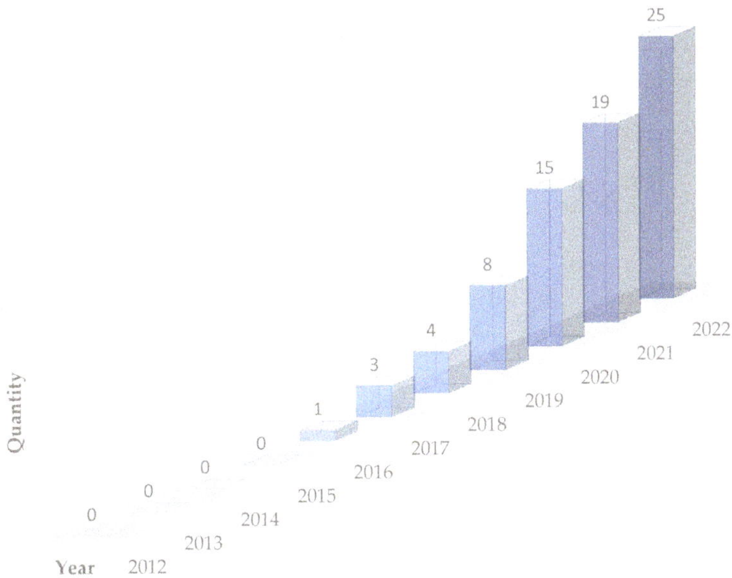

Figure 3. The number of publications published each year between 2012 and 2022.

Figure 4 depicts the relationship between the keywords of the studies chosen from the systematic literature review. Blockchain, business models, and similar terms are the most frequently used keywords. According to the analysis, the following words were frequently used: business process, innovation, business development, the Internet of Things, smart contract, supply chain management, sustainability, and many more.

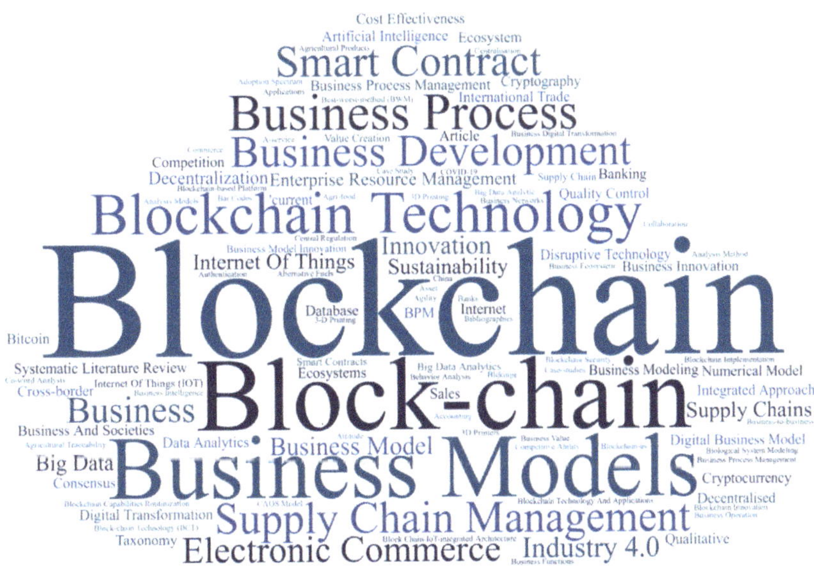

Figure 4. The primary keywords used in the articles.

Figure 5 depicts the included studies ordered by nationality and ranked according to the number of writers from each included nation. It is linked to population and development. China (both highly populated and developed) has the most writers (14), followed by India (highly populated), the United States (highly populated and developed), and Germany (developed), among others. Blockchain projects are being developed all over the world, with many of them focusing on new business models.

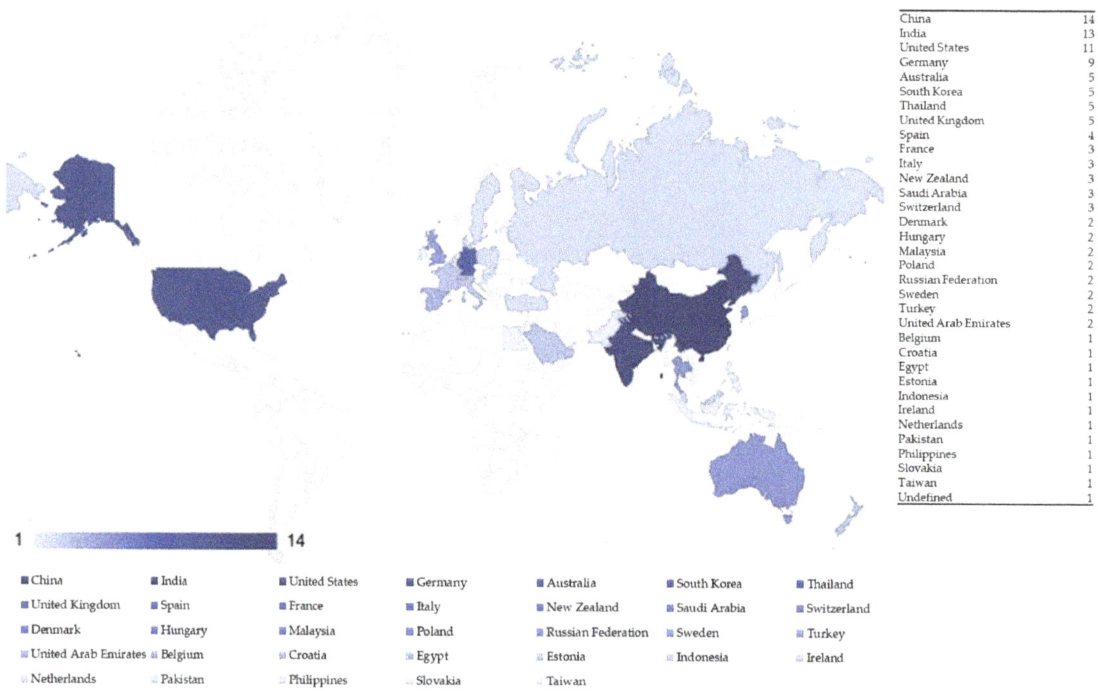

Figure 5. The distribution of authors by country.

Table 1 displays the most-cited articles, including the type of study conducted, on business models and their outcomes from 2012 to 2022. In short, a business model describes a company's plan or strategy for selling a product or service and profiting from it. Each company will develop its own business practices. There is, however, a centralized model that includes the owners or shareholders, the employees, the customers, and the organization. A blockchain business model possesses all three of blockchain technology's main characteristics: it is decentralized, based on peer-to-peer transactions, and operates within a trusted and reliable network.

Table 1. The most-cited articles (2012–2022) on business models and their outcomes.

Type of Study on Business Models	Year	Cited by	Outcome	Reference
Internet-of-Things (IoT) electric business model	2017	362	Design and implementation of IoT-based E-business models using a strong framework.	[62]
Impact of blockchain on the analyzed business model	2019	243	Business gains a competitive advantage, whether new or old, through business model innovation and blockchain.	[63]
Disruption of existing business models by blockchain	2017	136	Blockchain technology can affect and disrupt business models in three key ways: by authenticating traded goods, through disintermediation, and by lowering transaction costs.	[26]

Table 1. *Cont.*

Type of Study on Business Models	Year	Cited by	Outcome	Reference
The development of decentralized business models	2020	131	Decentralized finance may alter the way that modern finance is organized and open new opportunities for innovation and entrepreneurship while highlighting the benefits and drawbacks of decentralized business models.	[34]
Framework for blockchain-based business process management in industry 4.0 service environment	2020	111	To demonstrate how blockchain can be integrated to support quick, accurate, and reasonably priced evaluation and transfer of Quality of Services in the workflow's structure and management, a business process management framework can be used.	[64]
Online business fraud detection	2016	105	Blockchain technology is very effective for stopping objective information fraud, such as the falsification of loan application information. However, with respect to subjective information fraud, such as rating fraud, where the ground truth is difficult to verify, its effectiveness is constrained.	[65]
Blockchain's future applications in business and management	2017	93	To gain and maintain a competitive edge, business managers should comprehend the possible impacts of, and threats posed by, blockchain applications.	[4]
A decentralized token economy	2019	86	It is anticipated that future token economies will be established using new protocols enabled by blockchain technology, thereby creating a new economic framework.	[33]
Use of blockchain and IoT to develop new business processes in the digital economy	2019	71	To identify new areas of study, difficulties, and potentially useful applications when incorporating blockchain into strategies for the growth of business process management, the state of the art was presented.	[35]
Using digital innovations to transform business: data analytics, AI, cloud, and blockchain	2022	61	Wide-reaching and diversified applications across a range of vertical areas were addressed, providing exploratory study options for further inquiry.	[66]

Blockchain's universal technology may be used by numerous disciplines and presented to diverse audiences. The companies Banks and Fintech have embraced it. This explains why applications and new business models are constantly being developed.

4.2. Implications

RQ2: What implications will blockchain have for new business models?

Using blockchain technology, business models may be modified in several ways. Additionally, blockchain restricts the development of new business models. As a positive implication, blockchains provide clients with a variety of reasons to adopt blockchain-based business models [67,68]. Depending on how a blockchain is implemented, the benefits can include significant cost savings due to faster transaction times [69], disintermediation [70,71], less record-keeping with respect to customers due to distributed ledger technology, and improved data traceability and verification.

4.2.1. Theoretical Implications

With the rise of blockchain, the digital revolution, whose origins lie in the growth of the internet, is now approaching a new stage of development. A new, blockchain-based internet will usher in a period of the internet of value, which will reshape current business models through the increased reliability and transparency of information, whereas the pre-blockchain internet was dedicated to its role as the internet of information, which merely connects information providers with the consumers who use it [72]. In this paradigm, users assess goods by exchanging data they have generated as buyers rather than solely relying on data provided by service suppliers. Through a decentralized procedure that stops one organization from monopolizing information, the development of blockchain will support the achievement of a more objective and equitable consensus.

Blockchain's rules are among its most crucial and significant aspects. There are a variety of protocols with a set of conditions that are being applied to various industrial areas and objectives. These protocols' crucial components are the algorithms that create reliable tools and auxiliary technologies. These algorithms create confidence services that fall into different evidence-type categories. These three kinds of proof are as follows: evidence in an agreement, evidence as a service, and evidence in a service [73]. One of the most well-known protocols is proof of work (PoW), a cryptocurrency-specific method that is based on the proof-of-state agreement protocol. To use the PoW method, multiple miners work together to solve an issue. Although it uses a great deal of energy resources, it ensures stability and offers security from forgery in the absence of reliable intermediaries [74]. PoW guarantees that every transaction is replicated exactly in every network. Everyone can participate in the ecosystem and decide whether each transaction should be evaluated using the PoW consensus algorithm, especially in a public database. Although the enablers are anonymous, all interactions are visible [75]. The crucial aspect of PoW, however, is the steadily increasing cost and time per block and transaction [76]. Proof of stake, a suggested substitute for the PoW consensus algorithm, is cheaper and consumes less computing power. Each stake receives compensation or a penalty based on the success or failure of the deal [77]. A proof-of-value (PoV) procedure is another method of reaching an agreement. The worth of each node's input is established by this kind of consensus. In addition, the system assesses each input and its reputation in the system; then, it ascribes the impact appropriately. The proof-of-a-majority type also includes evidence of the power and preexisting procedures.

4.2.2. Practical Implications

Blockchain verifies assets in a manner distinct from centralized transaction systems, which depend on a single entity [26]. Blockchains replace centralized transaction systems to establish trust between parties. Specifically, blockchain technology enables tiny, scattered parties to manage transactions and conceal their identities [78]. All transactions are secure because of their encryption. Blockchains maintain a system's security and promote individual confidence when paired with decentralization and complex validation techniques [79]. Blockchains contribute to business models and organizational concept of a distributed autonomous organization (DAO) by lowering expenses, facilitating item tracking, and enhancing security [80–82].

A blockchain's assets affect a company's business model and how its practices are conducted [83]. Real, digital, monetary, or user-unique assets may be transferred over blockchain [83,84]. Using blockchain technology for a variety of assets creates several opportunities for altering and enhancing a company's interactions with its customers, rivals, and suppliers.

When blockchain is utilized to eliminate middlemen and provide consumers with access to, and the traceability of, their data, new business models are established by restructuring the status quo [27,85]. Therefore, this form of business model is often used by value networks [27]. When using public and consortium blockchains, which provide rigorous data validation and network access, one may collaborate with confidence. When individuals have mutual trust, data exchange among stakeholders is more effective. For instance, the democratization of access to financial resources minimizes consumer anxiety and preserves the privacy of transactions in the financial sector that lack dependable third-party guarantors [86]. These connections also provide supplementary services to clients by rendering transaction data transparent [87,88]. When you purchase an item, the ability to trace its origin increases your trust in the product's origins and manufacturing standards [87]. Service delivery is an advantage that is compatible with all blockchain network types. There are three ramifications of delivering services through digitally intermediated networks with robust data validation methods. It increases the efficacy of transactions between parties, who may freely exchange resources and data in an environment facilitated by blockchain technology. It also keeps users on the network, thereby reducing switching expenses [29,89]. Parties may collaborate to create new services, such as online-learning and on-demand services, using a blockchain-based peer-to-peer network [90]. By using blockchain technologies, transaction costs will decrease, security and financial fraud will be mitigated, and energy consumption will decrease. This will precipitate cost-effectiveness [91]. When a public blockchain makes business models conceivable, network effects occur. This kind of technology simplifies the socialization process between individuals without requiring them to exercise authority over one another. Network effects may improve efficiency by expanding the number of participants and dramatically increasing sales [92].

Finally, a blockchain-based business model enables the application of tokenization and cryptography. Cryptography can significantly alter the value proposition of a company model since it ensures that all network interactions are genuine [93]. Tokenization often refers to replacing a secret data component with a non-confidential data component [94]. The value of the business model may increase if tokens are distributed to stakeholders or if third-party tokens are accepted [78]. Tokens on the blockchain ledger may also serve as proof that a firm and its stakeholders are the legitimate owners of certain assets [83]. Various requirements, such as platform openness; the integration of numerous characteristics, such as identification, privacy [95], and interoperability; stability; scalability; and performance, pose challenges for blockchain technology [69]. In conclusion, blockchain technologies provide firms with several opportunities to alter and create new business models. However, there is insufficient research demonstrating how blockchain technology influences business models.

RQ3: What are the applications of blockchain-based business models?

The scope of the data is greater than ever, and the physical bounds are expanding. The platforms that have been developed are also communicating with outside parties. An ecosystem is created by the relationships between the systems, which resemble a symbiotic sort of dependence between external and internal companies. With these new technological foundations, lean and agile types of structures open new possibilities for enterprises to capture and generate new distributed and decentralized values [96]. To create a value-driven dynamic model, the entire system and its contributors should be considered. A structure with a more dynamic network is replacing traditional and linear business models as a result of recent technological advancements that enable hypoconnectivity. The creation of backup plans to incorporate corporate strategies with dynamic business models that account for digitalization is simplified by developing dynamic capabilities. To create digital models, new business model innovation employs sensing, seizing, and transformation skills. New approaches to business planning and design will be developed using digital business models with these dynamic characteristics as well as by employing clear business models that effectively capture a competitive edge. The capacity for sensing enables the discovery of opportunities within the external ecosystem to create value for digital business models [97].

Blockchain's capabilities have the potential to change established business models [98]. Blockchain innovation is creating a new approach to business model innovation by altering the traditional framework. There are case studies [27,30] that outline existing business models and anticipated blockchain business models in various industries, but they do not all share a methodology that examines the model holistically.

4.3. E-Business

In a corporate setting, the integration of IoT and blockchain with business process management will be crucial, especially in the context of intra/inter-organizational data systems and their various design possibilities [99]. The rapid advancement of Internet technology has improved the global economy's integration. The rapid growth of international e-commerce has been facilitated by the constant improvement of technology and business structures for international trade (CBE) [100]. Hu and Xu [100] explored the causes of the aforementioned issues in the development of CBE and addressed the creation of CBE business models based on blockchains developed according to research on the state of CBE development. The aim of such models was to research the big data and blockchain-based CBE business model. Additionally, they employed blockchain to address issues regarding cross-border trust, cross-border logistics, cross-border payment, and cross-border data flow.

The reputation framework has been developed as a powerful tool to help clients reduce the risks involved in online shopping; however, it is susceptible to rating fraud [65]. The study by Cai and Zhu [65] examined rating fraud by distinguishing objective from subjective fraud. Then, the efficacy of blockchain in preventing objective fraud and its shortcomings in preventing subjective fraud, particularly rating fraud, were covered. Finally, the study systematically examined how robust blockchain-based reputation systems are against various forms of rating fraud. As they might act strategically to conceal themselves, it is difficult to catch fake raters. They also studied the possible benefits and drawbacks of blockchain-based reputation frameworks under the two attack guises of "badmouthing" and "vote-stuffing", as well as several attack models, such as a "Sybil attack", a "whitewash attack", a "camouflage attack", and a "continuous attack". Vote-stuffing fraud is more resistant to badmouthing than blockchain-based reputation frameworks. The IoT e-business model presented by Zhang and Wen [62] reimagined many classic e-business model components, enabled P2P trade based on the blockchain and smart contracts, and realized the transaction of smart property and paid data on the IoT. Rane and Narvel [101] sensorized and IoTized an industrial pump to enable real-time operation monitoring and the use of predictive maintenance to manage these assets more quickly. The well-known properties of blockchain, such as boosting decentralization potential; enabling secure, trust-free transactions; and providing autonomous device coordination, together with the advantages of IoT, will aid in achieving Industry 4.0's stated goal of enhancing agility.

4.4. Digital Business

Digital transformation in the corporate world refers to the incorporation of digital technologies across all functional divisions, from product development to customer service. This idea is crucial for a company's and its economy's overall sustainable growth [102]. The study by Bhatti et al. [102] was carried out based on this reality, wherein the main objective was to investigate the significance of digital transformation within an organization through big data, the IoT, and blockchain-based abilities for strategic performance within the Chinese telecom industry. The findings showed a significant correlation between strategic performance and technical competence and between data quality and strategic performance. Moreover, the IoT and big data analytics played a crucial mediating role between the dependent and independent variables. Using the lenses of four emergent technology fields, namely, artificial intelligence, blockchain, cloud computing, and data analytics (ABCD), Akter et al. [66] investigated digital business transformation. The study specifically examined the workings and value propositions of these various but progressively convergent technologies. The potential of ABCD hybridization, integration, recombination, and convergence has not yet been considered due to the dynamic nature of innovation. The study's results, which were obtained via a multidisciplinary approach, demonstrated extensive and varied applications across a range of vertical sectors, thus opening potential areas for further research. The paper also emphasized how these new technologies have real-world applications. To improve business processes and preserve secure client interactions, Wang et al. [103] presented a business innovation strategy based on blockchain and artificial intelligence. There were only a few primary respondents from which the qualitative empirical data were collected, and they stemmed from two different business sectors. By comparing and contrasting how digitalization has affected value development, proposals, and business capture, blockchain and artificial intelligence were analyzed. Moreover, blockchain can help address problems regarding employee interaction and organizational capabilities. The outcome of the experiment reveals that digital transformation is typically viewed as crucial and enhances business innovation efforts. The numerical outcome suggested by the business innovation strategy based on blockchain and artificial intelligence enhances the demand forecast ratio (97.1%), business development ratio (98.9%), product quality ratio (98.3%), customer satisfaction ratio (97.2%), and customer behavior analysis ratio (96.3%).

In the article by Kifokeris and Koch [104], a brand-new digital business model for independent logistics consultants in the construction industry was proposed. It included the design of a socio-material blockchain solution for coordinated information, material, and financial flows. A permissioned and private proof-of-authority blockchain system integrating the supply chain flowing in a general socio-material environment was conceptualized by fusing academic research and empirical findings. The value proposition of a digital business strategy for an independent construction logistics consultant then incorporates this solution. The proposal calls for, among other things, increased output and better process management, while also supporting the consultants' ability to innovate and gain a competitive edge. While some business model sections, such as channels, are not considerably impacted, others, such as essential resources, are updated via blockchain. Issues such as the lack of knowledge about blockchain and the power imbalances within socio-material constellations should be resolved to prevent obstacles from obstructing the implementation of this digital business model.

Gimerská and Šoltés [105] minutely explained how blockchain may be used to digitize a purchasing group's processes. The analyzed company's core business, a financial service called central regulation, and other services were the key areas of focus. Following a review of the literature, the most well-known blockchain projects in big businesses over the past few years were examined to identify successful adoptions. The new blockchain extension was used to explain the purchasing group's processes. The findings may prove useful for the administration of purchasing groups because they indicate a rise in supply chain transparency and, concurrently, an improvement in payment-processing efficiency. For buying clubs that use a centralized payment system, the combination of permissionless and permissioned blockchains might be a workable approach.

The use of blockchain technology and the efficiency of supply chains were investigated in the study by Elrefae and Nuseir [106] regarding digital business strategy, information sharing, and trading partner pressure. A cross-sectional study design was used. The study's conclusions showed that the adoption of blockchain technology is significantly influenced by digital business strategy, information sharing, and trade partner pressure. In addition, implementing blockchain technology is essential for enhancing supply chain efficiency. The effectiveness of blockchain as a mediating factor was ultimately proven via the conducted analysis. Ivaninskiy and Ivashkovskaya [107] demonstrated that agency conflict is generally mitigated by digitization. Even while shareholders were not more hostile toward management, they did become more engaged. The authors determined that industries such as healthcare, banking, communications, and information technology have the most influence. These are the industries that ecosystem-based business model innovation has the largest impact on. The authors concluded that ecosystem-based business models and digitalization work in tandem to reduce principal–agent conflict.

4.5. Adoption and Industries

Blockchain implementation offers an organization many advantages that can lead to changes in its business model. However, it can be difficult to pinpoint how a blockchain contributes to the innovation of company models [108]. The research by Purusottama et al. [108] identified the use of blockchain as a new technology sub-element with respect to business model improvements precipitated by value creation. This form of technological adoption impacts value capture and value proposition in varying ways. Additionally, using the developed model, this study categorized the complexity of blockchain adoption and the degree of business model innovation to classify the adoption of blockchain in business model innovation. The results demonstrated that the cases in this study are scattered over the conceptual model's four quadrants.

Three new decentralized platform archetypes—hosted, shared, and federated platform models—were discovered through the use of cluster analysis [109]. The study by Lage et al. [109] advanced the understanding of newly emerging decentralized business systems. The results showed that shared and federated archetypes, which make up two-thirds of the platforms under study, do not adhere to conventional paradigms. Instead, they sought to forge new connections inside the community and in business. Moreover, the shared platform archetype is the most disruptive because it exhibits a greater degree of business-model-related and decentralization shift.

Blockchain's adoption in and application to the hospitality and tourism industries were identified by Kizildag et al. [110]. These areas include smart tourism, due diligence, the creation of loyalty programs, collaborative initiatives, integrated property management systems, verified review and rating systems, smart contracts, the de-intermediation of hospitality and tourism, tracking and service customization, and payments and cryptocurrencies. The adoption of blockchain-based systems may encourage the emergence of a weak intermediary (such as loyalty programs and/or review and rating systems) and multi-center (such as guest operations and customer service) business sectors in this sector. Table 2 summarizes some industries where blockchain-based business models play a role.

Table 2. Role of blockchain-based business models in some industries.

Ref	Agri-Food	Sustainable Business Models	Supply Chain	Transportation	Smart Contracts	Smart City	Sports	Music	Healthcare
[101]	√	√							
[105]			√						
[104]			√						
[111]				√	√				
[112]			√						
[113]						√			
[114]		√		√					
[115]							√		
[116]								√	
[117]		√							
[118]		√							
[119]			√						
[120]			√						√
[121]			√						
[122]	√	√							
[123]		√							
[124]		√							
[125]		√							
[126]		√							

4.6. Perspectives

RQ4: What new business models based on blockchain will emerge in the coming years?

The creation of new company initiatives and the alteration of existing business models are increasingly being combined with the enabling potential of blockchain. The study area is still relatively fresh, which provides researchers with new opportunities to observe.

Although managers are aware of and use blockchain to enhance the value of all business operations, further efforts are required for successful company management. The sophisticated aspects of distributed ledger technology are being tested and used, with industry professionals and developers introducing and experimenting with methods of integrating blockchain into regular corporate operations. Globally, researchers and academics are analyzing the costs and benefits of businesses employing blockchain technology. In the future, blockchain technology will be crucial in all areas, including the placing and hiring of staff, the management and organization of financial and accounting tasks, the implementation of marketing plans, and improvement of the cash or production cycle. The blockchain community is educating staff and managers about integrating technology into their daily operations. In the coming years, start-ups and small businesses will profit from blockchain's cheap cost. Businesses will be able to handle all the crucial data of company operations thanks to blockchain's simple method for tracing records that are permanent and irrevocable. Using a permissioned or private blockchain, this may be distributed to key parties such as suppliers, clients, investors, and staff members. The network's data will become more trustworthy and secure for all participants as more genuine transactions are added using encryption. The removal of middlemen will improve lead times for firms and shorten their operational and cash cycles. The business community's confidence will increase when internal and external company transactions are more transparent.

Due to the relative speed of developments in the private sector, it is even possible that enterprises' use of blockchain technology may become widespread and accepted before cryptocurrencies are more generally utilized by the general public and governments. Most people today may not even be aware of how blockchain technology impacts their dealings with major corporations. Blockchains could soon become as commonplace as internet connection. Although blockchain technology is very promising, it is still rather difficult to implement. As technology develops, it will have an influence on businesses at their core as opposed to solely goods and applications. In addition to merely their business models, corporations' modes of functioning have changed. The following sections present NFTs and play-to-earn (P2E) as emerging technologies in this field.

4.6.1. NFT

A digital commodity based on blockchain is called a non-fungible token (NFT). Cryptocurrencies and tokens can be used as blockchain-based digital commodities. Typically, smart contracts are used on the blockchain network to generate tokens [127]. NFTs were developed as a result of years of research and advancement regarding blockchain [128]. Digital assets known as NFTs are used to symbolize the possession of a variety of distinctive, substantial, and occasionally abstract but frequently concrete digital products [129]. The data unit used to symbolize these things on a blockchain digital record is called an NFT [130]. NFTs are non-exchangeable, unlike tradable tokens, which makes them one of the finest methods for individually identifying a commodity [131]. A digital asset's uniqueness or non-interchangeability is guaranteed by an NFT [130].

An NFT, as originally described in the literature [132], enables users to purchase, own, and exchange distinctive virtual objects that are recognized using blockchain. By utilizing NFTs' capabilities, businesses can expand their product offerings through virtual deals and boost exchanges between the virtual and physical worlds. For instance, Nike recently achieved remarkable price points for its exclusive, virtual, NFT-based goods. The value of NFTs can reach astounding sums thanks to their unique identifiers, whereas conventional virtual items (such as virtual clothing or virtual artwork) typically have less value than their physical equivalents [133].

Even in crowded and competitive marketplaces, NFTs provide businesses and creators with a way to accentuate themselves and offer distinctive experiences that can help them forge deeper bonds with their audiences. NFTs may serve as digital assets used to create special experiences for viewers by granting them exclusive access to information, products, and other items and services. Fans may feel more invested in and linked to a company or artist as a result, thereby helping to foster a feeling of exclusivity and community. The same can be said for NFT initiatives. Early users of social media platforms were able to develop larger followings and acquire more influence as the platforms increased in prominence than their competitors who were slower to engage in such platforms. Early adopters of NFT initiatives, such as companies and creators, stand to gain from the technology's development and adoption as more people join the NFT environment and it becomes more widely used.

4.6.2. P2E

Concerns about the rising price of games are not the only ones shared by the gaming community; in fact, according to certain studies, generally, playing games is not a rewarding activity [134]. The majority of people who participate in intense (and frequently competitive) gaming typically have little to show for all their effort, despite the emergence of a burgeoning, multi-billion-dollar esports business that benefits a small, elite proportion of players. Although they may spend 30 h per week playing games, such gamers rarely receive any real money for their efforts. Thus, interest in the possible benefits of so-called "Play-to-Earn" or P2E games, which may pay players for their gaming activity, has increased [135]. The concept behind this is that players may obtain both in-game tokens or incentives for their participation as well as tangible assets that they can change into fiat money. In other words, this model transcends the currently dominant in-game currencies, point systems, and assets and progresses towards one that more closely mimics an open trading market wherein in-game success can be converted into real-world financial results. The idea of P2E games has been around for a while in the form of virtual in-game currencies (such as those used in the *Diablo* series) and the trading of in-game assets for real money [136], which consists of skin trading, wherein players can sell and buy cosmetic features offered in games through trading systems or other third-party websites. Recently, this technology has advanced such that it incorporates blockchain.

The P2E business strategy enables players to gather and cultivate cryptocurrencies and NFTs, which may be exchanged for cash. In the "crypto gaming industry", where blockchain-based games allow token economics to thrive as a rewards system at scale for users to play and be involved in, this model has already established itself as a standard. There are three major kinds of agents in the gaming business, which is a subset of the entertainment industry. Such agents range from game system companies, which support the creation of games by creators, to those who produce games and video game devices. While offering the console at cost and making money from games has been the standard business strategy for decades, the advent of digital games has altered how games are promoted and sold, thus paving the way for free-to-play business methods. The P2E business strategy enables players to gather and cultivate cryptocurrencies and NFTs, which may be exchanged for cash. This strategy represents a new approach in the gaming world because players are monetarily rewarded for playing games. Figure 6 depicts various kinds of gaming company models.

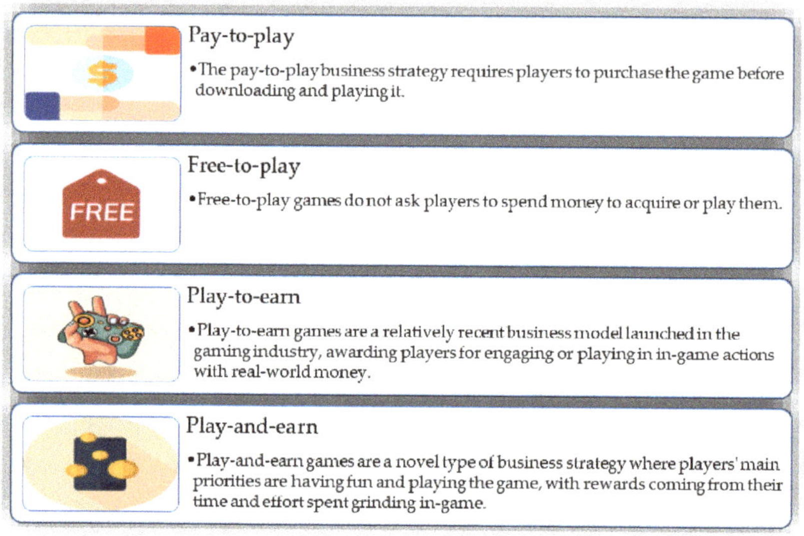

Figure 6. Various kinds of gaming business models.

5. Conclusions

In conclusion, a business model is a plan or strategy used by a company to provide goods or services and profit from them. Each company will develop its own way of conducting business. However, there is a centralized model that consists of the business, its clients, its employees, and its owners or shareholders. A blockchain-based business model is decentralized, runs on a secure network, and relies on peer-to-peer transactions, which are the three main characteristics of blockchain technology. Adopting blockchain-based technology may cause businesses to reevaluate their current business models, which could boost their profitability, productivity, and efficiency. By using blockchain, forecasting, optimization, scheduling, planning, management, and resource allocation can all be improved.

Since the technology is still in its infancy and there has been little research on the effects of the various blockchain networks, their role in new business model creation needs to be given more attention (such as consortium, private, and public). A thorough investigation of potential blockchain-based business models developed between 2012 and 2022 was sparked by this discovery. This study examined the state of blockchain-based new business models, their applications, and the revolutionary potential of their distinctive features. A total of 75 distinct publications on this topic were considered for this evaluation. As with any other business model, there is no predetermined blueprint for how every blockchain company model must operate. Thus, the objectives and business model of the company will determine which strategy is best. Business managers need to be aware of the risks and potential effects of blockchain applications to gain and maintain a competitive edge. Blockchain-based applications appear to have great potential in terms of improving performance and generating income. Thus far, the corresponding business models consist of decentralized apps that enable peer-to-peer interactions without the use of central authority or middlemen.

The present status of blockchain-based business strategies is characterized by a rapid upward rise. Blockchain technologies offer businesses numerous opportunities to change and create new business models. Blockchain technology can disrupt established company structures. By altering the conventional framework, blockchain innovation is developing a new method of business model innovation. New business endeavors and changes to established business models are increasingly being coupled with the facilitating potential of blockchain technologies such as NFT and P2E. Since this field is still relatively new, academics will have novel opportunities to analyze this field.

Author Contributions: Conceptualization, H.T.; methodology, H.T.; validation, H.T.; formal analysis, H.T. and M.M.; resources, H.T. and M.M.; data curation, H.T.; writing—original draft preparation, M.M. and H.T.; writing—review and editing, M.M.; visualization, H.T. and M.M.; supervision, H.T. All authors have read and agreed to the published version of the manuscript.

Funding: This research received no external funding.

Data Availability Statement: Data sharing not applicable.

Conflicts of Interest: The authors declare no conflict of interest.

References

1. Manski, S. Building the blockchain world: Technological commonwealth or just more of the same? *Strateg. Chang.* **2017**, *26*, 511–522. [CrossRef]
2. White, G.R. Future applications of blockchain in business and management: A Delphi study. *Strateg. Chang.* **2017**, *26*, 439–451. [CrossRef]
3. Sharma, J.; Taherdoost, H. Impact of Blockchain Technology on the Development of E-Businesses. In Proceedings of the International Conference on Advances in Data Computing, Communication and Security (I3CS), National Institute of Technology, Kurukshetra, India, 8–10 September 2021; Springer: Singapore, 2021.
4. Taherdoost, H. A Critical Review of Blockchain Acceptance Models—Blockchain Technology Adoption Frameworks and Applications. *Computers* **2022**, *11*, 24. [CrossRef]
5. Kewell, B.; Adams, R.; Parry, G. Blockchain for good? *Strateg. Chang.* **2017**, *26*, 429–437. [CrossRef]
6. Zamani, E.D.; Giaglis, G.M. With a little help from the miners: Distributed ledger technology and market disintermediation. *Ind. Manag. Data Syst.* **2018**, *118*, 637–652. [CrossRef]
7. Firica, O. Blockchain technology: Promises and realities of the year 2017. *Qual. Access Success* **2017**, *18*, 51.
8. Johnson, M.W.; Christensen, C.M.; Kagermann, H. Reinventing your business model. *Harv. Bus. Rev.* **2008**, *86*, 57–68.
9. Li, F. The digital transformation of business models in the creative industries: A holistic framework and emerging trends. *Technovation* **2020**, *92*, 102012. [CrossRef]
10. Schallmo, D.; Williams, C.A.; Boardman, L. Digital transformation of business models—Best practice, enablers, and roadmap. *Int. J. Innov. Manag.* **2017**, *21*, 1740014. [CrossRef]
11. Frizzo-Barker, J.; Chow-White, P.A.; Adams, P.R.; Mentanko, J.; Ha, D.; Green, S. Blockchain as a disruptive technology for business: A systematic review. *Int. J. Inf. Manag.* **2020**, *51*, 102029. [CrossRef]
12. Upadhyay, N. Demystifying blockchain: A critical analysis of challenges, applications and opportunities. *Int. J. Inf. Manag.* **2020**, *54*, 102120. [CrossRef]
13. He, Q.; Meadows, M.; Angwin, D.; Gomes, E.; Child, J. Strategic alliance research in the era of digital transformation: Perspectives on future research. *Br. J. Manag.* **2020**, *31*, 589–617. [CrossRef]
14. Abbas, Y.; Martinetti, A.; Moerman, J.-J.; Hamberg, T.; van Dongen, L.A. Do you have confidence in how your rolling stock has been maintained? A blockchain-led knowledge-sharing platform for building trust between stakeholders. *Int. J. Inf. Manag.* **2020**, *55*, 102228. [CrossRef]
15. Dwivedi, A.D.; Srivastava, G.; Dhar, S.; Singh, R. A decentralized privacy-preserving healthcare blockchain for IoT. *Sensors* **2019**, *19*, 326. [CrossRef] [PubMed]
16. Rimba, P.; Tran, A.B.; Weber, I.; Staples, M.; Ponomarev, A.; Xu, X. Quantifying the cost of distrust: Comparing blockchain and cloud services for business process execution. *Inf. Syst. Front.* **2020**, *22*, 489–507. [CrossRef]
17. Fotaki, M.; Voudouris, I.; Lioukas, S.; Zyglidopoulos, S. More accountable, more ethical, yet less trusted: Misplaced corporate governance in the era of big data. *Br. J. Manag.* **2021**, *32*, 947–968. [CrossRef]
18. Kumar, V.; Ramachandran, D.; Kumar, B. Influence of new-age technologies on marketing: A research agenda. *J. Bus. Res.* **2021**, *125*, 864–877. [CrossRef]
19. Lacity, M.C. Addressing key challenges to making enterprise blockchain applications a reality. *MIS Q. Exec.* **2018**, *17*, 201–222.
20. Kavanagh, D.; Ennis, P.J. Cryptocurrencies and the emergence of blockocracy. *Inf. Soc.* **2020**, *36*, 290–300. [CrossRef]
21. Pereira, J.; Tavalaei, M.M.; Ozalp, H. Blockchain-based platforms: Decentralized infrastructures and its boundary conditions. *Technol. Forecast. Soc. Chang.* **2019**, *146*, 94–102. [CrossRef]
22. Valtanen, K.; Backman, J.; Yrjölä, S. Creating value through blockchain powered resource configurations: Analysis of 5G network slice brokering case. In Proceedings of the 2018 IEEE Wireless Communications and Networking Conference Workshops (WCNCW), Barcelona, Spain, 15–18 April 2018; pp. 185–190.
23. Wang, Y.; Chen, C.H.; Zghari-Sales, A. Designing a blockchain enabled supply chain. *Int. J. Prod. Res.* **2021**, *59*, 1450–1475. [CrossRef]

24. Erceg, A.; Damoska Sekuloska, J.; Kelić, I. Blockchain in the tourism industry—A Review of the situation in Croatia and Macedonia. *Informatics* **2020**, *7*, 5. [CrossRef]
25. Schneider, S.; Leyer, M.; Tate, M. The transformational impact of blockchain technology on business models and ecosystems: A symbiosis of human and technology agents. *IEEE Trans. Eng. Manag.* **2020**, *67*, 1184–1195. [CrossRef]
26. Nowiński, W.; Kozma, M. How can blockchain technology disrupt the existing business models? *Entrep. Bus. Econ. Rev.* **2017**, *5*, 173–188. [CrossRef]
27. Chong, A.Y.L.; Lim, E.T.K.; Hua, X.; Zheng, S.; Tan, C.W. Business on chain: A comparative case study of five blockchain-inspired business models. *J. Assoc. Inf. Syst.* **2019**, *20*, 1308–1337. [CrossRef]
28. Mariappan, S. Blockchain technology: Disrupting the current business and governance model. *Int. J. Recent Technol. Eng.* **2019**, *8*, 6285–6292. [CrossRef]
29. Morkunas, V.J.; Paschen, J.; Boon, E. How blockchain technologies impact your business model. *Bus. Horiz.* **2019**, *62*, 295–306. [CrossRef]
30. Oh, J.; Shong, I. A case study on business model innovations using Blockchain: Focusing on financial institutions. *Asia Pac. J. Innov. Entrep.* **2017**, *11*, 335–344. [CrossRef]
31. Aydiner, A.S. New Approach to A Disruptive Business Model with Dynamic Capability under the Blockchain Technology. In *Management Strategies to Survive in a Competitive Environment: How to Improve Company Performance*; Springer: Cham, Switzerland, 2021; pp. 17–32.
32. Marikyan, D.; Papagiannidis, S.; Rana, O.F.; Ranjan, R. Blockchain: A business model innovation analysis. *Digit. Bus.* **2022**, *2*, 100033. [CrossRef]
33. Lee, J.Y. A decentralized token economy: How blockchain and cryptocurrency can revolutionize business. *Bus. Horiz.* **2019**, *62*, 773–784. [CrossRef]
34. Chen, Y.; Bellavitis, C. Blockchain disruption and decentralized finance: The rise of decentralized business models. *J. Bus. Ventur. Insights* **2020**, *13*, e00151. [CrossRef]
35. Viriyasitavat, W.; Xu, L.D.; Bi, Z.; Pungpapong, V. Blockchain and Internet of Things for Modern Business Process in Digital Economy—The State of the Art. *IEEE Trans. Comput. Soc. Syst.* **2019**, *6*, 1420–1432. [CrossRef]
36. Bürer, M.J.; de Lapparent, M.; Pallotta, V.; Capezzali, M.; Carpita, M. Use cases for Blockchain in the Energy Industry Opportunities of emerging business models and related risks. *Comput. Ind. Eng.* **2019**, *137*, 106002. [CrossRef]
37. Kimani, D.; Adams, K.; Attah-Boakye, R.; Ullah, S.; Frecknall-Hughes, J.; Kim, J. Blockchain, business and the fourth industrial revolution: Whence, whither, wherefore and how? *Technol. Forecast. Soc. Chang.* **2020**, *161*, 120254. [CrossRef]
38. Fielt, E. Conceptualising business models: Definitions, frameworks and classifications. *J. Bus. Model.* **2013**, *1*, 85–105.
39. Teece, D.J. Business models, business strategy and innovation. *Long Range Plan.* **2010**, *43*, 172–194. [CrossRef]
40. Kinder, T. Emerging e-commerce business models: An analysis of case studies from West Lothian, Scotland. *Eur. J. Innov. Manag.* **2002**, *5*, 130–151. [CrossRef]
41. Taran, Y.; Boer, H.; Lindgren, P. A business model innovation typology. *Decis. Sci.* **2015**, *46*, 301–331. [CrossRef]
42. Simmons, G.; Palmer, M.; Truong, Y. Inscribing value on business model innovations: Insights from industrial projects commercializing disruptive digital innovations. *Ind. Mark. Manag.* **2013**, *42*, 744–754. [CrossRef]
43. George, G.; Bock, A.J. The business model in practice and its implications for entrepreneurship research. *Entrep. Theory Pract.* **2011**, *35*, 83–111. [CrossRef]
44. Sabatier, V.; Mangematin, V.; Rousselle, T. From recipe to dinner: Business model portfolios in the European biopharmaceutical industry. *Long Range Plan.* **2010**, *43*, 431–447. [CrossRef]
45. Di Carlo, E.; Fortuna, F.; Testarmata, S. Boundaries of the business model within business groups. *J. Manag. Gov.* **2016**, *20*, 321–362. [CrossRef]
46. Zott, C.; Amit, R.; Massa, L. The business model: Recent developments and future research. *J. Manag.* **2011**, *37*, 1019–1042.
47. Chesbrough, H. Business model innovation: It's not just about technology anymore. *Strategy Leadersh.* **2007**, *35*, 12–16. [CrossRef]
48. Taran, Y.; Nielsen, C.; Montemari, M.; Thomsen, P.; Paolone, F. Business model configurations: A five-V framework to map out potential innovation routes. *Eur. J. Innov. Manag.* **2016**, *19*, 492–527. [CrossRef]
49. Heikkilä, M.; Bouwman, H.; Heikkilä, J.; Solaimani, S.; Janssen, W. Business model metrics: An open repository. *Inf. Syst. e-Bus. Manag.* **2016**, *14*, 337–366. [CrossRef]
50. Ugray, Z.; Paper, D.; Johnson, J. How Business Value Is Extracted from Operational Data: A Case Study. In *Digital Business Models: Driving Transformation and Innovation*; Springer: Cham, Switzerland, 2019; pp. 117–145.
51. Wrigley, C.; Straker, K. Designing innovative business models with a framework that promotes experimentation. *Strategy Leadersh.* **2016**, *44*, 11–19. [CrossRef]
52. Keane, S.F.; Cormican, K.T.; Sheahan, J.N. Comparing how entrepreneurs and managers represent the elements of the business model canvas. *J. Bus. Ventur. Insights* **2018**, *9*, 65–74. [CrossRef]
53. Aagaard, A. The concept and frameworks of digital business models. In *Digital Business Models: Driving Transformation and Innovation*; Springer: Cham, Switzerland, 2019; pp. 1–26.
54. Böhm, M.; Weking, J.; Fortunat, F.; Müller, S.; Welpe, I.; Krcmar, H. The business model DNA: Towards an approach for predicting business model success. In Proceedings of the 13th International Conference on Wirtschaftsinformatik, St. Gallen, Switzerland, 12–15 February 2017.

55. Sun, Y.; Yan, H.; Lu, C.; Bie, R.; Thomas, P. A holistic approach to visualizing business models for the internet of things. *Commun. Mob. Comput.* **2012**, *1*, 4. [CrossRef]
56. Wang, Y.; Han, J.H.; Beynon-Davies, P. Understanding blockchain technology for future supply chains: A systematic literature review and research agenda. *Supply Chain. Manag. Int. J.* **2019**, *24*, 62–84. [CrossRef]
57. Nakamoto, S. Bitcoin: A Peer-to-Peer Electronic Cash System. *Decentralized Bus. Rev.* **2008**, 21260. Available online: https://bitcoin.org/bitcoin.pdf (accessed on 14 March 2023).
58. Angelis, J.; Da Silva, E.R. Blockchain adoption: A value driver perspective. *Bus. Horiz.* **2019**, *62*, 307–314. [CrossRef]
59. Tushar, W.; Saha, T.K.; Yuen, C.; Smith, D.; Poor, H.V. Peer-to-peer trading in electricity networks: An overview. *IEEE Trans. Smart Grid* **2020**, *11*, 3185–3200. [CrossRef]
60. Tripoli, M.; Schmidhuber, J. *Emerging Opportunities for the Application of Blockchain in the Agri-Food Industry*; FAO: Rome, Italy, 2018.
61. Taherdoost, H. Non-Fungible Tokens (NFT): A Systematic Review. *Information* **2023**, *14*, 26. [CrossRef]
62. Zhang, Y.; Wen, J. The IoT electric business model: Using blockchain technology for the internet of things. *Peer Peer Netw. Appl.* **2017**, *10*, 983–994. [CrossRef]
63. Weking, J.; Mandalenakis, M.; Hein, A.; Hermes, S.; Böhm, M.; Krcmar, H. The impact of blockchain technology on business models—A taxonomy and archetypal patterns. *Electron. Mark.* **2020**, *30*, 285–305. [CrossRef]
64. Viriyasitavat, W.; Da Xu, L.; Bi, Z.; Sapsomboon, A. Blockchain-based business process management (BPM) framework for service composition in industry 4.0. *J. Intell. Manuf.* **2020**, *31*, 1737–1748. [CrossRef]
65. Cai, Y.; Zhu, D. Fraud detections for online businesses: A perspective from blockchain technology. *Financ. Innov.* **2016**, *2*, 20. [CrossRef]
66. Akter, S.; Michael, K.; Uddin, M.R.; McCarthy, G.; Rahman, M. Transforming business using digital innovations: The application of AI, blockchain, cloud and data analytics. *Ann. Oper. Res.* **2020**, *308*, 7–39. [CrossRef]
67. Wörner, D.; Von Bomhard, T.; Schreier, Y.-P.; Bilgeri, D. The Bitcoin Ecosystem: Disruption beyond Financial Services? In Proceedings of the 24th European Conference on Information Systems, Istanbul, Turkey, 12–15 June 2016.
68. Taherdoost, H. Blockchain Technology and Artificial Intelligence Together: A Critical Review on Applications. *Appl. Sci.* **2022**, *12*, 12948. [CrossRef]
69. Underwood, S. Blockchain beyond bitcoin. *Commun. ACM* **2016**, *59*, 15–17. [CrossRef]
70. Ying, W.; Jia, S.; Du, W. Digital enablement of blockchain: Evidence from HNA group. *Int. J. Inf. Manag.* **2018**, *39*, 1–4. [CrossRef]
71. Xu, L.; Shah, N.; Chen, L.; Diallo, N.; Gao, Z.; Lu, Y.; Shi, W. Enabling the sharing economy: Privacy respecting contract based on public blockchain. In Proceedings of the ACM Workshop on Blockchain, Cryptocurrencies and Contracts, Abu Dhabi, United Arab Emirates, 2 April 2017; pp. 15–21.
72. Tapscott, D.; Tapscott, A. *Blockchain Revolution: How the Technology behind Bitcoin is Changing Money, Business, and the World*; Penguin: London, UK, 2016.
73. Mougayar, W. *The Business Blockchain: Promise, Practice, and Application of the Next Internet Technology*; John Wiley & Sons: New York, NY, USA, 2016.
74. Risius, M.; Spohrer, K. A blockchain research framework: What we (don't) know, where we go from here, and how we will get there. *Bus. Inf. Syst. Eng.* **2017**, *59*, 385–409. [CrossRef]
75. Filipova, N. Blockchain–an opportunity for developing new business models. Бизнес Управление **2018**, *28*, 75–92.
76. Conte de Leon, D.; Stalick, A.Q.; Jillepalli, A.A.; Haney, M.A.; Sheldon, F.T. Blockchain: Properties and misconceptions. *Asia Pac. J. Innov. Entrep.* **2017**, *11*, 286–300. [CrossRef]
77. Puthal, D.; Mohanty, S.P. Proof of authentication: IoT-friendly blockchains. *IEEE Potentials* **2018**, *38*, 26–29. [CrossRef]
78. Subramanian, H. Decentralized blockchain-based electronic marketplaces. *Commun. ACM* **2017**, *61*, 78–84. [CrossRef]
79. Zhu, H.; Zhou, Z.Z. Analysis and outlook of applications of blockchain technology to equity crowdfunding in China. *Financ. Innov.* **2016**, *2*, 1–11. [CrossRef]
80. Adams, R.; Parry, G.; Godsiff, P.; Ward, P. The future of money and further applications of the blockchain. *Strateg. Chang.* **2017**, *26*, 417–422. [CrossRef]
81. Elsden, C.; Manohar, A.; Briggs, J.; Harding, M.; Speed, C.; Vines, J. Making Sense of Blockchain Applications: A typology for HCI. In Proceedings of the 2018 Chi Conference on Human Factors in Computing Systems, Montreal, QC, Canada, 21–26 April 2018; pp. 1–14.
82. Shermin, V. Disrupting governance with blockchains and smart contracts. *Strateg. Chang.* **2017**, *26*, 499–509. [CrossRef]
83. Dai, J.; Vasarhelyi, M.A. Toward blockchain-based accounting and assurance. *J. Inf. Syst.* **2017**, *31*, 5–21. [CrossRef]
84. Smith, K.; Dhillon, G. Blockchain for Digital Crime Prevention: The Case of Health Informatics. In Proceedings of the Americas Conference on Information Systems (AMCIS 2017), Boston, MA, USA, 10–12 August 2017.
85. Bauer, I.; Zavolokina, L.; Leisibach, F.; Schwabe, G. Exploring Blockchain Value Creation: The Case of the Car Ecosystem. In Proceedings of the 52nd Hawaii International Conference on System Sciences, Maui, HI, USA, 8–11 January 2019.
86. Sydow, A.; Sunny, S.A.; Coffman, C.D. Leveraging blockchain's potential–The paradox of centrally legitimate, decentralized solutions to institutional challenges in Kenya. *J. Bus. Ventur. Insights* **2020**, *14*, e00170. [CrossRef]
87. Caro, M.P.; Ali, M.S.; Vecchio, M.; Giaffreda, R. Blockchain-based traceability in Agri-Food supply chain management: A practical implementation. In Proceedings of the 2018 IoT Vertical and Topical Summit on Agriculture-Tuscany (IOT Tuscany), Tuscany, Italy, 8 May 2018; pp. 1–4.

88. Zavolokina, L.; Ziolkowski, R.; Bauer, I.; Schwabe, G. Management, governance and value creation in a blockchain consortium. *MIS Q. Exec.* **2020**, *19*, 1–17. [CrossRef]
89. Gerth, S.; Heim, L. Trust through digital technologies: Blockchain in online consultancy services. In Proceedings of the 2020 The 2nd International Conference on Blockchain Technology, Hilo, HI, USA, 12–14 March 2020; pp. 150–154.
90. Sun, H.; Wang, X.; Wang, X. Application of blockchain technology in online education. *Int. J. Emerg. Technol. Learn.* **2018**, *13*. [CrossRef]
91. Brilliantova, V.; Thurner, T.W. Blockchain and the future of energy. *Technol. Soc.* **2019**, *57*, 38–45. [CrossRef]
92. Kundu, D. Blockchain and trust in a smart city. *Environ. Urban. ASIA* **2019**, *10*, 31–43. [CrossRef]
93. Christidis, K.; Devetsikiotis, M. Blockchains and smart contracts for the internet of things. *IEEE Access* **2016**, *4*, 2292–2303. [CrossRef]
94. Panarello, A.; Tapas, N.; Merlino, G.; Longo, F.; Puliafito, A. Blockchain and iot integration: A systematic survey. *Sensors* **2018**, *18*, 2575. [CrossRef]
95. Chen, L.; Xu, L.; Shah, N.; Diallo, N.; Gao, Z.; Lu, Y.; Shi, W. Unraveling Blockchain Based Crypto-Currency System Supporting Oblivious Transactions: A Formalized Approach. In Proceedings of the ACM Workshop on Blockchain, Cryptocurrencies and Contracts, New York, NY, USA, 2–6 April 2017; pp. 23–28.
96. Krčo, S.; van Kranenburg, R.; Lončar, M.; Ziouvelou, X.; McGroarty, F. Digitization of Value Chains and Ecosystems. In *Digital Business Models: Driving Transformation and Innovation*; Springer: Cham, Switzerland, 2019; pp. 81–116.
97. Warner, K.S.; Wäger, M. Building dynamic capabilities for digital transformation: An ongoing process of strategic renewal. *Long Range Plan.* **2019**, *52*, 326–349. [CrossRef]
98. Taherdoost, H. Blockchain: A catalyst in fintech future revolution. *Future Technol. (FUTECH)* **2023**, *2*, 25–31. [CrossRef]
99. Al-Rakhami, M.S.; Al-Mashari, M. Blockchain and internet of things for business process management: Theory, challenges, and key success factors. *Int. J. Adv. Comput. Sci. Appl.* **2020**, *11*, 552–562. [CrossRef]
100. Hu, J.; Xu, L. Cross-Border E-Commerce Business Model Based on Big Data and Blockchain. *Mob. Inf. Syst.* **2022**, *2022*, 9986371. [CrossRef]
101. Rane, S.B.; Narvel, Y.A.M. Re-designing the business organization using disruptive innovations based on blockchain-IoT integrated architecture for improving agility in future Industry 4.0. *Benchmarking* **2019**, *28*, 1883–1908. [CrossRef]
102. Bhatti, A.; Malik, H.; Kamal, A.Z.; Aamir, A.; Alaali, L.A.; Ullah, Z. Much-needed business digital transformation through big data, internet of things and blockchain capabilities: Implications for strategic performance in telecommunication sector. *Bus. Process Manag. J.* **2021**, *27*, 1854–1873. [CrossRef]
103. Wang, Z.; Li, M.; Lu, J.; Cheng, X. Business Innovation based on artificial intelligence and Blockchain technology. *Inf. Process. Manag.* **2022**, *59*, 102759. [CrossRef]
104. Kifokeris, D.; Koch, C. A conceptual digital business model for construction logistics consultants, featuring a sociomaterial blockchain solution for integrated economic, material and information flows. *J. Inf. Technol. Constr.* **2020**, *25*, 500–521. [CrossRef]
105. Gimerská, V.; Šoltés, M. Analysis and Possibilities of Innovation of the Business Model Called Central Regulation Using Blockchain Technology. *Qual. Innov. Prosper.* **2022**, *26*, 55–71. [CrossRef]
106. Elrefae, G.; Nuseir, M.T. The relationship among digital business strategy, knowledge sharing and supply chain: Exploring mediating effect of blockchain adoption. *Uncertain Supply Chain. Manag.* **2021**, *9*, 1027–1036. [CrossRef]
107. Ivaninskiy, I.; Ivashkovskaya, I. Are blockchain-based digital transformation and ecosystem-based business models mutually reinforcing? The principal-agent conflict perspective. *Eurasian Bus. Rev.* **2022**, *12*, 643–670. [CrossRef]
108. Purusottama, A.; Simatupang, T.M.; Sunitiyoso, Y. The spectrum of blockchain adoption for developing business model innovation. *Bus. Process Manag. J.* **2022**, *28*, 834–855. [CrossRef]
109. Lage, O.; Saiz-Santos, M.; Zarzuelo, J.M. Decentralized platform economy: Emerging blockchain-based decentralized platform business models. *Electron. Mark.* **2022**, *32*, 1707–1723. [CrossRef]
110. Kizildag, M.; Dogru, T.; Zhang, T.; Mody, M.A.; Altin, M.; Ozturk, A.B.; Ozdemir, O. Blockchain: A paradigm shift in business practices. *Int. J. Contemp. Hosp. Manag.* **2020**, *32*, 953–975. [CrossRef]
111. Tan, W.K.A.; Sundarakani, B. Assessing Blockchain Technology application for freight booking business: A case study from Technology Acceptance Model perspective. *J. Glob. Oper. Strateg. Sourc.* **2021**, *14*, 202–223. [CrossRef]
112. Li, G.; Xue, J.; Li, N.; Ivanov, D. Blockchain-supported business model design, supply chain resilience, and firm performance. *Transp. Res. Part E Logist. Transp. Rev.* **2022**, *163*, 102773. [CrossRef]
113. Hussein, D.M.E.D.M.; Taha, M.H.N.; Khalifa, N.E.M. A Blockchain technology evolution between Business Process Management (BPM) and Internet-of-Things (IoT). *Int. J. Adv. Comput. Sci. Appl.* **2018**, *9*, 442–450. [CrossRef]
114. Philipp, R. Blockchain for LBG Maritime Energy Contracting and Value Chain Management: A Green Shipping Business Model for Seaports. *Environ. Clim. Technol.* **2020**, *24*, 329–349. [CrossRef]
115. Lv, C.; Wang, Y.; Jin, C. The possibility of sports industry business model innovation based on blockchain technology: Evaluation of the innovation efficiency of listed sports companies. *PLoS ONE* **2022**, *17*, e0262035. [CrossRef] [PubMed]
116. Owen, R.; O'Dair, M. How blockchain technology can monetize new music ventures: An examination of new business models. *J. Risk Financ.* **2020**, *21*, 333–353. [CrossRef]
117. Chin, T.; Shi, Y.; Singh, S.K.; Agbanyo, G.K.; Ferraris, A. Leveraging blockchain technology for green innovation in ecosystem-based business models: A dynamic capability of values appropriation. *Technol. Forecast. Soc. Chang.* **2022**, *183*, 121908. [CrossRef]

118. Son-Turan, S. Fostering Equality in Education: The Blockchain Business Model for Higher Education (BBM-HE). *Sustainability* **2022**, *14*, 2955. [CrossRef]
119. Hu, S.; Huang, S.; Qin, X. Exploring blockchain-supported authentication based on online and offline business in organic agricultural supply chain. *Comput. Ind. Eng.* **2022**, *173*, 108738. [CrossRef]
120. Jung, D.H. Enhancing Competitive Capabilities of Healthcare SCM through the Blockchain: Big Data Business Model's Viewpoint. *Sustainability* **2022**, *14*, 4815. [CrossRef]
121. Huang, L.; Han, Y.; Yuan, A.; Xiao, T.; Wang, L.; Yu, Y.; Zhang, X.; Zhan, H.; Zhu, H. New Business Form of Smart Supply Chain Management Based on "internet of Things + Blockchain". *Mob. Inf. Syst.* **2022**, *2022*, 1724029. [CrossRef]
122. Mercuri, F.; della Corte, G.; Ricci, F. Blockchain technology and sustainable business models: A case study of devoleum. *Sustainability* **2021**, *13*, 5619. [CrossRef]
123. Leelasantitham, A. A Business Model Guideline of Electricity Utility Systems Based on Blockchain Technology in Thailand: A Case Study of Consumers, Prosumers and SMEs. *Wirel. Pers. Commun.* **2020**, *115*, 3123–3136. [CrossRef]
124. Büyüközkan, G.; Tüfekçi, G. A decision-making framework for evaluating appropriate business blockchain platforms using multiple preference formats and VIKOR. *Inf. Sci.* **2021**, *571*, 337–357. [CrossRef]
125. Liu, Y.; Li, Z.; Huang, L. The application of blockchain technology in smart sustainable energy business model. *Energy Rep.* **2022**, *8*, 7063–7070. [CrossRef]
126. de Villiers, C.; Kuruppu, S.; Dissanayake, D. A (new) role for business—Promoting the United Nations' Sustainable Development Goals through the internet-of-things and blockchain technology. *J. Bus. Res.* **2021**, *131*, 598–609. [CrossRef]
127. Bouraga, S. On the popularity of non-fungible tokens: Preliminary results. In Proceedings of the 2021 3rd Conference on Blockchain Research & Applications for Innovative Networks and Services (BRAINS), Paris, France, 27–30 September 2021; pp. 49–50.
128. Rehman, W.; e Zainab, H.; Imran, J.; Bawany, N.Z. NFTs: Applications and challenges. In Proceedings of the 2021 22nd International Arab Conference on Information Technology (ACIT), Muscat, Oman, 21–23 December 2021; pp. 1–7.
129. Karapapas, C.; Pittaras, I.; Polyzos, G.C. Fully decentralized trading games with evolvable characters using NFTs and IPFS. In Proceedings of the 2021 IFIP Networking Conference (IFIP Networking), Espoo, Finland, 21–24 June 2021; pp. 1–2.
130. Takahashi, H.; Lakhani, U. Voting blockchain for high security NFT. In Proceedings of the 2021 IEEE 10th Global Conference on Consumer Electronics (GCCE), Kyoto, Japan, 12–15 October 2021; pp. 358–361.
131. Ertürk, E.; Doğan, M.; Kadiroğlu, Ü.; Karaarslan, E. NFT based fundraising system for preserving cultural heritage: Heirloom. In Proceedings of the 2021 6th International Conference on Computer Science and Engineering (UBMK), Ankara, Turkey, 15–17 September 2021; pp. 699–702.
132. Chohan, R.; Paschen, J. What marketers need to know about non-fungible tokens (NFTs). *Bus. Horiz.* **2021**, *18*.
133. Hofstetter, R.; de Bellis, E.; Brandes, L.; Clegg, M.; Lamberton, C.; Reibstein, D.; Rohlfsen, F.; Schmitt, B.; Zhang, J.Z. Crypto-marketing: How non-fungible tokens (NFTs) challenge traditional marketing. *Mark. Lett.* **2022**, *33*, 705–711. [CrossRef]
134. Francisco, R.; Rodelas, N.; Ubaldo, J.E. The perception of Filipinos on the advent of cryptocurrency and non-fungible token (NFT) games. *arXiv* **2022**, arXiv:2202.07467. [CrossRef]
135. Serada, A.; Sihvonen, T.; Harviainen, J.T. CryptoKitties and the new ludic economy: How blockchain introduces value, ownership, and scarcity in digital gaming. *Games Cult.* **2021**, *16*, 457–480. [CrossRef]
136. Davidovici-Nora, M. Innovation in business models in the video game industry: Free-To-Play or the gaming experience as a service. *Comput. Games J.* **2013**, *2*, 22–51. [CrossRef]

MDPI
St. Alban-Anlage 66
4052 Basel
Switzerland
www.mdpi.com

Electronics Editorial Office
E-mail: electronics@mdpi.com
www.mdpi.com/journal/electronics

www.ingramcontent.com/pod-product-compliance
Lightning Source LLC
Chambersburg PA
CBHW040817250926
51850CB00007BA/1196
9783725811939